ACTA BIOLOGICA ET MEDICA GERMANICA

Zeitschrift für funktionelle Biowissenschaften

l[th] Symposium
ucture and Function
Erythrocytes
rt I

Band 36
Heft 3/4 · 1977
Seite 291—610

48, – M

IGA J 36 (3/4) 291 –610 (1977)

2

AKADEMIE-VERLAG
BERLIN

Aufnahmebedingungen

1. Die ACTA BIOLOGICA ET MEDICA GERMANICA, Zeitschrift für funktionelle Biowissenschaften, publiziert Arbeiten aus den Fachgebieten Biochemie, Molekular- und Zellbiologie, Physiologie (einschließlich Pathophysiologie), Pharmakologie und Immunbiologie. Es werden nur Arbeiten angenommen, die nicht an anderer Stelle mit demselben Inhalt veröffentlicht oder zur Veröffentlichung angeboten werden. Der Autor verpflichtet sich nach Annahme, die Arbeiten an keiner anderen Stelle zu veröffentlichen.

2. Die Arbeit muß wissenschaftlich wertvoll sein. Bestätigungen bekannter Tatsachen, Versuche und Beobachtungen ohne positives Ergebnis werden, wenn überhaupt, nur in kürzester Form aufgenommen. Nicht aufgenommen werden Polemiken und spekulative Arbeiten, falls sie nicht wesentlich neue Gesichtspunkte enthalten.
Die Arbeiten sollen den Charakter wissenschaftlicher Originalarbeiten haben. Als solche gelten alle Mitteilungen, die zur vorwärtsführenden Erweiterung des Erkenntnisstandes auf den genannten Fachgebieten führen. Originalarbeiten sollen 20 Manuskriptseiten nicht überschreiten. Kurzmitteilungen werden bei der Drucklegung zeitlich bevorzugt; sie dürfen 5 Manuskriptseiten nicht überschreiten. Als Kurzmitteilung gelten solche Arbeiten, in denen über neue Ergebnisse berichtet wird, ohne Details einer Originalarbeit zu enthalten.
Besonders aktuelle Untersuchungsergebnisse können in kurzer Form (bis 4 Seiten) im Offsetverfahren publiziert werden, wofür reproduktionsreife Manuskripte erforderlich sind. In Form von Übersichtsarbeiten (Reviews) werden Artikel entgegengenommen, die zu aktuellen Gebieten einen Überblick geben, in dem Fakten dargestellt, besprochen und kritisch bewertet werden.

3. Die Arbeiten müssen so kurz als möglich abgefaßt werden und in einem druckreifen Zustand geschrieben sein. Einleitung (Problematik), Methodik, Befunde und Diskussion sollen deutlich in Erscheinung treten. Der Arbeit ist eine Zusammenfassung der wesentlichsten Ergebnisse voranzustellen, wovon bei deutschsprachigen Manuskripten auch eine englische Übersetzung notwendig ist.
Arbeiten werden in Deutsch, Englisch und Russisch angenommen. Die Manuskripte sind in zweifacher Ausfertigung einzureichen. Bei Manuskripten in deutscher Sprache ist die Schreibweise des „Duden“ verbindlich; bei eingedeutschten Wörtern ist die „K-Z“-Schreibweise anzuwenden.
Von den Abbildungen sind 2 Kopien sowie 1 Satz reproduktionsreife Vorlagen beizufügen. Genaue Hinweise zur Manuskriptgestaltung sind von der Redaktion der Zeitschrift anzufordern und unbedingt einzuhalten. Manuskripte, die diesen Bedingungen nicht entsprechen, gehen unbearbeitet zur Revision an den Autor zurück.

4. Die Arbeiten werden im Sofortumbruch gesetzt; Korrekturen in Form von Streichungen bzw. Zusätzen sind daher in der Umbruchkorrektur nicht zulässig.

5. Manuskripte sind an die Redaktion der ACTA BIOLOGICA ET MEDICA GERMANICA, DDR-1115 Berlin-Buch, Lindenberger Weg 70, zu senden.

6. Von jeder Originalarbeit werden kostenlos 80 Sonderdrucke geliefert; weitere Sonderdrucke können gegen Bezahlung durch den Autor erworben werden.

Chefredaktion Herausgeber

Zeitschrift ACTA BIOLOGICA ET MEDICA GERMANICA

Herausgeber: Prof. Dr. R. Baumann, Prof. Dr. H. Dutz, Prof. Dr. A. Graffi, Prof. Dr. F. Jung, Prof. Dr. O. Prokop, Prof. Dr. S. M. Rapoport.

Verlag: Akademie-Verlag, DDR-108 Berlin, Leipziger Straße 3–4; Fernruf 2236221 oder 2236229; Telex-Nr. 114420; Postscheckkonto: Berlin 35021. Bank: Staatsbank der DDR, Berlin, Kto.-Nr.: 6836-26-20712.

Chefredaktion: Prof. Dr. Heinz Bielka, Prof. Dr. Werner Scheler.

Anschrift der Redaktion: DDR-1115 Berlin-Buch, Lindenberger Weg 70. Fernruf: 5697851, App. 222.
Veröffentlicht unter der Lizenznummer 1318 des Presseamtes beim Vorsitzenden des Ministerrates der Deutschen Demokratischen Republik.

Gesamtherstellung: VEB Druckerei „Thomas Müntzer“, DDR-582 Bad Langensalza.

Erscheinungsweise: Die Zeitschrift erscheint monatlich. Die 12 Hefte eines Jahrganges bilden einen Band. Bezugspreis je Heft 35,— M (Preis für die DDR: 24,— M); Bandpreis 420,— M zuzüglich Versandspesen (Preis für die DDR 288,— M).

Bestellnummer dieses Heftes: 1053/XXXVI/3/4.

AN (EDV) 50117

Zeitschrift ACTA BIOLOGICA ET MEDICA GERMANICA

Herausgeber: Prof. Dr. R. Baumann, Prof. Dr. H. Dutz, Prof. Dr. A. Graffi, Prof. Dr. F. Jung, Prof. Dr. O. Prokop, Prof. Dr. S. M. Rapoport.

Verlag: Akademie-Verlag, DDR-108 Berlin, Leipziger Straße 3—4; Fernruf 2236221 oder 2236229; Telex-Nr. 114420; Postscheckkonto: Berlin 5021. Bank: Staatsbank der DDR, Berlin, Kto.-Nr. 6836-26-20712.

Chefredaktion: Prof. Dr. Heinz Bielka, Prof. Dr. Werner Scheler.

Anschrift der Redaktion: DDR-115 Berlin-Buch, Lindenberger Weg 70. Fernruf: 569781, App. 222.

Veröffentlicht unter der Lizenznummer 1318 des Presseamtes beim Vorsitzenden des Ministerrates der Deutschen Demokratischen Republik.

Gesamtherstellung: VEB Druckerei „Thomas Müntzer", DDR-582 Bad Langensalza.

Erscheinungsweise: Die Zeitschrift erscheint monatlich. Die 12 Hefte eines Jahrganges bilden einen Band. Bezugspreis je Heft 35,— (Preis für die DDR; 24,— M): Bandpreis 420,— M zuzüglich Versandspesen (Preis für die DDR 288,— M).

Bestellnummer dieses Bandes: 1053/XXXV.

AN (EDV) 50 117

INHALTSVERZEICHNIS

Heft 1

Biochemie

Physiologie/Pathophysiologie

Pharmakologie

Immunbiologie

Virologie

Kurzmitteilungen

Heft 2

Biochemie

Physiologie/Pathophysiologie

Pharmakologie

Zellbiologie.

Immunologie

Kurzmitteilungen

Heft 3/4

Biochemie

Physiologie/Pathophysiologie

Toxikologie

Kurzmitteilungen

Heft 5

Biochemie

Physiologie

Pharmakologie

Immunologie

Kurzmitteilungen

Heft 6

Biochemie

Physiologie

Pharmakologie

Kurzmitteilungen

Heft 7

Biochemie

Physiologie/Pathophysiologie

Kurzmitteilungen

Heft 8/9

Prostaglandinbiosynthese

Kardiovaskuläre Wirkungen von Prostaglandinen und Fettsäuren

Prostaglandine, Hypertonie und Niere

Prostaglandinantagonisten

Klinische Anwendung der Prostaglandine in der Gynäkologie und kardiovaskuläre Nebeneffekte

Sonstige Beiträge

Heft 10

Biochemie

Physiologie/Pathophysiologie

Pharmakologie

Immunologie

Kurzmitteilungen

Heft 11

Biochemie

Zellbiologie

Physiologie/Pathophysiologie

Pharmakologie

Immunologie

Heft 12

Biochemie

Physiologie/Pathophysiologie

Pharmakologie

Immunologie

Kurzmitteilungen

SACHWORTVERZEICHNIS

AUTORENVERZEICHNIS

ACTA BIOLOGICA ET MEDICA GERMANICA

Herausgeber:

R. Baumann, H. Dutz,
A. Graffi, F. Jung,
O. Prokop, S. M. Rapoport

Chefredaktion:

H. Bielka, W. Scheler

Band 36 1977 Heft 3—4

Zeitschrift für funktionelle Biowissenschaften

VIII. Internationales Berliner Symposium über Struktur und Funktion der Erythrozyten

1.—5. August 1976 Berlin, DDR

Herausgeber

Prof. Dr. Dr. S. M. Rapoport

Institut für Physiologische und Biologische Chemie
der Humboldt-Universität zu Berlin

Prof. Dr. F. Jung

Zentralinstitut für Molekularbiologie der
Akademie der Wissenschaften der DDR

Dear Colleagues:

In opening the VIIIth Symposium on "Structure and Function of Erythrocytes" I should like to recall that the first symposium was held 24 years ago. What are the reasons for this continuing tradition? For one it is the well-chosen object of research, the simple red cell which has remained a focus of a variety of both practical and theoretical interests, which are partly reflected in the program of the present symposium. A second reason is that the red cell has attracted strongly motivated people who devote a good part of their life to it. This dedication has been the framework for cooperation and companionship and has greatly suppressed feelings of rivalry which are not unknown to scientists. A third reason is the continous participation of young people which has given continuity to the work without a generation gap. An important part is played by the stability of support which has been given to the organizers of these symposium by the Minister of Health and by the Rector of Humboldt-University over the entire period, for which I should like to express our great thanks.

S. M. Rapoport

The papers of the symposium are published in two parts. This issue (3/4) contains part I and part II will be published in the following issue (5/6).
Contents of both parts are indicated at the end of issue 5/6.

Section I

Molecular Biology of the Erythron

Acta biol. med. germ., Band 35, Seite 295—303 (1977)

Moscow State University, Moscow, USSR

Avian erythropoiesis: Cellular and molecular aspects

K. G. Gazaryan

Erythropoiesis represents a highly suitable model for the study of gene functioning during terminal cell differentiation. From this point of view avian erythropoiesis is preferable since here the maturing erythroid cell does not lose its nucleus, which is preserved, though in largely inactivated form. Thus, using this system one can investigate not only translation problems, as in mammals, but also factors determining a change in the level of transcription, in particular, mechanisms of gene inactivation.

Such studies with this model have been carried out in several laboratories [1—4]; however, avian erythropoiesis in general has not as thoroughly been investigated as in mammals. Main processes in the terminal period of erythropoiesis are essentially similar in both cases, namely: gradual transition to the accumulation of a limited protein set, mainly haemoglobin and connected with this, specialization of nuclear and cytoplasmic functions. Morphologically this tendency is expressed in a change of basophilic cytoplasm into an orthochromic one and conversion of the rounded nucleus with relatively diffused chromatin to elongated forms with more compact chromatin.

The changes in cytoplasm reflect the accumulation of haemoglobin and specialization of translation, and those in the nucleus reflect the specialization of the genome transcriptional sphere. Discussing avian erythropoiesis as a genetic model it should be noted that haemoglobin is not the only marker of interest. Besides, the serine-rich histone may be interesting as a marker. At present many investigators are paying special attention to this histone.

Anaemia is provoked to obtain a great number of immature erythroid cells for studying the early stages of erythrocyte maturation. This stimulates erythropoiesis, but at the same time as is shown in mammals its picture is rather changed. Taking pigeon erythropoiesis as a model for the study of the molecular mechanisms of cell differentiation we found it desirable to compare the course of erythroid cell maturation under normal and anaemic conditions [3, 4].

To provoke acute anaemia, pigeons were injected daily with 15 mg of phenylhydrazine per kg of body weight for 4 days. Fig. 1 shows the changes in relative number of bone marrow erythroblasts in the course of anaemization and restoration of normal state. It is seen that soon after the treatment with the drug the latest forms of erythroblasts (i. e. polychromatic and orthochromatic) disappear. At the same time the portion of early erythroblasts, mainly basophilic ones, increases considerably reaching its peak on the 5th day. On the 7th—8th day the latest erythroblasts appear again and the picture is normalized. In peripheral blood immature forms begin to appear rather soon, but their quantity increases after the 5th day and reaches the maximum on the 7th day. Correlating these pictures we have supposed [4] that the early forms rather than the latest erythroblasts were the precursors of most of the immature forms found in peripheral blood. To test this possibility, we injected [^{3}H]-thymidine on the 6th day of anaemization (before the period when in peripheral blood immature forms are accumulating). In erythropoiesis the ability of [^{3}H]-thymidine incorporation gradually decreases in the maturing erythroblasts and reticulocytes do not incorporate it at all. It was found that the labeled cells in normal pigeons appear in peripheral blood only after 48 hrs of [^{3}H]-thymidine injection, the mean label number in their nuclei being lower than in the latest bone marrow erythroblasts [4]. In anaemized animals the labeled cells appear already within 6 hrs and the label number in their nuclei is much higher than in the latest erythroblasts [4]. Hence, these immature forms of peripheral blood cells could hardly been considered as arising from the latest bone marrow erythroblasts. Fig. 2 shows the histograms of distribution of bone marrow erythroblast populations subject to the amount of [^{3}H]-thymidine label in their nuclei. Here, we also see the distribution of immature forms of peripheral blood. It is clearly seen that in normal conditions the histo-

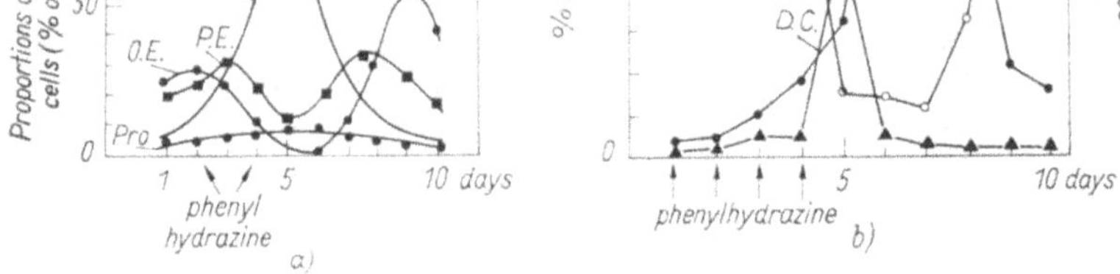

Fig. 1. Changes in the number of different forms of erythroid cells in bone marrow (a) and peripheral blood (b) during anaemization.

Pro. = proerythroblast; B.E. = basophilic erythroblast; P.E = polychromatic erythroblast; O.E. = orthochromatic erythroblast; E = erythrocyte; D.C. = degenerating cells.

gram of immature forms of peripheral blood corresponds to that of the late erythroblasts. There is no such correspondence under anaemia, where many cells in peripheral blood with high numbers of label are observed. Cells with such an amount of label are absent among the latest erythroblasts but are present in the early ones. These data support the conclusion that immature forms of peripheral blood arise predominantly from the early erythroblasts but not from the latest.

Up to now we have considered immature forms of peripheral blood on the whole. But it is known that under anaemia provoked in birds with phenylhydrazine, forms differing in the degree of maturation including erythroblast-like cells appear in the peripheral blood [1, 2].

Fig. 3a depicts immature forms which we have observed in pigeons. Only the last one (orthochromatic reticulocyte) is present in normal animals, the rest being characteristic of anaemia. The forms designated as polychromatic and basophilic reticulocytes have cytoplasms resembling those of bone marrow polychromatic and basophilic erythroblasts, respectively. But their nuclei are somewhat more elongated as in reticulocytes. The form termed basophilic erythroblast is hardly distinguishable from the corresponding bone marrow erythroblasts. It seemed interesting to know in what relations such a "precursor-product" these forms consist. One can imagine two ways of their appearance: the first — everyone appears independently of the corresponding erythroblast form of bone marrow, the second — they are generated from each other in the order displayed in the figure 3b. Table 1 gives the time-course of these forms labeling after [^{3}H]-thymidine injection.

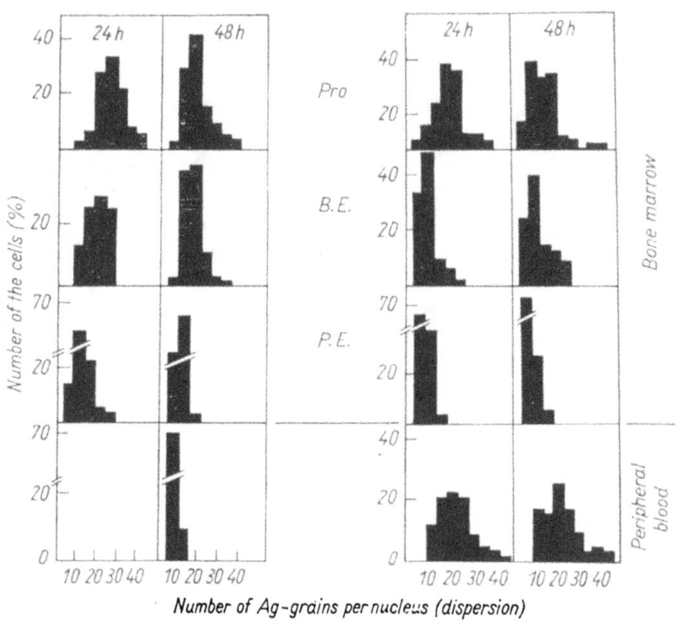

Fig. 2. Histograms showing distribution of different erythroid cell forms subject to the amount of [^{3}H]-thymidine label in the nuclei

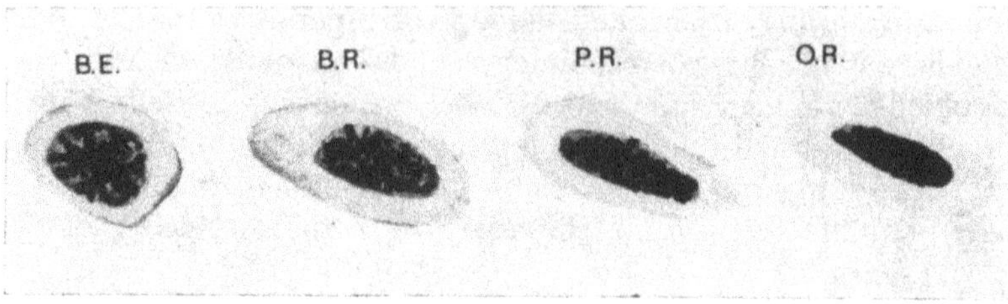

Fig. 3a. Circulating immature erythroid forms of anaemized pigeon.
B.E. = basophilic erythroblast; B.R. = basophilic reticulocyte; P.R. = polychromatic reticulocyte; O.R. = orthochromatic reticulocyte

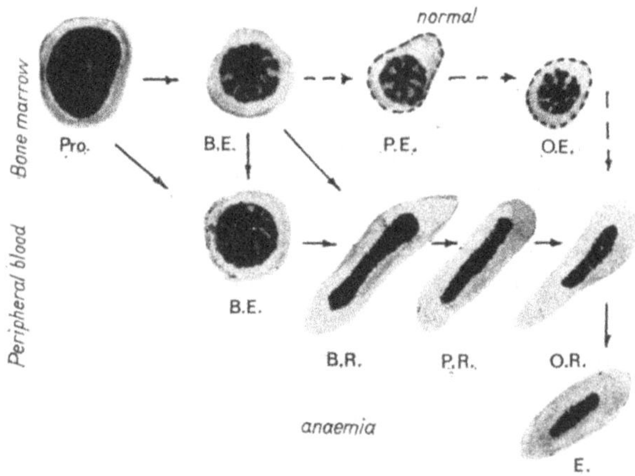

Fig. 3b. The pathways of erythrocyte maturation in pigeon

Table 1

Appearence of different immature cell forms, labeled with [^{3}H]-thymidine in the peripheral blood.

1 mC [^{3}H]-thymidine was injected to each animal on the 6th day of anaemization. Smears of the blood cells have been prepared and used for autoradiography

Hrs after [^{3}H]-thymidine injection	% of labeled cells (a) and mean grain count per nucleus (b)							
	B.E.		B.R.		P.R.		O.R.	
	a	b	a	b	a	b	a	b
2	25	38	—	—	—	—	—	—
6	55	45	21	38	—	—	—	—
12	64	31	20	40	—	—	—	—
24	89	17	58	26	20	35	—	—
48	87	8	95	14	45	32	9	37
72	98	7	91	8	49	14	22	22

B.E. = basophilic erythroblasts; B.R. = basophilic reticulocytes; P.R. = polychromatic reticulocytes; O.R. = orthochromatic reticulocytes.

Analysis of these data shows the following: (1) the more mature a form is, the later it appears as labeled, (2) with time, the mean label number in the nuclei of each form decreases since probably the cells with lower label come from the bone marrow (as a result of cell division and dilution of label). Meanwhile more mature forms appearing in peripheral blood (P.E. and O.E.) contain as much label as less mature forms we have at the beginning (before label dilution). These facts demonstrate directly that every more mature form of peripheral blood arises from the less mature form of peripheral blood but not from the corresponding bone marrow erythroblast. Bearing all this in mind one can propose the pathway of erythroid cell development under anaemia as shown in Fig. 3b. This way of development ocours only during anaemia and then erythropoiesis proceeding in normal animals is restored. We have investigated some aspects of erythropoiesis under anaemia and came to the conclusion that it hardly represents a pathological process and mature erythrocytes developing here do not basically differ in their characteristics from those arising in normal conditions.

It can be supposed that in the precursor-cells giving erythroid population two developmental programs exist: one is repressed in normal and induced only under anaemic conditions. However, an alternative cannot be excluded, namely, two kinds of the precursor-cells exist but, in normal conditions, one is developing while the development of another is induced only under acute anaemia.

As a result of bone marrow erythroblast conversion into mature erythrocytes, the transcriptional level considerably decreases. As estimated by [^{3}H]-uridine incorporation in intact cells the RNA synthesis decreases 20 fold or more [3].

The decrease in the amount or activity of nuclear RNA polymerases might be a possible cause of the RNA synthesis decline in the cell. Fig. 4 shows that the total RNA-polymerase activity in the nuclei isolated from the mature erythrocytes is several fold lower than in the nuclei of bone marrow cell population taken from anaemized pigeons. The nuclei of immature cells of peripheral blood occupy the intermediate state in this respect. It was demonstrated using α-amanitine that the nucleolar RNA polymerase activity decreases more prominently than the activity of nucleoplasmic polymerases. As a result we have found in the mature erythrocyte nuclei practically only nucleoplasmic polymerase activity [5].

Several facts mentioned below show that the only decrease of RNA polymerase activity cannot explain the process of transcription inactivation taking place in erythropoiesis. This is supported by the data obtained in the analysis of non-

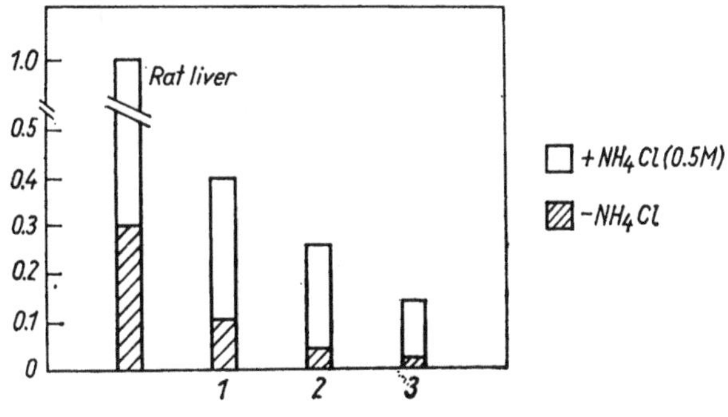

Fig. 4. Endogeneous RNA-polyuerase activity of isolated nuclei.
1 = bone marrow (anaemized); 2 = peripheral blood (anaemized); 3 = peripheral blood (normal)

histone proteins and histones, especially serine-rich histone. Comparison of electrophoretic spectra of the total non-histone proteins extracted from immature cells of peripheral blood or from populations of bone marrow cells and erythrocytes shows that erythrocyte protein contains considerably less components [6—8]. An interesting difference between these proteins has been revealed in the experiments on testing their ability to activate transcription in an in vitro system. The system contained erythrocyte chromatin as a template, non-histone proteins and E. coli RNA polymerase. It turns out that the proteins of bone marrow cells activate transcription like non-histone proteins from other cells which actively synthesize RNA [9], but erythrocyte proteins have no such ability.

When non-histone proteins were divided into fractions weakly-bound and tightly-bound with chromatin it appears that these fractions are present in erythroblast chromatin in a 1:1 ratio, with both fractions stimulating transcription (Fig. 5). In the erythrocyte chromatin only tightly-bound proteins are found. It is interesting that the spectra of the total erythrocyte proteins and the tightly-bound erythroblast fraction are similar [8]. They are most likely the same proteins. Therefore during erythrocyte maturation the weakly-bound fraction of nonhistone proteins disappears or its quantity is substantially reduced. The tightly-bound proteins are retained in the erythrocyte chromatin but lose their ability to activate transciption in an in vitro test-system.

It is still unclear how these changes in the properties of non-histone proteins might be connected with gene inactivation in the maturing erythrocytes since the function of these proteins has not yet been elucidated. These events seem to reflect a decrease of the number of regulatory proteins activating transcription in chromatin but such an assumption being in accordance with the modern hypothesis about the role of non-histone proteins requires confirmation.

Another group of questions centers around the role of histones, first of all serine-rich histone in gene inactivation during avian erythropoiesis. It has been shown earlier [10] that if histone fractions are gradually removed from mature erytrocytes with increasing ionic strength, the difference in their template activity remains so long as histone H_1 is extracted. But this difference disappears completely as soon as histone H_5 is removed.

H_5-histone starts to be extracted by the treatment with 0.6 M NaCl and is completely removed by 0.7 M NaCl. After 0.7 M NaCl treatment two chromatins do not differ in their transcriptional level [10]. Fig. 6 demonstrates the same results and also gives the data on chromatin reconstruction. When proteins removed by 0.7 M NaCl are used to reconstitute the chromatins, their template activity decreases but the decrease is greater in the case when the proteins extracted from the erythrocyte chromatin are added. It is important that the decrease in the template activity was always greater irrespective of whether the erythrocyte proteins were added to the chromatin from erythrocytes or erythroblasts. Therefore one can conclude that proteins extracted from the erythrocyte chromatin with 0.7 M NaCl contain factors responsible for inactivation of the genome during erythrocyte maturation. H_5 histone is the main component of these proteins. Here is also present some quantity of H_1 histone but the data available [10] prove that the latter is of less importance in creation of the difference between template activities

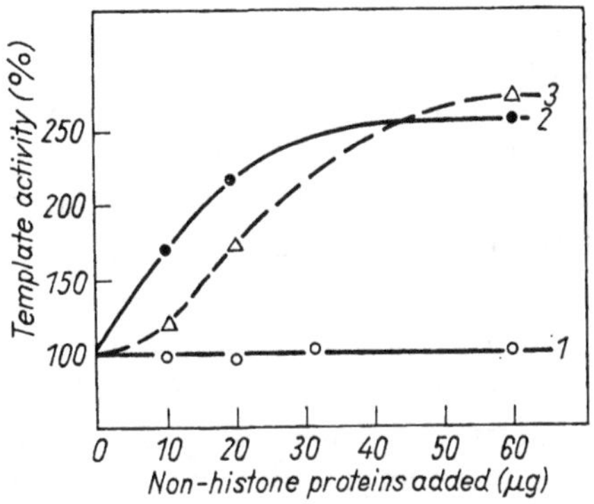

Fig. 5

Fig. 6 ▶

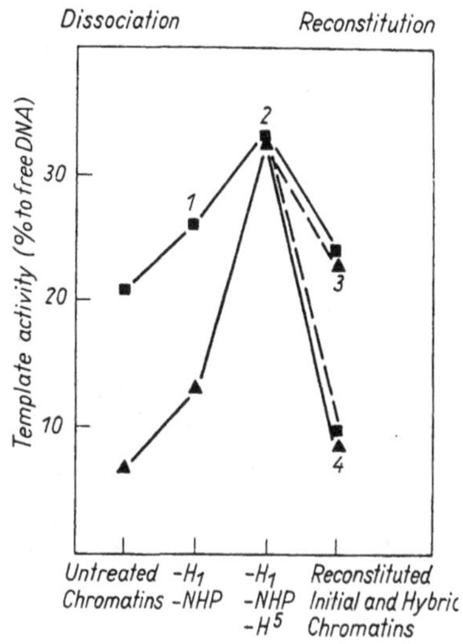

Fig. 5. Effect of non-histone proteins (NHP) on template activity of erythrocyte chromatin assayed with E. coli RNA-polymerase.

1 = total (extracted with 5 M NaCl-8 M urea) NHP from erythrocyte chromatin; 2 = weakly-bound (extracted with 0.35 M NaCl) NHP from bone marrow chromatin; 3 = tightly-bound (extracted with 5 M NaCl-8 M urea) NHP from bone marrow chromatin pretreated with 0.35 M NaCl. 100% = without adding NHP

Fig. 6. Dissociation and reconstitution of chromatins. Change in template activity assayed with E. coli RNA-polymerase.

Dissociation: 1 = 0.5 M NaCl; 2 = 0.7 M NaCl. Reconstitution: 3 = bone marrow proteins extracted with 0.7 M NaCl + $chromatin_{0.7}$; 4 = erytrocyte proteins extracted with 0.7 M NaCl + $chromatin_{0.7}$; (chromatin pretreated with 0.7 M NaCl).
■ = bone marrow chromatin; ▲ = erythrocyte chromatin

of erythroblast and erythrocyte chromatins. All evidence supports the idea that H_4-histone is the factor which restricts transcription in maturing erythrocytes [11].

It is assumed that H_5-histone is more effective as a repressor ("superrepressor") than H_1-histone and this causes the necessity of H_5-histone appearance [12]. However, several reasons prevent this problem from being completely settled:

1. Bolund and Johns [13] in their experiments have not revealed any essential difference between histone H_5 from chicken erythrocytes and H_1-histone from calf thymus.

2. Histone H_5 is also present in bone marrow cell populations, but when it is removed from the chromatin the template activity does not increase so substantially as it does after removal of the same histone from erythrocyte chromatin [10].

In this connection we have compared the repressor efficiencies of histones H_1 and H_5 obtained from erythroblasts and erythrocytes in chromatographically homogeneous state. The results showed that (Fig. 7a, b): (1) histones H_1 both from erythroblasts and erythrocytes do not differ from one another by their repressor ability; (2) the efficiency of H_5-histone from erythroblasts is comparable to the efficiency of both histones H_1; (3) histone H_5 from erythrocytes possesses more pronounced repressor ("superrepressor") ability. In other words, H_5-histone appearing in erythroblasts is not yet "superrepressor" which it becomes later in mature erythrocytes. This difference in repressor strengths of serine-rich histones isolated from the bone marrow and erythrocytes could be caused by a different degree of their phosphorylation.

Fig. 7c gives the data on the influence of serine-rich histone subfractions extracted from reticulocyte nuclei differentially by 0.1 M citric acid with 0.125 M NaCl and then using 5% $HClO_4$. The subfraction extracted at first (i.e. most phosphorylated) is less effective as a repressor and the last one (less phosphorylated) is the strongest repressor. These results support the conclusion that a lower efficiency of H_5-histone from erythroblasts is caused by a high degree of its phosphorylation.

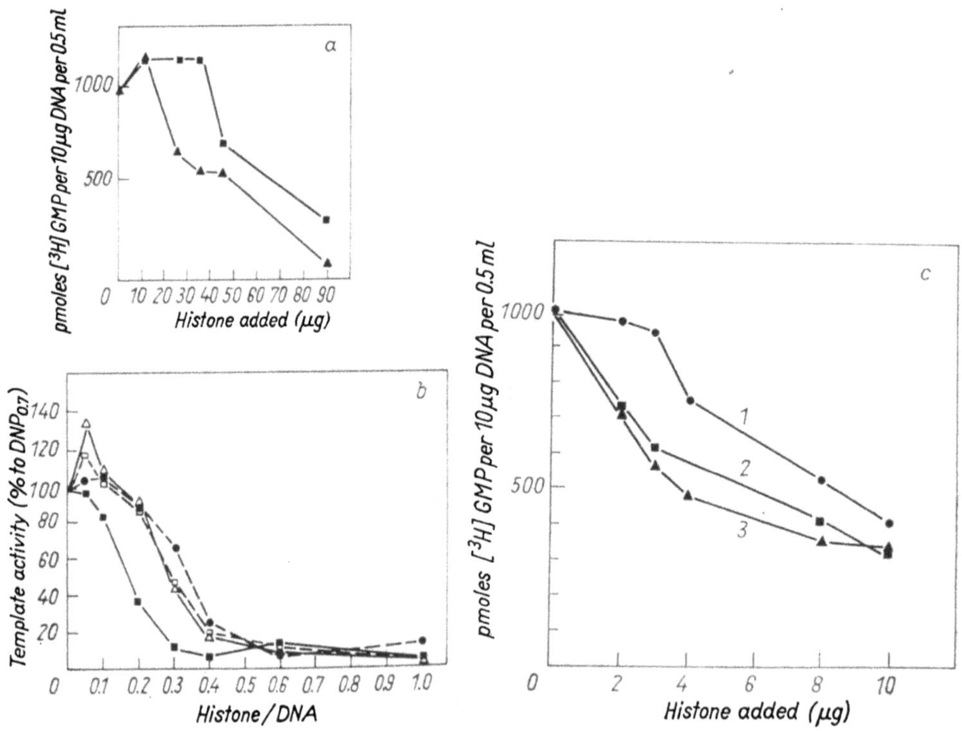

Fig. 7. The effect of purified H_5- and H_1-histones on template activity of chromatin$_{0.7}$.

a) ■ = H_1 from erythrocytes; ▲ = H_5 from erythrocytes;

b) △ = H_1 from bone marrow; □ = H_1 from erythrocytes; ● = H_5 from bone marrow; ■ = H_5 from erythrocytes;

c) subfractions with different degree of phosphorylation: 1 = extracted with 0.1 M citric acid — 0.125 M NaCl (the 1st extract, highly phosphorylated); 2 = the same (the 2nd extract, moderately phosphorylated; 3 = extracted with 5% $HClO_4$ (less phosphorylated). H_5-histones of the subfractions were purified by chromatography on CM-cellulose.

Our results are consistent with the idea that it is H_5-histone which plays the main role in the process of erythrocyte genome inactivation. The process apparently proceeds at least in two stages: the first step is the accumulation of phosphorylated form of the histone H_5 which is yet unable to associate tightly with chromatin and condense it. In some moment these phosphorylated molecules of H_5-histone are substituted for non-phosphorylated ones. It remains unclear why H_5-histone firstly appears in less active form. It seems that a definite amount of this histone should be accumulated on the chromatin at first and only after that it must start working as a "superrepressor". As demonstrated above the development process in erythropoiesis can proceed in different ways. Now, this can be related to H_5-histone behaviour. Suggesting that the process of nucleus inactivation may begin in principle at the stage of early erythroblasts as well as at the stage of the latest ones, depending on the state of erythropoietic activity. Such a situation is possible if we may accept that the factor responsible for inactivation of the nucleus is present already in early erythroblasts and it is ready to play its role. However, it does not act until the way of cell development is "chosen". In normal conditions, the action of the repressor is triggered only after the cell has passed through all stages in bone marrow. Under anaemic conditions, the repressor begins to work at the earlier erythroblasts stages. In other words, the fact that H_5-histone appears initially in early erythroblasts in weakly active form and then as "superrepressor" may be a device which gives a possibility to choose the moment of nucleus inactivation depending on the way of cell development.

The facts available at present do not permit to speak of a real mechanism of genome inactivation in erythroid cells. One can only suppose now that nuclear protein factors considered here in total participate in this mechanism but it is unclear in what manner. In particular, serine-rich histone can probably be considered only as a non-specific factor which acts together with more specific ones.

References

[1] SCHERRER, K., L. MARKOUD, F. ZAJDELA, I. M. LONDON, and F. GROSS: Proc. natn. Acad. Sci. U.S.A. **56**, 1571 (1966)

[2] MODAK, S. P., D. COMMELIN, L. GROSSET, M.-T. IMAIZUMI, M. MONNAI, and K. SCHERRER: Eur. J. Biochem. **60**, 407 (1975)

[3] GASARYAN, K. G., A. S. KULMINSKAYA, T. G. ANANJANZ and G. I. KIRYANOV: Ontogenes SSSR **2**, 263 (1971)

[4] GASARYAN, K. G., and A. S. KULMINSKAYA: Ontogenes SSSR **6**, 31 (1975)

[5] GASARYAN, K. G., T. G. ANANYAN, A. B. FEDINA and N. B. ANDREEVA: Molek. Biol. USSR **7**, 73 (1973)

[6] VIDALI, G., L. C. BOFFA, V. C. LITTAU, K. M. ALLFREY, and V. G. ALLFREY: J. biol. Chem. **248**, 4065 (1973)

[7] SANDERS, L. A.: Biochemistry **13**, 527 (1974)

[8] GASARYAN, K. G., and A. B. FEDINA: Differentiation (in press)

[9] KAMIJAMA, M., and T. Y. WANG: Biochim. biophys. Acta **228**, 563 (1971)

[10] GASARYAN, K. G., and N. B. ANDREEVA: FEBS Lett. **27**, 263 (1972)

[11] GASARYAN, K. G., N. B. ANDREEVA, and V. I. PENKINA: Differentiation **5**, 21 (1976)

[12] DICK, C., and E. W. JOHNS: Biochim. biophys. Acta **175**, 414 (1969)

[13] BOLUND, L., and E. W. JOHNS: Eur. J. Biochem. **40**, 591 (1973)

Acta biol. med. germ., Band 36, Seite 305--313 (1977)

[1] Abteilung für Elektronenmikroskopie der Charité
am Pathologischen Institut der Humboldt-Universität, 104 Berlin, DDR
[2] Institut für Physiologische und Biologische Chemie
der Humboldt-Universität, 104 Berlin, DDR
[3] Bereich Bioregulation des Zentralinstituts für Molekularbiologie
der Akademie der Wissenschaften der DDR, 1115 Berlin-Buch

The structure of pre-messenger RNA and messenger RNA from erythroid cells

A.-M. Ladhoff[1], B. J. Thiele[2], Ch. Coutelle[3], and S. Rosenthal[3]

Summary

Pre-mRNA fractions (>45 S) were characterized by electron microscopy. High salt concentrations (0.2 M ammonium acetate, *p*H 8) yield linear molecules of different length (0.5—17 µm). In 10% of the molecules a compact-nonlinear contour (cn-contour) is detectable at one end. A significant enhancement of the number of cn-contour carrying molecules is observed after binding pre-mRNA to poly(U)-sepharose. The terminal cn-contour could be the depiction of a secondary and/or tertiary structure including the poly(A)-tail. 9 S globin mRNA appear in 80% with virtually the same cn-contour as detected in pre-mRNA molecules. After denaturing the mRNA in 80% formamide — 4 M urea in connection with heating to 90 °C for 10min, a percentage of 77% of stretched, linear molecules results. This structural transformation is reversible when the denatured RNA is precipitated and redissolved in 0.2 M ammonium acetate. 73% of the stretched molecules are characterized by a mean length of 0.44 µm. This value is twice as high as commonly assumed for a globin mRNA chain.

Introduction

The recent experimental proofs [1—4] for the suggested precursor product relationship between the nuclear pre-mRNA and the cytoplasmic mRNA [5, 6] have created increased interest also in the structure of these RNA species.

One of the possible methods for the study of RNA structure is electron microscopy. Previously we have published electron micrographs of individual pre-mRNA and mRNA molecules from erythroid cells [7, 8]. An interesting observation was the appearance of a contour, probably corresponding to a higher ordered structure, on one end of 10% of the pre-mRNA molecules from rabbit bone marrow cells. A virtually similar contour was observed in molecules of 9 S globin mRNA fractions from rabbit reticulocytes. The same contour was also described by Dubochet et al. [9] for rabbit and duck globin mRNA molecules.

In the present paper we show that it is possible to enhance the percentage of cn-contour carrying pre-mRNA-molecules by binding to poly(U)-sepharose and to transform the cn-contour of globin mRNA to a linear form by use of extensive denaturation conditions.

Material and methods

RNA preparation

Pre-mRNA (>45 S) from rabbit bone marrow cells and 9 S globin mRNA from rabbit reticulocytes were prepared as described in detail in a previous paper [8].

Electron microscopy

For electron microscopic investigation a simple protein-free preparation method was used. Methodical manipulations and techniques of staining were the same as described [8].
Using standard conditions RNA was dissolved in 1 M ammonium acetate and layered in drops of about 10 μl on a hypophase of 10 mM ammonium acetate up to a final concentration of ammonium acetate of 0.2 M. In several experiments the RNA was simply dissolved in 0.2 M ammonium acetate and layered in drops of 20 or 100 μl (200 μg RNA/ml) on a siliconized Petri-dish. After 20 min the grids were layered on the surface of the drops. Details of variations of the media used for induction of conformational changes of the globin mRNA are given in "Results and discussion".

Results und discussion

Pre-messenger RNA

DNase treated, protein-free fractions of nuclear RNA from erythroid enriched bone marrow cells, sedimenting higher than 45 S, appear in the presence of 0.2 M ammonium acetate as a bulk of linear molecules of different contour length. The frequency distribution of the length of 98 molecules is shown in Fig. 1. The distribution ranges from a length of 0.5 μm up to 17 μm. It is necessary to point out that the percentage of longer molecules is probably higher because some of them could not be measured due to ambiguity on contour length resulting from background contaminations. Assuming a base distance of about 0.3 nm the longest molecules found (17 μm) should consist of about 57.000 nucleotides. Electron micrographs of molecules of such a length were published in a previous paper [8]. The structural resolution of the micrographs of 1.0 to 1.5 nm permits the detection of end-to-end hybridization and the proof of longer base paired regions. In 90% of the stained pre-mRNA molecules, no indications exist that the polynucleotide chain is of higher structural order, while in about 10% of mainly the longer molecules a quite peculiar compact non-linear contour (in the following named cn-contour) is detectable at one end. Fig. 2 shows the depiction of 3 pre-mRNA molecules carrying a terminal cn-contour. A considerable enhancement of the number

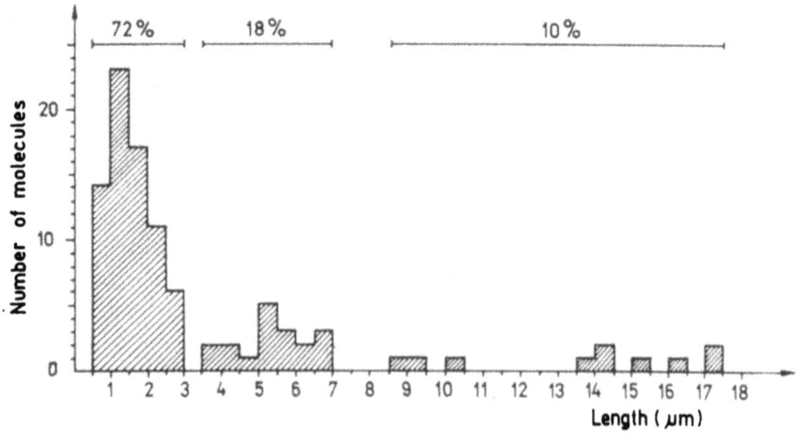

Fig. 1. Frequency distribution of the length of 98 molecules of nuclear RNA fractions (>45 S) from enriched erythroid bone marrow cells of the rabbit

of cn-contour carrying molecules was registered after binding the pre-mRNA fractions to poly(U)-sepharose. The length of the terminal cn-section is characterized by a mean value of 0.40 μm (Fig. 3).

At present we cannot give a convincing explanation for this peculiar molecule contour. Because protein could never be detected in the examined RNA samples and ribosomal RNA appears completely linear, the most likely explanation of the terminal compact structure is that it is the depiction of a secondary and/or tertiary structure specific for erythroid pre-mRNA. It is likewise possible that the cn-contoured section represents a region consisting of double stranded regions functioning as processing marker proposed by Ryskov et al. [10].

Globin messenger RNA

Molecules from the 9 S region of a sucrose density gradient after binding of polysomal RNA to poly(U)-sepharose appear with virtually the same cn-contour as

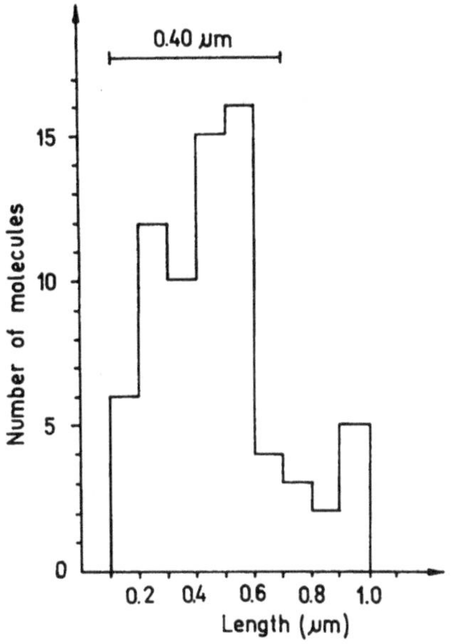

Fig. 3. Frequency distribution of the length of the terminal cn-contour of poly(A)+pre-mRNA molecules

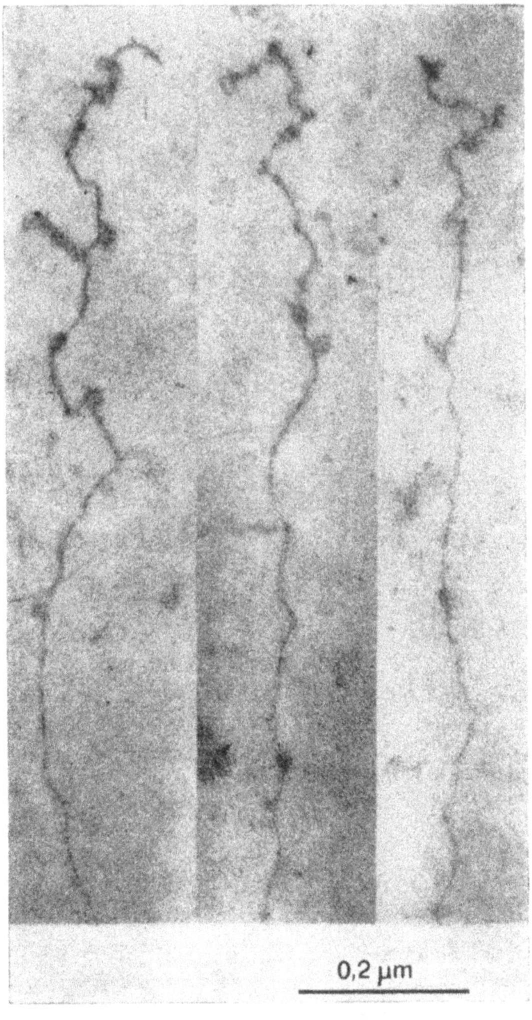

Fig. 2. Terminal sections of poly(A)+ pre-mRNA molecules from rabbit bone marrow cells. Uranyl acetate staining ▶

the terminal section of erythroid pre-mRNA. Typical contour forms of 9 S globin mRNA fractions are shown in Fig. 4. In the presence of 0.2 M ammonium acetate 80% of the molecules are characterized by a cn-contour (Fig. 4c). The cn-contour of the platinium shadowed samples corresponds exactly to that published by DUBOCHET et al. [9] if uranyl acetate staining is used [7]. 4% of the molecules are of linear character (Fig. 4a) and the remaining percentages correspond to molecules showing contours which could be discussed as intermediate forms (Fig. 4b).

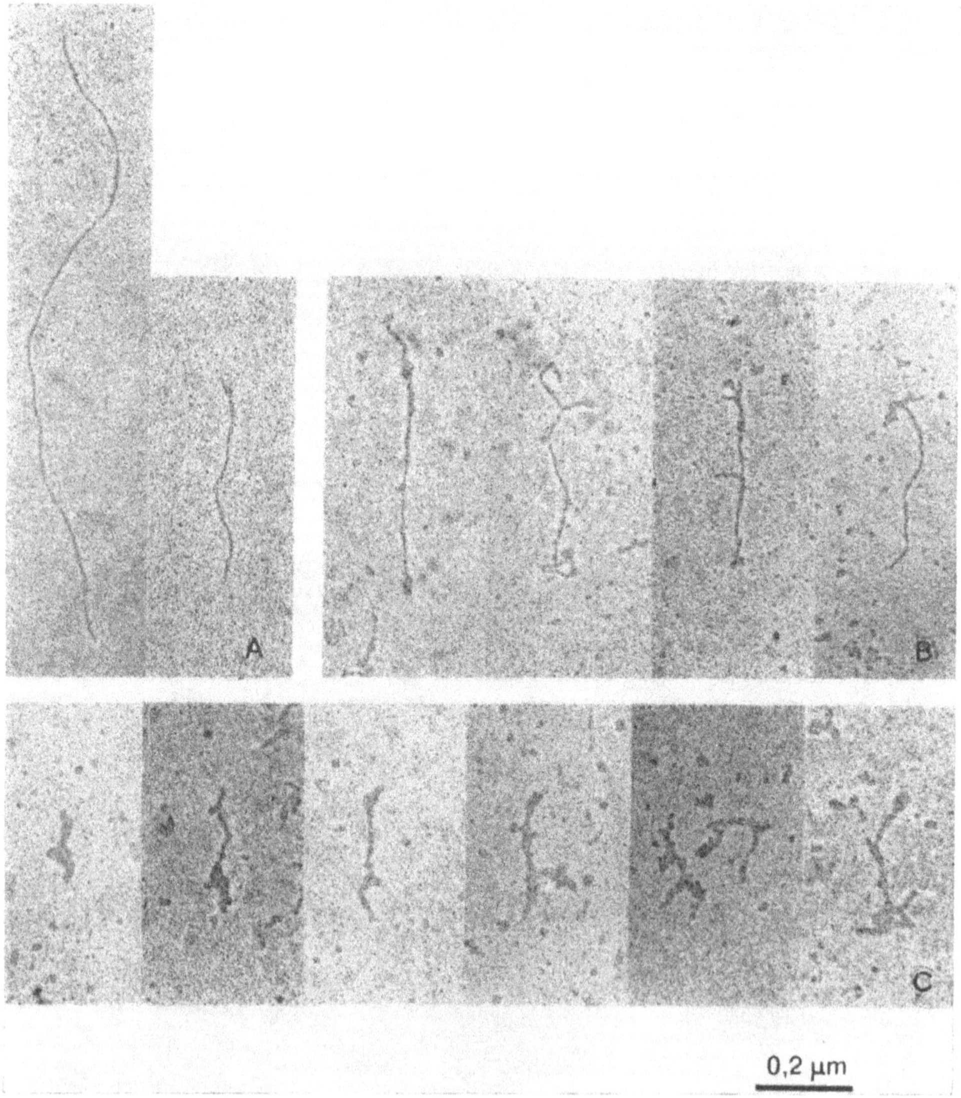

Fig. 4. 9 S globin mRNA from rabbit reticulocytes prepared for electron microscopy by adsorption at the liquid-air-interphase in the presence of 0.2 M ammonium acetate. Contrasting was done by staining with 10^{-3} M uranyl acetate in ethanol followed by platinium rotary shadowing.
A) linear contour (4%); B) intermediate form (16%); C) compact-nonlinear contour (80%)

Because in fractions of 9 S mRNA no protein contamination could be detected one must assume that the cn-contour is most likely the result of a structure of higher order within the mRNA molecule. The share of linear molecules could be due to a contamination of ribosomal RNA which remains stretched under these conditions. A length distribution of molecules with a cn-contour is shown in Fig. 5. The distribution ranged from 0.05 μm up to 0.35 μm (mean value 0.14 μm).
In the presence of 0.2 M ammonium acetate a partial structural transformation occurs if the RNA sample is heated for 10 min to 68 °C. The percentage of cn-contours shifts from 80% down to 63% and that of the intermediate forms from 16% up to 35%. The linear form remains constant. Only when the RNA was dissolved in 80% formamide — 4 M urea (0.04 M Tris, 0.001 M EDTA, *p*H 7.5) and heated for 10 min to 90 °C a substantial change of linear molecules occurred. The percentage of linear molecules rises up to 77% (electron micrographs of typical linear molecules are shown in Fig. 6). 12% of the molecules remain in the cn-state while 11% show different degrees of transformation. The results of the temperature denaturing experiments are summarized in Table 1.
A distribution of the length of molecules stretched in the denaturation medium of 80% formamide and 4 M urea is shown in Fig. 7. 73% of the linear molecules are in the range between 0.1 and 0.9 μm (mean value 0.44 μm). To investigate the reversibility of this structural transformation, the denatured RNA was precipitated with ethanol and resuspended in 0.2 M ammonium acetate. Typical contour forms of renatured molecules are shown in Fig. 8. Only about half this material was

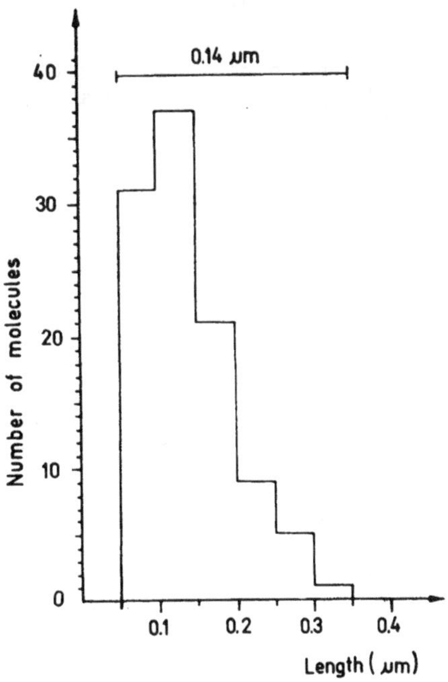

Fig. 5. Length distribution of globin mRNA molecules from rabbit reticulocytes characterized by a compact-nonlinear contour

Table 1

Influence of the denaturation conditions on the percentage of typical contour forms of globin mRNA from rabbit reticulocytes

contour form	 linear	 intermediate	 compact-nonlinear
room temperature	4	16	80
10 min 68°C	2	35	63
10 min 90°C 4M urea 80% formamide	77	11	12

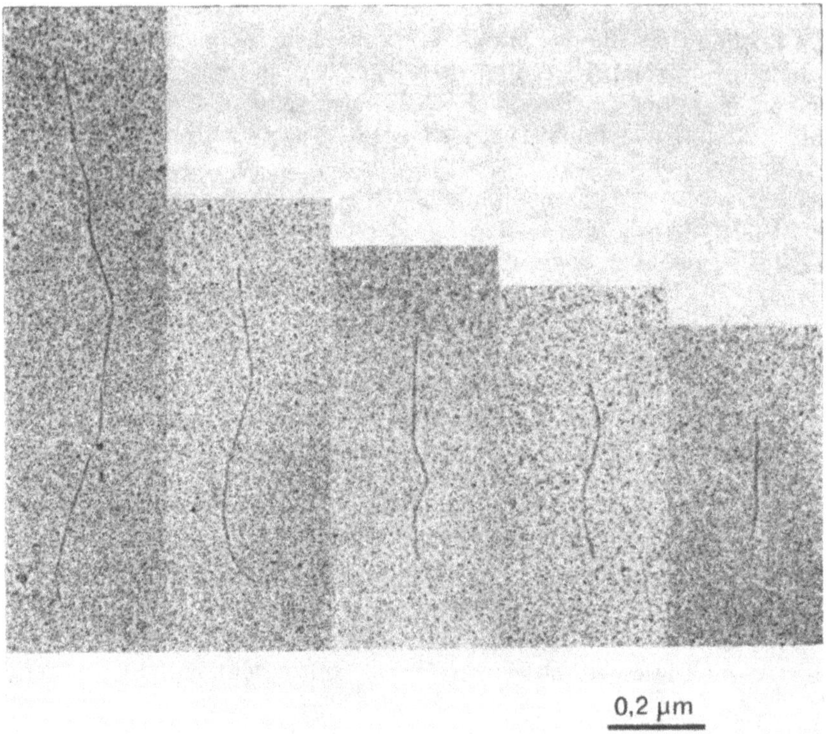

Fig. 6. Linear contour form of 9 S globin mRNA after 10 min heating to 90 °C in the presence of 80% formamide, 4 M urea, 0.04 M Tris, 0.001 M EDTA, *p*H 7.5. Double contrasting as described in legend to Fig. 4

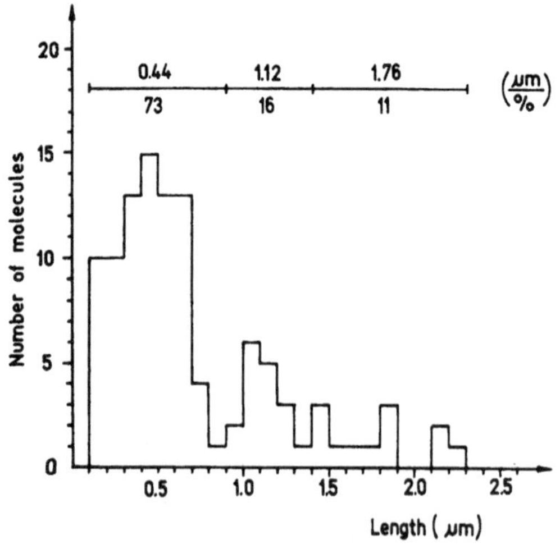

Fig. 7. Frequency distribution of globin mRNA denatured by 10 min heating to 90 °C in 80% formamide and 4M urea (0.04 M Tris, 0.001 M EDTA, *p*H 7.5)

transformed originally to cn-contours while the other half shows different degrees of transformation. The length distribution of the renatured forms corresponding to cn-contour (Fig. 9) is nearly identical with the frequency distribution of the length of cn-contour in the presence of 0.2 M ammonium acetate (Fig. 5). The percentage of molecules which are not completely retransformed to cn-state suggests a certain degree of specificity in the formation of a structure corresponding to the cn-contour. Compact contour forms, similar to those demonstrated here were also observed by DUBOCHET et al. [9] for messenger ribonucleoprotein particles (mRNP). If we assume that the structure of the mRNP particles is the naturally occurring structure, this similarity suggests a high degree of stability

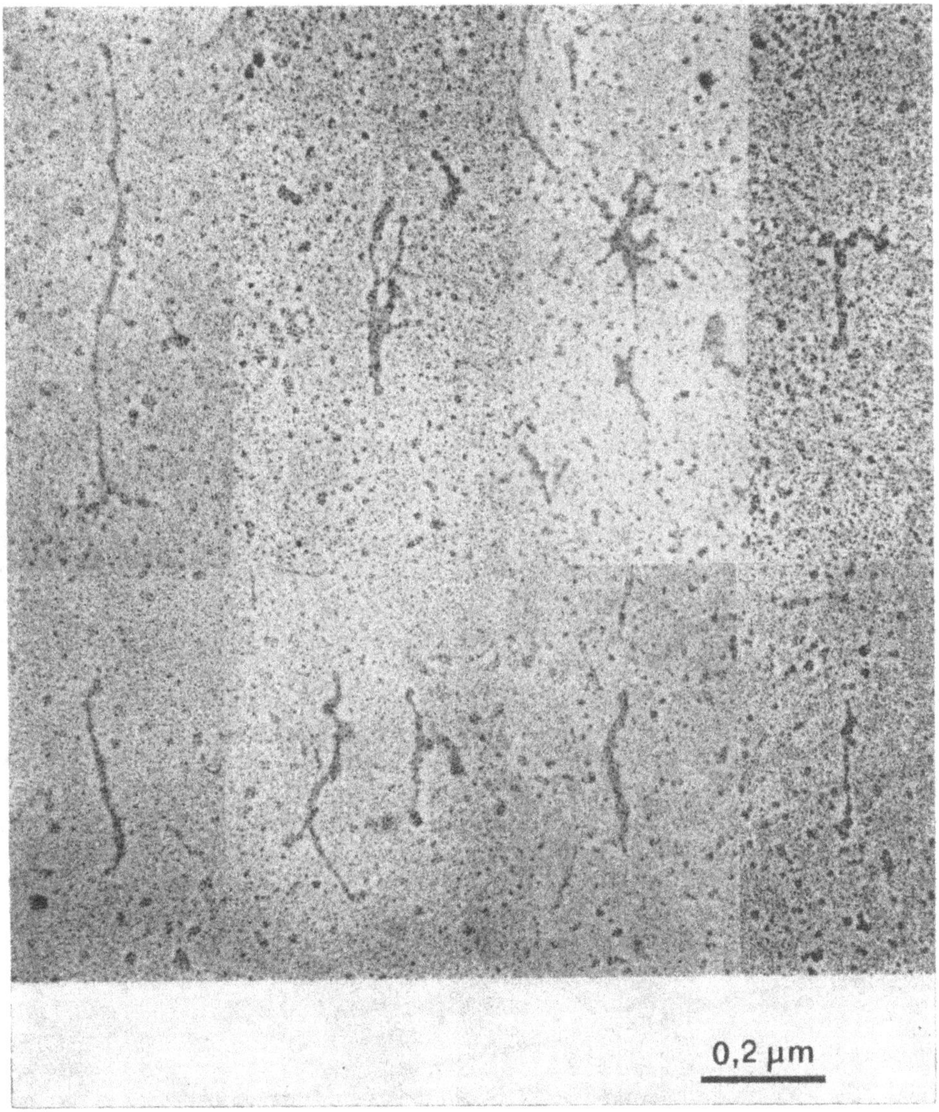

Fig. 8. Compact-nonlinear contour forms of globin mRNA after denaturation (compare legend to Fig. 6), precipitation by ethanol and solution in 0.2 M ammonium acetate. Doublo contrasting as described in legend to Fig. 4.

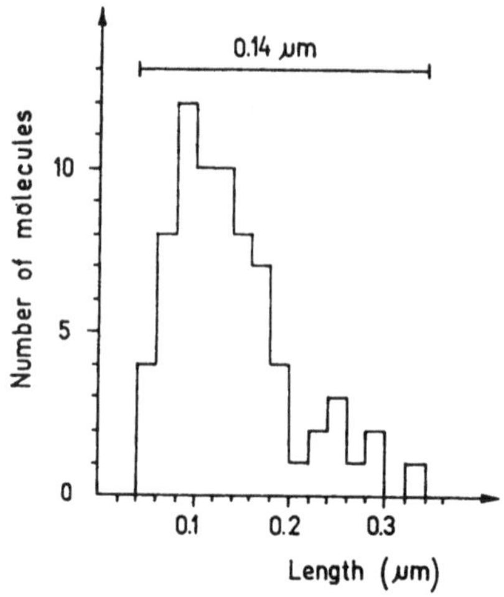

Fig. 9. Frequency distribution of the length of renatured molecules of globin mRNA characterized by a compact-nonlinear contour

of the higher order structure of the mRNA molecule, resisting even the phenol-SDS-deproteinization procedure. This suggestion is supported by the temperature denaturation experiments. Taking into account these results, one must assume that globin mRNA contains base paired sections with different degree of stability. Some could be melted at low temperature (up to 68 °C), others (most of them) need strong denaturation conditions.

Another surprising result is the length of the denatured 9 S globin mRNA molecules. Assuming a nucleotide distance of about 0.3 nm the 650 nucleotides of the globin mRNA correspond to a length of 0.2 µm. This value is found for the longest compact contours, but the mean value of the bulk of the denatured molecules is 0.44 µm, that means more than 100% higher. Although this value corresponds to the theoretical maximal length of a globin mRNA molecule stretched to Van der Waals radii [11], it should kept in mind that more than 60% of the measured molecules reached a length above 0.44 µm.

The main suggestions resulting from these investigations of the globin mRNA are: Firstly, either the nucleotide distance in RNA of about 0.3 nm is too small and should be at least twice as big, or the noncoding sequence of the globin mRNA is much larger than commonly assumed. Secondly, the very broad size distribution and especially the appearance of very long mRNA molecules could be due to the contamination of 9 S reticulocyte RNA with other nonglobin mRNA. The longer molecules could also be interpreted as a sign of further processing of mRNA in cytoplasm. This might be a peculiar event under conditions of extreme anaemia but could also be a general mechanism of post-transcriptional regulation of gene expression.

References

[1] Georgiev, G. P., A. P. Ryskov, Ch. Coutelle, V. L. Mantieva, and E. R. Avakyan: Biochim. biophys. Acta **259**, 259 (1972)

[2] Melli, M., and R. E. Pemberton: Nature New Biol. **236**, 172 (1972)

[3] Imaizumi, T., H. Diggelmann, and K. Scherrer: Proc. natn. Acad. Sci. U.S.A. **70**, 1122 (1973)

[4] WILLIAMSON, R., C. R. DREWIENKIEWICZ, and J. PAUL: Nature New Biol. **241**, 66 (1973)
[5] SCHERRER, K., and L. MARCAUD: J. Cell Physiol. **72**, 81 (1968)
[6] GEORGIEV, G. P.: J. theor. Biol. **25**, 473 (1969)
[7] LADHOFF, A.-M., B. J. THIELE, CH. COUTELLE, and S. ROSENTHAL: Acta biol. med. germ. **33**, 771 (1974)
[8] LADHOFF, A.-M., B. J. THIELE, and CH. COUELTLE: Eur. J. Biochem. **58**, 431 (1975)
[9] DUBOCHET, J., C. MOREL, B. LEBLEU, and M. HERZBERG: Eur. J. Biochem. **36**, 465 (1973)
[10] RYSKOV, A. P., D. A. KRAMEROV, S. A. LIMBORSKAJA, and G. P. GEORGIEV: Molek. Biol. USSR **9**, 6 (1975)
[11] FAVRE, A., C. MOREL, and K. SCHERRER: Eur. J. Biochem. **57**, 147 (1975)

Acta biol. med. germ., Band 36, Seite 315—317 (1977)

[1] Institut für Physiologische und Biologische Chemie
der Humboldt-Universität, 104 Berlin, DDR
[2] Zentralinstitut für Molekularbiologie der
Akademie der Wissenschaften der DDR, 1115 Berlin-Buch
[3] Institute of Molecular Biology of the Academy of Sciences of the USSR, Moscow

Specific sequences in pre-mRNA from erythroid bone marrow cells

B. J. Thiele[1], Ch. Coutelle[2], H.-D. Hunger[2], and A. P. Ryskov[3]

Since it could be shown that the mRNA of erythroid cells contains molecules carrying covalently bound globin coding sequences, it is well established that globin mRNA is synthesized as a high molecular weight precursor [1—3]. The conversion of the large primary transcripts to functional cytoplasmic mRNA ("processing") is one of the main problems in posttranscriptional regulation of gene expression [4].

We have investigated sequence elements of pre-mRNA which are possibly related to the processing of the large precursor molecules (oligo(U)- and double stranded sequences) and to the conversion of the translatable mature mRNA (poly(A)-sequence).

For our investigation on the structure and fate of the precursors of mRNA we have used rabbit bone marrow enriched in erythroid cells up to 80% of the nucleated cells by phenylhydrazine treatment [5].

Pre-mRNA fractions were obtained either by the hot phenol fractionation procedure introduced by Georgiev and Mantieva [6] or by the isolation of poly(A)-carrying molecules or molecules >45 S from pulse labelled total cellular RNA [7, 8].

Poly(A)-, oligo(U)- and double stranded sequences were isolated by T1- and pancreatic RNase treatment and separation of the sequences by poly(A)- and poly(U)-sepharose chromatography respectively [9].

To compare the poly(A)- and oligo(U)-sequences isolated from erythroid bone marrow cells with sequences of pre-mRNA from other origin, base analysis and electrophoresis were performed [9]. The poly(A)-segment consists at least of 96.5% A. Poly(A) from poly(A)$^+$ pre-mRNA (isolated from total cellular RNA extracted with phenol/SDS/urea at room temperature) shows a size distribution between 40 and 130 nucleotides with an average of 75. Hot phenol extraction of pre-mRNA however leads to a degradation to a size of about 25 nucleotides.

Sheiness et al. [10] have shown that poly(A) appears in the nucleus of HeLa cells as a relatively homogeneous unit of about 200 nucleotides. Our experiments suggest that this initial poly(A) in pre-mRNA of erythroblasts is not longer than 130 nucleotides.

For the determination of the molecular weights standards of definite length were used, because it could be shown that the use of heteropolymers (like tRNA

Abbrevations: hnRNA = heterogeneous nuclear RNA; pre-mRNA = precursor of messenger RNA; ds-RNA = double stranded RNA and 5 S RNA) as markers leads to an overestimation of the length of the poly(A)-sequence [11]. We obtained quite similar results, also for oligo(U). These differences in migration could be due to the more compact secondary structure of heteropolymers rather than to differences in the charge of the bases.

The oligo(U)-segment of erythroblasts pre-mRNA consists of 80% of U and is about 25 nucleotides long. Poly(A)$^+$ pre-mRNA of about 12,000—16,000 nucleotides contains 1—2 oligo(U)-sequences. No oligo(U) was detectable in poly(A)$^+$ pre-mRNA of about 1700 nucleotides as well as in mature globin mRNA. This suggests its localization to be at least 1700 nucleotides away from the 3'end of the long pre-mRNA molecules.

The occurrence of oligo(U)- sequences in mRNA of HeLa and hepatoma cells were first described by MOLLOY et al. [12] and by SHENKIN and BURDON [13]. The oligo(U) sequence obtained from the pre-mRNA of our more specialized cells shows no significant differences neither in the percentage of the total RNA nor in the base composition or the length of the oligonucleotide in comparison to their investigations. This suggests a universal signal function of the oligo(U)-segment.

Since the discovery of poly(A)-sequences in pre-mRNA and mRNA a possible regulatory function related to transport, translation or stability has been discussed. Experiments of MARBAIX et al. [14] have ensured that the poly(A)-segment is responsible for mRNA stability.

We have carried out some experiments regarding the functional meaning of the double stranded sequences in pre-mRNA [15]. For this purpose we studied the

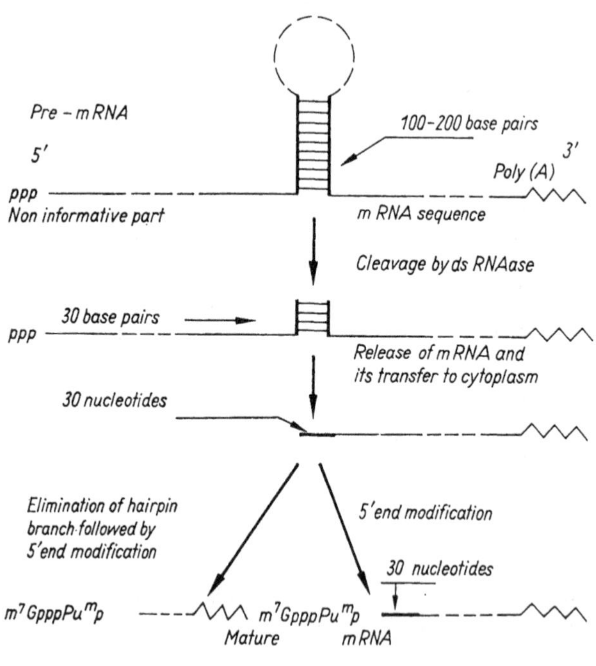

Fig. 1. A hypothetical model for the localization and functioning of double-stranded regions of pre-mRNA as a separator of mRNA from non informative part of the precursor

hybridization reaction of unlabelled globin mRNA from rabbit reticulocytes with a belled double strands from pre-mRNA of erythroid precursor cells.

lIt was previously shown that the cytoplasmic mRNA from mouse liver or Ehrlich carcinoma cells could form RNase-stable hybrids with melted ds-RNA sequences prepared from a nuclear pre-mRNA [16].

Long ds-sequences from erythroblast pre-mRNA were shown by electrophoresis to contain about 100 nucleotide pairs. Melted double strands were hybridized with an excess of globin mRNA. 7—9% of the double strands became RNase stable and about 30% of the sequences could be bound to poly(U)-sepharose through poly(A) of the mRNA. This shows that globin mRNA contains a sequence complementary to the double stranded region of the pre-mRNA molecule. The RNase stable hybrid is about 30 nucleotides long, i.e. it corresponds to one third of one strand of the ds RNA.

It is suggested that the double stranded hairpin-like structures formed the border lines between mRNA sequences and the noninformative part of the pre-mRNA. In the course of processing a part of the ds-region would be destroyed resulting in the separation of the mRNA (see Fig. 1). A part of one branch of the ds-RNA would remain bound to mRNA and could hybridizide with the ds-RNA from pre-mRNA.

References

[1] Imaizumi, T., D. Diggelmann, and K. Scherrer: Proc. natn. Acad. Sci. U.S.A. **70**, 1122 (1973)

[2] Williamson, R., C. E. Drewienkiewicz, and J. Paul: Nature New Biol. **241**, 66 (1973)

[3] Macnaughton, M., K. B. Freeman, and J. O. Bishop: Cell **1**, 117 (1974)

[4] Scherrer, K. in: Protein Synthesis in Reproductive Tissue. 6. Symposium, Geneva 1973 (= Karolinska Symposia on Research Methods in Reproductive Endocrinology). E. Diczfalusy (ed.). Acta Endocrinol. Suppl. **180**, 95 (1973)

[5] Coutelle, Ch., H. H. Reinecke, E. Steindamm, W. Meurer, M. Grieger, and S. Rosenthal: Haematologia **9**, 195 (1975)

[6] Georgiev, G. P., and V. L. Mantieva: Biokhimiya **28**, 949 (1962)

[7] Holmes, D. S., and J. Bonner: Biochemistry **12**, 2330 (1973)

[8] Thiele, B. J., B. Papies, and Ch. Coutelle: In Vorbereitung

[9] Thiele, B. J., Ch. Coutelle, H.-D. Hunger and A. P. Ryskov: In Vorbereitung

[10] Sheiness, D., L. Puckett, and J. E. Darnell: Proc. natn. Acad. Sci. U.S.A. **72**, 1077 (1975)

[11] Burness, A. T. H., I. U. Pardoe and N. O. Goldstein: Biochem. Biophys. Res. Commun. **67**, 1408 (1975)

[12] Molloy, G. R., W. Jelinek, M. Salditt and J. E. Darnell: Cell **1**, 43 (1974)

[13] Shenkin, A., and R. H. Burdon: FEBS Lett. **22**, 157 (1972)

[14] Marbaix, G., G. Huez, A. Burny, Y. Cleuter, E. Hubert, M. Leclercq, H. Chantrenne, H. Soreq, U. Nudel and U. Z. Littauer: Proc. natn. Acad. Sci. U.S.A. **72**, 3085 (1975)

[15] Ryskov, A. P., O. V. Tokarskaya, G. P. Georgiev, Ch. Coutelle and B. J. Thiele: Nucl. Acids Res. **3**, 1487 (1976)

[16] Ryskov, A. P., S. A. Limborska and G. P. Georgiev: Molec. biol. Rep. **1**, 215 (1973)

Acta biol. med. germ., Band 36, Seite 319—321 (1977)

[1] Department of Molecular Biology, University of Brussels, Belgium and
[2] Neurobiology Unit, Weizmann Institute of Science, Rehovot, Israel

Role of the polyadenylate sequence in the stability of globin messenger RNA injected into Xenopus oocytes

G. Marbaix[1], G. Huez[1], A. Burny[1], E. Hubert[1], M. Leclercq[2], Y. Cleuter[1], H. Chantrenne[1], H. Soreq[2], U. Nudel[2], and U. Littauer[2].

Summary

It is shown that the polyadenylate sequence covalently bound to the 3'-OH end of globin messenger RNA is responsible for the stability of the message when translated in Xenopus oocytes.

The existence of polyadenylate sequences (poly(A)) covalently bound at the 3'-OH end of most eukaryotic messenger RNAs (mRNAs) is well documented [1]. It seems that these segments become shorter with ageing of mRNA and this was clearly proven in the case of mouse [2, 3] and rabbit [4, 5] globin mRNAs. However, their precise role remained so far obscure. Since the mRNAs coding for histones do not appear to contain poly(A), it seems unlikely that the presence of these sequences is required for translation.

We considered the hypothesis that the length of the poly(A) segment is related to the half-life of the mRNA in a given system. One would therefore predict that poly(A)-lacking mRNA molecules would have a much shorter half-life than those containing a poly(A) stretch. A specific method was developed for removal of the 3'-OH poly(A) sequence from globin mRNA. It is based on synchronous processive phosphorolysis of mRNA using molar excess of E. coli polynucleotide phosphorylase at 0 °C in high ionic strength conditions [6]. Rabbit poly(A)-free globin mRNA obtained by this method could still be translated in ascites cell-free extracts but after long periods of incubation, the template activity of poly(A)-free mRNA leveled off earlier than that of native mRNA. This result supported the assumption that poly(A) is not required for translation but could ensure the functional stability of the message.

Translation of globin mRNA in the ascites extract was active only for less than 90 minutes. To overcome the time limitation of this in vitro protein synthesizing system, we decided to compare the translation of intact and poly(A)-free globin mRNA in the in vivo system constituted by the oocytes of the frog Xenopus laevis [7]. Microinjection into oocytes permitted the use of a living cell translation machinery to compare the translational efficiency of the two globin mRNA preparations for periods of time as long as several days [8]. In this system, the initial rate of translation of poly(A)-free mRNA was close to that found for intact mRNA. However, after long incubation periods of injected oocytes (two days), the rate of globin synthesis with poly(A)-free mRNA was more than ten fold lower than with native mRNA [9]. One could thus conclude that the presence of the 3'-OH poly(A) sequence in globin mRNA is required to ensure its high functional stability.

One could however argue that the enzymatic treatment which was used for poly(A) removal had perhaps also degraded some other structure in the message molecule and that this hypothetical structure was in fact responsible for the stability of globin mRNA. To test this possibility, a poly(A) sequence was resynthesized onto globin mRNA from which the poly(A) had been previously removed. This was achieved using an ATP/RNA adenyltransferase from E. coli [10]. The functional stability of this "reconstituted" globin mRNA was compared to that of poly(A)-free mRNA in Xenopus oocytes. One clearly observed that the functional stability of the message was recovered when a poly(A) segment was readded to poly(A)-free mRNA [11]. This experiment definitely proved that the poly(A) sequence itself is responsible for the functional stability of globin mRNA in Xenopus oocytes. However, how translation of poly(A)-free globin mRNA was prevented after long periods of time was unknown. One possibility was that removal of the poly(A) segment reduced the stability of the message molecule and led to its degradation. Alternatively, some other process might have inactivated poly(A)-free mRNA and have interfered with its translation.

To answer this question, we quantitated the amount of globin mRNA remaining in oocytes incubated after micro-injection of poly(A)-free or native mRNA preparations. This was performed by molecular hybridization with a radioactive complementary DNA probe synthesized onto globin mRNA by purified reverse transcriptase from avian myeloblastosis virus. The results of this experiment showed that 56 hrs after injection into oocytes, about 85% of poly(A)-free mRNA molecules were degraded while native, poly(A)-containing mRNA chains were almost completely preserved during the same period of time [12]. This proves that the poly(A) stretch is required to ensure the stability of the mRNA molecule itself in this cytoplasmic environment.

We also compared the functional stabilities in Xenopus oocytes of globin mRNA preparations having well-defined, decreasing sizes of poly(A) segments. For the considered period of incubation of oocytes (two days) we observed that the stability remained constant for all mRNA preparations having a poly(A) stretch of more than 30 nucleotides. However, below a threshold which lies around 15 nucleotides, the poly(A) sequence did not ensure the stability of the message molecule any more [13]. A minimum length of the 3'-OH poly(A) segment is thus required to ensure its protective function.

The above described experiments have been performed with haemin injected at the same time as globin mRNA since this compound is required for an efficient translation of a message [14]. This fact offered the opportunity to decide if poly(A)-free globin mRNA degradation in Xenopus oocytes depended on its translation.

For that purpose, we compared the extent of degradation of poly(A)-free globin mRNA enriched in a message if injected with or without haemin. The amount of a enriched globin mRNA remaining after 72 hrs in incubated oocytes was estimated by molecular hybridization as previously described. We observed that 87% of poly(A)-free enriched mRNA was degraded at the end of the incubation period if haemin was present whilst in the absence of haemin, but 40% of poly(A)-free mRNA had disappeared [15].

As haemin per se has no degradative effect on injected native globin mRNA, the most straightforward interpretation of this result is that, in Xenopus oocytes, poly(A)-free globin mRNA must be translated for its physiological degradation to occur.

Acknowledgements

This work has been supported by grants of the "Fonds Cancérologique de la Caisse Générale d'Epargne et de Retraite", of the "US National Cancer Institute, Contract NO1CP 33220" and of the "US-Israel Binational Science Foundation".

References

[1] Darnell, J. E., W. R. Jelinek, and G. R. Molloy: Science, N.Y. **181**, 1215 (1973)
[2] Gorski, J., M. Morrison, C. Merkel, and J. Lingrel: Nature, Lond. **253**, 749 (1975)
[3] Merkel, C., S. Kwan, and J. Lingrel; J. biol. Chem. **250**, 3725 (1975)
[4] Nokin, P., G. Huez, G. Marbaix, A. Burny, and H. Chantrenne: Eur. J. Biochem. **62**, 509 (1976)
[5] Nokin, P., A. Burny, H. Huez, and G. Marbaix: Eur. J. Biochem. (in press)
[6] Soreq, H., U. Nudel, R. Salomon, M. Revel, and U. Z. Littauer: J. molec. Biol. **88**, 233 (1974)
[7] Gurdon, J. B., C. D. Lane, H. R. Woodland, and G. Marbaix: Nature, Lond. **233**, 177 (1971)
[8] Gurdon, J. B., J. B. Lingrel, and G. Marbaix: J. molec. Biol. **80**, 539 (1973)
[9] Huez, G., G. Marbaix, E. Hubert, M. Leclercq, U. Nudel, H. Soreq, R. Salomon, B. Lebleu, M. Revel, and U. Z. Littauer: Proc. natn. Acad. Sci. U.S.A. **71**, 3143 (1974)
[10] Sippel, A.: Eur. J. Biochem. **37**, 31 (1973)
[11] Huez, G., G. Marbaix, E. Hubert, Y. Cleuter, M. Leclercq, H. Chantrenne, R. Devos, H. Soreq, U. Nudel, and U. Z. Littauer: Eur. J. Biochem. **59**, 589 (1975)
[12] Marbaix, G., G. Huez, A. Burny, Y. Cleuter, E. Hubert, M. Leclercq, H. Chantrenne, H. Soreq, U. Nudel, and U. Z. Littauer: Proc. natn. Acad. Sci. U.S.A. **72**, 3085 (1975)
[13] Nudel, U., H. Soreq, U. Z. Littauer, G. Marbaix, G. Huez, M. Leclercq, E. Hubert, and H. Chantrenne: Eur. J. Biochem. **64**, 115 (1976)
[14] Giglioni, B., A. M. Gianni, P. Comi, S. Ottolenghi, and D. Rungger: Nature New Biol. **246**, 99 (1973)
[15] Huez, G., G. Marbaix, A. Burny, Y. Cleuter, M. Leclercq, E. Hubert, and H. Chantrenne: Archs int. Physiol. Biochim. (in press)

Address of the authors: 1. Department of Molecular Biology, University of Brussels. B-1640 Rhode St-Genèse, Belgium; 2. Neurobiology Unit, The Weizmann Institute of Science, P.O.B. 26, Rehovot, Israel

Acta biol. med. germ., Band 36, Seite 323—333 (1977)

[1] Institute of Molecular Biology, Academy of Sciences, Moscow, USSR
[2] Zentralinstitut für Molekularbiologie der Akademie der Wissenschaften der DDR, 1115 Berlin-Buch, DDR

Synthesis and characterization of globin DNAs

[1]L. Yu. Frolova, [1]A. V. Tennov, [1]N. A. Scobeleva, [1]L. L. Kisselev, [2]V. Hahn, [2]Ch. Coutelle, and [2]H. Grützmann

Introduction

A deep insight into the problem of gene expression and gene structure is intimately connected with the necessity to isolate and to investigate appropriate individual genes. The isolation of genes from mammalian and other higher organisms remained underdeveloped although tremendous success has been achieved for prokaryotic and simple eukaryotic systems due to cloning of recombinant DNA molecules.

An apparent paradox consists in the fact that at the moment it is more realistic to synthesize mammalian genes than to isolate them.

Red blood cells highly specialized in the production of one type of protein and hence one dominant type of mRNA, namely globin and globin mRNA, is a unique object for the attempts to synthesize genes by means of the gene product, mRNA. Since mRNA is not a transcriptional but a posttranscriptional product the gene to be synthesized should correspond to a structural gene, not to a complete one including regulatory regions.

The way from mRNA back to DNA was discovered by Temin and Mizutani [1] and Baltimore [2] when they found a new enzyme, RNA-dependent DNA-polymerase as an integral part of the oncogenic RNA-containing viruses.

This enzyme which we call after Engelhardt "revertase" is at the moment a key component for all enzymatic gene-synthesizing systems.

In 1972 it was discovered in three laboratories simultaneously that poly(A)-containing mRNA could be used for the synthesis of complementary DNA strand [3—5]. At that time as well as now the globin messenger was the best known mRNA obtained in a pure state and preparative quantities in a number of places. In these first experiments the DNA copies (called cDNA) were probably incomplete in most cases and only single-stranded DNA was obtained.

Later with the improvement of the technology for RNA-directed DNA synthesis it became possible to obtain full-length copies of globin and other mRNAs [6—11]. In essence, the most important observations here were the use of high deoxyribonucleoside triphosphate concentrations in incubation mixture, low ionic strength and the enzyme fully deprived of nuclease contaminants. Besides, the quality of mRNA preparation is of key importance. In addition to the absence of hidden breaks in polynucleotide chain and the presence of a sufficiently long poly(A) sequence at 3′ terminus a complete removal of potentional revertase inhibitors is necessary. Among them are polyvinylsulphate, heparin, dodecylsulphate, bentonite and diethylpyrocarbonate used as nuclease inhibitors during isolation of mRNA preparation and poly(U) which itself inhibits revertase [12] and covers also the poly(A) portion of mRNA preventing the effective binding of the oligo(dT) primer. Poly(U) appears in preparations after the passage of mRNA through the bioaffinity columns containing poly(U).

However, the problem of synthesis of double-stranded full-length DNAs which might be considered as structural genes was not solved at that time.

Only in 1976 a few laboratories succeeded in enzymatic synthesis of true structural genes. Again the first structural gene to be synthesized in a test tube was a globin gene.

In principle, various ways may be used for the enzymatic synthesis of genes based on the ability of revertase to catalyze DNA synthesis complementary to RNA sequence which serve as template.

Fig. 1 described schematically possible ways for gene synthesis beginning with pure mRNA.

In all the three versions (A, B and C) the first step is performed in the presence of revertase isolated from an oncorna virus and oligo(dT) which serves as a primer.

Among the revertase transcripts full-length copies were also obtained [6, 11]. It is noteworthy that these full-length single-stranded globin cDNAs contain a small fraction of nuclease S_1 resistant, i.e. double-helical sequences [6, 13], probably hairpins. Earlier, in 1972 Leis and Hurwitz [14] suggested that a 3'-terminal hairpin is generated in the course of reverse transcription. They assumed also that this short terminal hairpin might be further extended by revertase to form a double-stranded provirus DNA sequence in vivo.

Scheme A (Fig. 1) corresponds to what was proposed but not proved by Leis and Hurwitz. Scheme B is close to scheme A. The main difference is connected with the generation of the second DNA strand: in A it is made by revertase whereas in B the second enzyme, DNA-dependent DNA-polymerase I (Pol I) from E. coli fulfills the elongation of the second strand.

In both schemes a double-stranded full-length structural gene is generated with the two chains covalently linked at one side ("hairpin gene"). To convert the hairpin gene into the "open" form the short loop between two strands might be split by S_1 nuclease.

Scheme B was used very recently by Efstratiadis et al. [15] to synthesize globin gene.

Scheme C, at variance with A and B, includes generation of a new initiation point for the second strand synthesis. Terminal nucleotidyltransferase is used to make a short homopolymeric sequence at the 3'end of the cDNA molecules. After this terminal addition reaction a complementary primer is added and DNA synthesis is performed by means of DNA-polymerase I. In scheme C the third enzyme, terminal transferase, is engaged in the process of synthesis but the S_1 nuclease step is omitted.

Fig. 1. Principal pathways for in vitro enzymatic synthesis of structural genes ▶

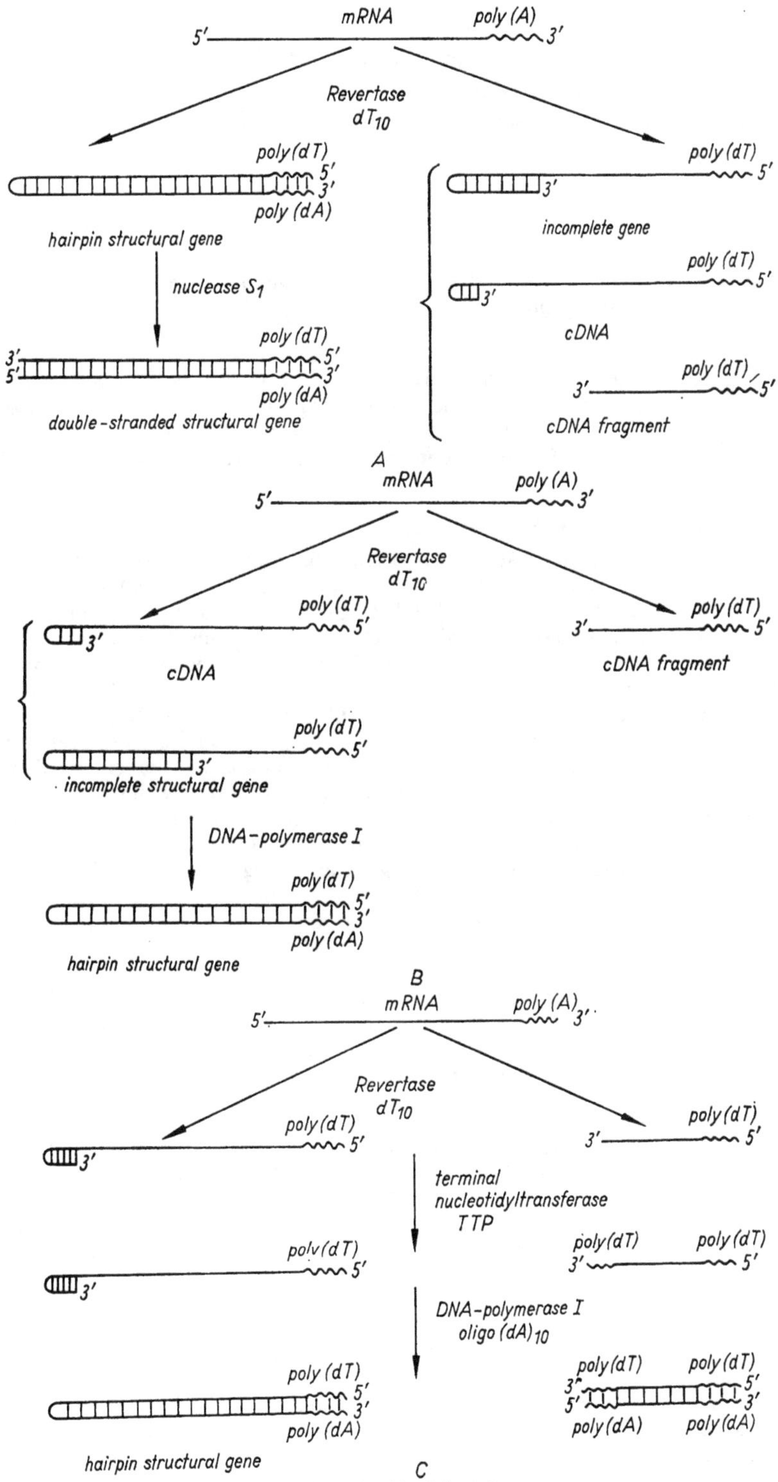

Legende siehe Seite 324

Scheme C has been utilized by Rougeon et al. [16] for insertion of a portion of mammalian gene into plasmid, however, only short DNA pieces were successfully incorporated into recombinant DNA molecule.
The aim of the present work was to test experimentally all the above mentioned schemes (Fig. 1).

Experimental procedures

Purification of globin mRNA

Rabbit globin mRNA was isolated as described [21] by oligo(dT)-cellulose chromatography and sucrose gradient centrifugation.

Enzymes

Avian myeloblastosis virus (AMV) DNA-polymerase (revertase) was isolated and purified as described [22]. Single stranded specific nuclease-S_1 from takadiastase was purified according to the procedure described in [23]. DNA-dependent DNA-polymerase I from E. coli was obtained from Boehringer. Calf thymus deoxynucleotidyl terminal transferase (TdT) was purified according to [24].

Enzymatic synthesis of DNA on RNA template

The reaction mixture (100 µl) consisted of the following components: Incubation mixture 1: 50 mM Tris-HCl, *p*H 8.3, 6 mM $MgCl_2$, 4 mM dithiothreitol (DTT), 40 mM KCl, 0.1 mM (each) dATP, TTP, dGTP and 0.02 mM [^{3}H] dCTP (20—25 Ci/mmole), 100 µg/ml of actinomycin D, 5 µg/ml rabbit globin mRNA, 5 µg/ml of $(dT)_{10}$ and 30—40 units/ml of revertase. 1 unit of activity catalyzes the incorporation of 1 nmole TTP in TCA-precipitable fraction at 37 °C. Incubation mixture 2: the same as in incubation mixture 1, but mRNA = 20—30 µg/ml, oligo$(dT)_{10}$ = 25 µg/ml, 0.2 mM (each) of dATP, dGTP, TTP and [^{3}H]dCTP (1500 cpm/pmole) and in the absence of actinomycin, incubation at 37 °C for 1 hr. For preparative purposes the reaction mixture was scaled up to 5—10fold. Aliquots were precipitated with TCA and counted for determination of the yield. The mixture was treated with phenol and ethanol-precipitated without or after addition of tRNA-carrier in the case of incubation mixture 1. The ethanol-precipitated material was dissolved in water, hydrolyzed with 0.3 N KOH at 37 °C for 5 or 17 hrs, neutralized, passed through the Ultrogel column AcA34, equilibrated with 0.05 M triethyl-aminobicarbonate buffer, *p*H 6.7. The excluded material was evaporated and solubilized in deionized water.

Elongation of globin DNA

The reaction mixture (100 µl) contained: 0.1 M cacodylate, *p*H 7.2, 10 mM $ZnSO_4$, 1 mM $CoCl_2$, 0.2 mM [^{3}H]TTP (300—500 cpm/pmole), 60 units/ml of deoxynucleotidyl transferase, and DNA-template. The incubation was at 37 °C for 1 hr. The procedure for the purification and isolation of the product was the same as in the previous case.

Enzymatic synthesis of DNA complementary to cDNA

The reaction mixture (50 µl) consisted of the following components: 50 mM Tris-HCl, *p*H 8.0, 6 mM $MgCl_2$, 17 mM KCl, 4 mM DTT, 0.4 mM (each) dATP, dGTP, TTP and [^{14}C]dCTP (440 Ci/mole), 54 units/ml of E. coli DNA-polymerase, [^{3}H]dCMP DNA template or dT elongated [^{3}H]dCMP-DNA annealed with $(dA)_{10}$. The incubation was at 30 °C for 2 hrs. The isolation and purification of the product was the same as in the previous cases.

Polyacrylamide gel electrophoresis

Electrophoresis at 2.5 mA/tube was run on 4.5% polyacrylamide gels in 98% formamide according to [25]. Plastic tubes (10 × 0.5 cm) were used and gels were scanned by means of Joyce Loebl Scan 400. The samples were boiled for 2 min, cooled rapidly on ice and dissolved in deionised formamide before application. 4 S, 5 S and 16 S RNAs were used as markers. After scanning gels were frozen and cut into 2 mm slices. To each slice 0.5 ml of Protosol (NEN) was added, incubated for 1 hr at 55 °C, cooled, diluted with toluene scintillator and counted in an Intertechnique SL 30.

S_1 nuclease treatment

The reaction mixture (500 μl) consisted of 0.05 M NaCl, 0.03 M CH_3COONa, *p*H 4.6, 1 mM $ZnSO_4$, 5% glycerol, 20 μg/ml of denaturated DNA, and 600 units of activity of S_1 nuclease [26]; incubation was for 1 hr at 37 °C. The samples were precipitated with TCA, washed on millipore filters and counted.

Hydroxyapatite (HAP) chromatography

1 ml HAP (DNA grade, Biogel NTP, Bio-Rad) column was used. Fractions were eluted with 0.16 M and 0.4 M phosphate buffer, *p*H 6.8, at 60 °C. 1 ml fractions were collected on GF/C glass filters, dried, washed with TCA, and counted in scintillation spectrometer.

Results

Revertase catalyzed synthesis of globin hairpin gene

It has been demonstrated earlier in our laboratory [17] that revertase catalyzes the synthesis of the DNA fraction which contains poly(dA) and does not hybridize with mRNA. This observation was extended in this work via analysis of the length of the DNA products synthesized by revertase in the absence of actinomycin D.

To analyse the molecular weight of the fractions, polyacrylamide gel electrophoresis in 98% formamide has been used. It is known that under these conditions DNA and RNA chains are completely denatured [18].

Three main radioactive peaks are observed (Fig. 2). The first peak contains material the length of which is roughly equal to the double length of the mRNA used. The length of rabbit globin mRNA molecules was determined separately and found to be equal to 700 ± 50 nucleotides.

The second peak (Fig. 2) is rather heterogeneous and contains fractions with a length which corresponds to the full-length and incomplete cDNAs. The third peak is a relatively low-molecular one and contains short reverse transcripts.

This interpretation was strengthened further when the reaction products were separated on hydroxyapatite columns into two fractions: the first (single-stranded) fraction was eluted with 0.14—0.16 M phosphate buffer and the second (double-stranded, helical) fraction with 0.4 M phosphate. As shown in Fig. 3 the first

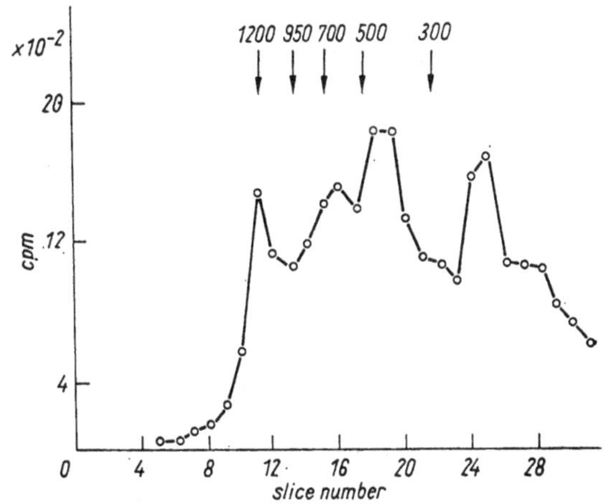

Fig. 2. Radioactivity of [³H]dCMP-cDNA electrophoresed on polyacrylamide gel.
DNA was synthesized by revertase with the rabbit globin mRNA as a template. Curve 1: incubation mixture 1; curve 2: incubation mixture 2 (see "Experimental procedures").

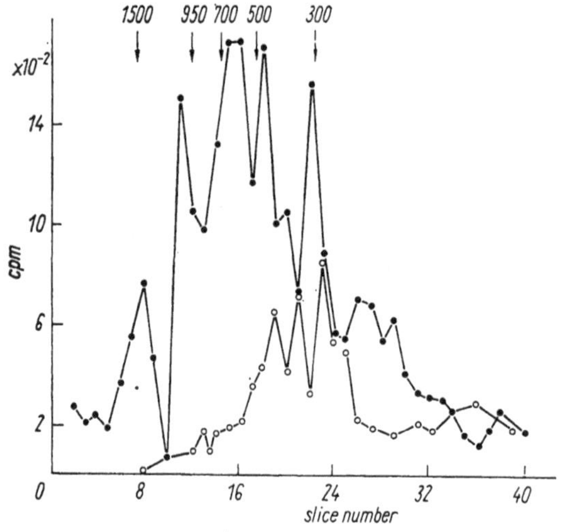

Fig. 3

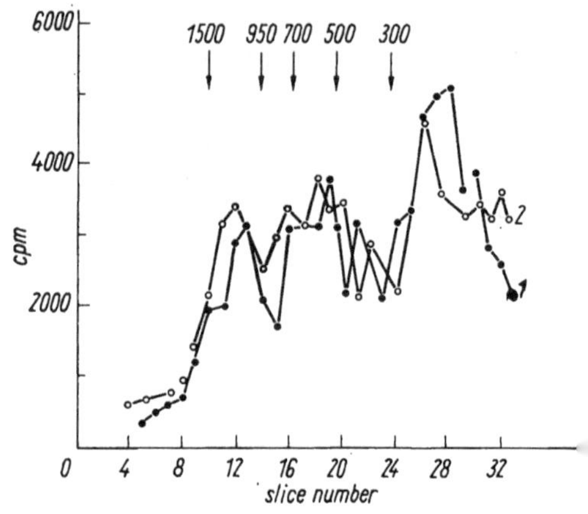

Fig. 4

Fig. 3. Radioactivity of [^{3}H]dCMP-DNA electrophoresed on polyacrylamide gel. DNA was synthesized by revertase in incubation mixture 2 and then eluted from HAP at 0.1M Na-phosphate (curve 1) and 0.4 M phosphate (curve 2)

Fig. 4. Distribution of [^{3}H]dCMP-DNA radioactivity in polyacrylamide gel. DNA was synthesized in incubation mixture 2

fraction contains no double-length transcripts and is presented mainly by relatively short chains. All the long transcripts are located in the "double-helical" fraction.

The conditions used in our work for DNA synthesis are not the same that were used by others [15]. The authors of this work were unable to reveal double-length transcripts with a single enzyme, revertase but only after Pol I-action on the cDNA. One may assume that in the experiments of Efstratiadis et al. [15] for reverse transcription, revertase is unable to catalyze the second strand formation. When we compared the products made at high and low triphosphate concentrations (Fig. 4, curves 1 and 2) the same double-length fraction was revealed and the whole pattern of radioactivity was quite similar. The only difference which we noticed was an increase in the proportion of high-molecular weight fractions accompanied by a decrease of the low-molecular weight peak.

Fig. 4 shows that DNA synthesis at high mRNA and nucleoside-triphosphate concentrations in the absence of actinomycin D proceeds in a step-wise manner. The first peak corresponds to a double length of mRNA, the second one to a single length, the third one to 2/3, and the last one to 1/3—1/4. We assume that the first peak is a hairpin gene whereas all the others reflect the cDNA synthesis of various length.

A discontinious character of DNA synthesis which was also observed by others [6] is probably caused by some structural features of the mRNA template. For example, long and GC-rich internal hairpins may produce "structural stops" where the rate of the synthesis is diminished. Although this question needs further clarification it might be a useful tool to study the secondary structure of mRNA.

To summarize, revertase alone is able to catalyze "hairpin gene" synthesis. The conversion of hairpin gene into the open form of the double-helix might be achieved as described by EFSTRATIADIS et al. [15].

DNA synthesis after B and C schemes

Scheme B (Fig. 1) realized for the first time by EFSTRATIADIS et al. [15] is based on a sequential action of two DNA-polymerases. The experiment was arranged in the following way. [^{3}H]dCMP-DNA was synthesized by revertase and isolated from incubation mixture. This preparation was used as a template-primer for DNA synthesis catalyzed by Pol I. In this case [^{14}C]dCTP was used as a labeled substrate.

The difference in distribution of [^{3}H] and [^{14}C] radioactivity after formamide gel electrophoresis is shown in Fig. 5. After the action of the second DNA-polymerase (curve 2) only one main peak of radioactivity is observed coinciding in its mobility with the double-length peak made by revertase. Incorporation into the shorter fractions is insignificant.

Two essential conclusions should be made from these data:

1) DNA chains synthesized by revertase can be elongated by Pol I. Since it is known that this enzyme catalyzes polymerization of nucleotides only in the presence of primer with a free 3'-OH group [19] it means that DNA fraction made by revertase contains internal primer, i.e. hairpins with unequal lengths where the longer side plays the role of template and the shorter one functions as a primer. This is in a good agreement with scheme B (Fig. 1).

2) The elongation of DNA chains proceeds nonrandomly since after Pol I action most of the chains, in spite of the initial heterogeneity, become equal to the double-length of globin mRNA. This observation proves also scheme A where it has been postulated that revertase acting alone produces those DNA molecules which differ in the length of the second DNA strand.

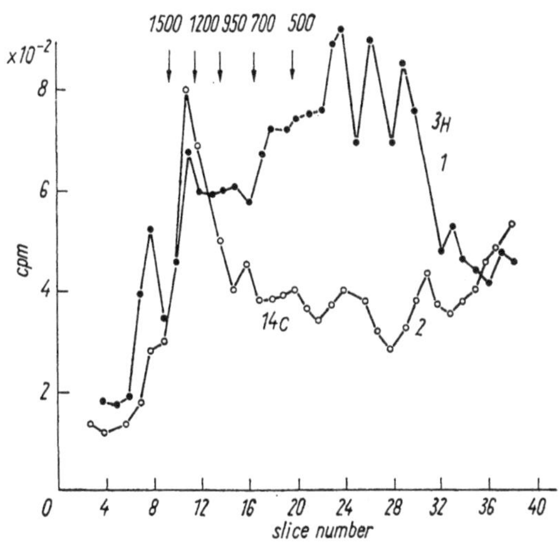

Fig. 5. Radioactivity profile of [^{3}H] dCMP-DNA fractions after gel electrophoresis. Curve 1 (control): the same as in Fig. 4.; curve 2: [^{14}C]dCMP-DNA synthesized by DNA-polymerase I in the presence of [^{3}H]dCMP-DNA (see curve 1)

From Fig. 5 it is clear that the role of Pol I consists mostly in elongation of those DNA chains which were not completed by revertase. It enhances the revertase action, and therefore improves the yield of the hairpin gene fraction.

Based on these data we synthesized DNA in a double polymerase system, then subjected the material to alkaline sucrose concentration gradient fractionation and recovered the peak from the region which corresponds according to the marker to the length of about 1200—1500 nucleotides. The isolated material was electrophoresed in gel in denaturating conditions (Fig. 6). It is obvious that the major fraction is about double-length of globin mRNA. The minor fraction (700 nucleotides) is presumably a cDNA fraction. After action of S_1 nuclease this fraction can be hydrolyzed. The resistance of the isolated product to S_1 nuclease was very high (85%). This is in agreement with the assumption that the major fraction is a self-complementary hairpin gene and the minor fraction is a single-stranded cDNA.

In conclusion, scheme B although more sophisticated methodologically than scheme A is efficient in producing the hairpin gene. The yield of the double-length product is much higher than after revertase action alone.

The common shortening of both schemes is connected with the necessity to use nuclease S_1, to split the loop and to open the hairpin. It may happen that during this procedure the hidden breaks are produced along the DNA helix.

Therefore it looks reasonable to try to eliminate this procedure due to its potentional hazard and to use the scheme C (Fig. 1) where nuclease action is avoided.

The initial step of the procedure was the same as in schemes A and B, namely DNA synthesis due to revertase action. After isolation of the product it was treated with deoxynucleotidyl terminal transferase in the presence of [^{3}H] TTP. The extent of the terminal addition reaction was followed after incorporation of [^{3}H]TTP into acid-precipitable material. The average chain length of the poly(T) at 3'ends was $\approx$50 nucleotides. After sequential action of revertase and terminal nucleotidyl transferase the preparation is heterogeneous (Fig. 7, curve 1) which is a super-

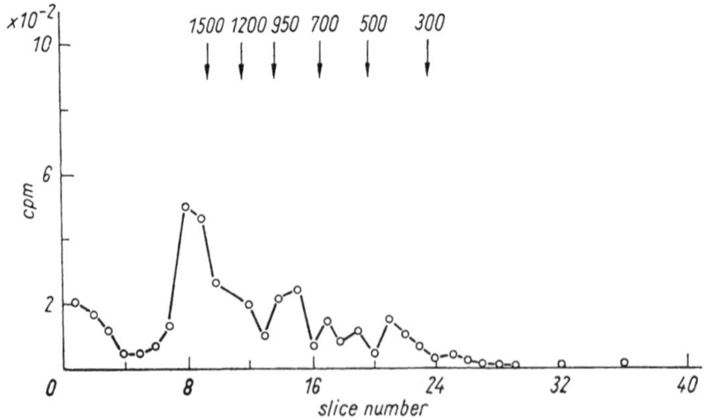

Fig. 6. Distribution of [^{3}H,^{14}C]dCMP-DNA radioactivity in polyacrylamide gel. [^{3}H,^{14}C]-dCMP-DNA fraction was synthesized by sequential action of revertase and DNA-polymerase I and isolated from a zone of alkaline sucrose centrifugation gradient containing polynucleotides of 1200—1400 nucleotides in length

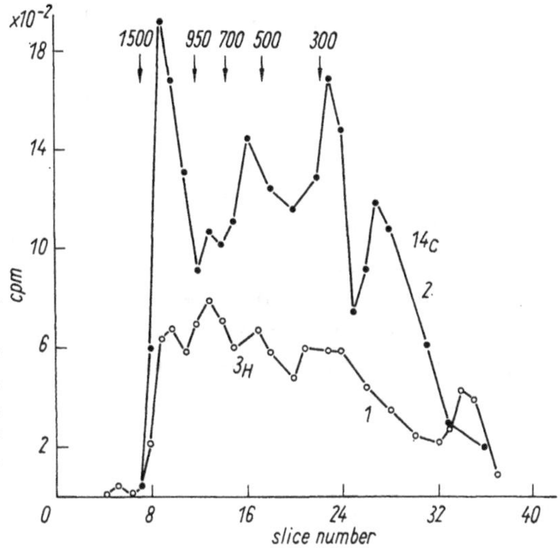

Fig. 7. Radioactivity profile for DNAs after gel electrophoresis.
Curve 1: globin [^{3}H dCMP-DNA] synthesized by revertase and elongated with terminal nucleotidyl transferase; curve 2: [^{14}C]dCMP-DNA synthesized by DNA-polymerase I in the presence of "elongated DNA" (curve 1)

position of heterogeneity produced by revertase and by transferase. Again as in all other previous cases, a heavy peak is seen (hairpin gene) and the fraction slightly longer than cDNA.

After addition of Pol I together with [^{14}C]dCTP, other triphosphates and oligo(dA) an interesting length distribution was observed (Fig. 7, curve 2). A large peak appears which corresponds to the double-length of mRNA. This is expected from scheme B, since elongation of DNA should be related to the presence of internal but not an external primer. A successful elongation of the fraction of DNA molecules after action of terminal nucleotidyltransferase means that this fraction does not serve as a primer for homopolymer elongation reaction catalyzed by terminal transferase. It is known [20] that this enzyme has a strong preference for single-stranded primers, whereas 3'-OH groups involved in the secondary structure are very poor primers. The inability of the terminal transferase to prevent the elongation of DNA chains by Pol I is in favour of the notion that the 3'-OH group of DNA is a part of a double-helical hairpin region. This statement has been already made on the basis of all other data mentioned above.

On the other hand significant incorporation of [^{14}C]-label into shorter fractions indicates that in this case the external primer was used since in its absence (Fig. 5) Pol I does not considerably synthesize short DNA chains, in other words the second strand of DNA was made in those cases when poly(T) appeared at 3'-end of cDNAs. In turn those cDNAs which serve as a primer for terminal transferase have single-stranded 3'-termini and therefore have no hairpin-like structure at the 3'-ends.

From comparison of Fig. 7 and Fig. 5 one may conclude that two simultaneous processes take place after addition of Pol I to the "elongated" DNA preparation: 1) elongation of preexisting hairpin-containing DNA fractions, 2) synthesis of separate strands of DNAs complementary to DNAs elongated with poly(T). In the first case a full-length hairpin gene is formed whereas in the second one the double-stranded pieces of globin structural gene are made varying in length in a wide range. The last statement is correlated with high S_1 nuclease resistance of these fractions, $\approx 70\%$.

Further treatment of the incubation mixture can be made in two different ways. After S_1 nuclease digestion all fractions should be converted into open (double-stranded) form and then used directly for producing recombinant DNA molecules or fractionated according to the chain length. Alternatively, the mixture can be separated according to the molecular weights and then used separately.

It should be stressed that scheme C is a rather complicated one since it is necessary to isolate the products three times: after revertase, terminal transferase and Pol I treatments leading to considerable loss of material. However, the advantage of this scheme consists in obtaining DNA products usefull for various types of further experimentation.

Conclusions

Three schemes for in vitro enzymatic synthesis of double-stranded full-length structural genes are considered. Each of the scheme was tested experimentally using rabbit globin mRNA as a starting material.

The common feature of all the protocols applied consists in exploiting the ability of RNA-directed DNA-polymerase (revertase) from avian myeloblastosis virus to copy mRNA chains in appropriate conditions via reverse transcription.

It is shown that (1) revertase catalyzes the mRNA-directed synthesis of DNA chains which are twice as long as mRNA template. This fraction is presumably a hairpin gene, i.e. full-length double-helical globin structural gene with a loop linking together the two strands. (2) The same fraction can be obtained by sequential action of revertase and DNA-dependent DNA-polymerase I (Kornberg enzyme) and the yield of the product is higher than after revertase alone. (3) A mixture of hairpin and double-stranded forms of globin structural gene is obtained after sequential action of revertase, terminal nucleotidyl transferase and DNA-polymerase I. In case of "open" forms a large portion of them is incomplete.

This work has been performed under the scope of "Revertase" project. The authors are very grateful to Professor J. Řiman for his kind gift of chicken plasma containing avian myeloblastosis virus, to Dr. V. Prasolov who provided us with terminal nucleotidyl transferase and to Dr. T. Avdonina for her participation in some experiments.

References

[1] Temin, H. M., and S. Mizutani: Nature, Lond. **226**, 1211 (1970)
[2] Baltimore, D.: Nature, Lond. **226**, 1209 (1970)
[3] Verma, I. M., G. F. Temple, H. Fan, and D. Baltimore: Nature New Biol. **235**, 163 (1972)
[4] Kacian, D. L., S. Spiegelmann, A. Bank, M. Terada, S. Metafora, and P. A. Marks: Nature New Biol. **235**, 167 (1972)
[5] Ross, J., H. Aviv, E. M. Scolnick, and P. Leder: Proc. natn. Acad. Sci. U.S.A. **69**, 264 (1973)
[6] Efstratiadis, A., T. Maniatis, F. C. Kafatos, A. Jeffrey, and J. N. Vournakis: Cell **4**, 367 (1975)
[7] Junghans, R. P., P. H. Duesberg, and C. A. Knight: Proc. natn. Acad. Sci. U.S.A. **72**, 4895 (1975)
[8] Kisselev, L. L., A.-L. Haenni, and F. Chapeville: FEBS Lett. **62**, 64 (1976)

[9] MONAHAN, J. J., S. E. HARRIS, S. L. WOO, D. L. ROBBERSON, and B. W. O'MALLEY: Biochemistry **15**, 223 (1976)
[10] ROTHENBERG, E., and D. BALTIMORE: J. Virol. **17**, 168 (1976)
[11] FROLOVA, L. YU., A. V. TENNOV, K. G. GASARYAN, V. Z. TARANTUL, YU. N. BARANOV, CH. COUTELLE, H. GRÜTZMANN, V. HAHN, F. HIEPE, N. A. GRAEVSKAJA, and L. L. KISSELEV: Molec. Biol. USSR **10**, 944 (1976)
[12] SRIVASTAVA, B. I. S.: Biochim. biophys. Acta **335**, 77 (1973)
[13] GASARYAN, K. G., V. Z. TARANTUL, YU. N. BARANOV, L. YU. FROLOVA, and L. L. KISSELEV: Molec. biol. Rep. **1**, 465 (1974)
[14] LEIS, J. P., and J. HURWITZ: J. Virol. **9**, 130 (1972)
[15] EFSTRATIADIS, A., F. C. KAFATOS, A. M. MAXAM, and T. MANIATIS: Cell **7**, 279 (1976)
[16] ROUGEON, F., P. KOURILSKY, and B. MACH: Nucl. Acids Res. **2**, 2365 (1975)
[17] FROLONA, L. YU., YA. MELDRAYS, and L. L. KISSELEV: Dokl. Akad. Nauk SSSR **225**, 1142 (1975)
[18] PINDER, J. C., D. Z. STAYNOV, and W. B. GRATZER: Biochemistry **13**, 5373 (1974)
[19] KORNBERG, A.: DNA synthesis. Freman and Company, San Francisco 1974
[20] BOLLUM, F. J. in: Enzymes. Vol. **10**, p. 145. P. D. BOYER (Ed.). Academic Press, New York, London 1974
[21] LADHOFF, A., B. THIELE, and CH. COUTELLE: Eur. J. Biochem. **58**. 431 (1975)
[22] KISSELEV, L. L., L. YU. FROLOVA, K. G. GASARYAN, V. Z. TARANTUL, and V. A. ENGELHARDT: Dokl. Akad. Nauk SSSR **213**, 1203 (1973)
[23] MECHALI, M., A. M. RECONDO, and M. GIRARD: Biochem. biophys. Res. Commun. **54**, 1306 (1973)
[24] BOLLUM, F. J., L. M. S. CHANG, C. M. TSIAPALIS and J. W. DORSON: Meth. Enzymol. **29**, 70 (1974)
[25] MANIATIS, T., A. JEFFREY, and H. VAN DE SANDE: Biochemistry **14**, 3787 (1975)
[26] VOGT, V. M.: Eur. J. Biochem. **33**, 192 (1973)

Acta biol. med. germ., Band 36, Seite 335—339 (1977)

Laboratory of Biophysics, Institute of Medical Genetics, Academy of Medical Sciences, Moscow, USSR, and
Laboratory of Protein Synthesis Enzymology, Institute of Molecular Biology, Academy of Science, Moscow, USSR

Isolation of human globin mRNA and synthesis of complementary DNA

S. A. Limborska and L. Yu. Frolova

There are two alternative approaches towards the preparation of single genes: gene isolation and gene synthesis. Cesium salt gradient centrifugation of DNA fragments and hybridization of DNA in excess of specific RNA are the methods that are customarily used for gene isolation. This technique was chosen for purification of reiterated genes, such as ribosomal and transfer RNA genes and histone genes [1—3]. Furthermore single-stranded DNA fragments enriched in globin genes were isolated using this technique [4]. On the other hand, individual genes can be obtained by the reverse transcription of corresponding mRNA samples. This method was used for preparation of globin, immunoglobulin, ovalbumin and some other genes [5—10].

The problem of human gene isolation is of particular interest. For example they may be useful either for studying the structure, function and regulation of the activity of human genes, or as probes for detecting and quantitating specific portions of the genome or mRNA in a variety of normal and abnormal conditions. Similar to animals, human globin genes are more available for analysis. Reverse transcription of globin mRNA is the main method for obtaining these genes (complementary DNA's). In most of the works already published, samples of cDNA with a size in the range of 300 to 600 nucleotides, that is from 40 to 90% of the length of mRNA, were used. This may depend on the conditions of the reaction and nativeness of mRNA samples.

The present paper describes the isolation and characteristics of human globin mRNA and the synthesis of complementary DNA, the greatest part of which is made of full length copies of mRNA.

Material and methods

Human globin mRNA was prepared from samples of normal and haemolytic anaemia blood with reticulocyte counts from 5 to 40% by a modified method described in [15]. Blood cells were centrifuged and washed three times with cold buffer, containing 0.13 M NaCl; 0.01 M Tris-HCl, *p*H 7.8. The washed cells were lysed at 0 °C by addition of 3 volumes of buffer containing 0.01 M Tris-HCl, *p*H 7.8, 1 mM $MgCl_2$, 6 mM β-mercaptoethanol and 100 µg/ml heparin. The cell membranes were sedimented at 15,000 rev./min in a Janetzki K-24 centrifuge for 20 min at 2 °C and discarded. The supernatant was carefully removed and brought to 1% sodum dodecyl sulfate. RNA was exctracted three times with phenol, *p*H 8.5, chloroform-isoamyl alcohol (50:50:1 by vol.) at room temperature. RNA was precipitated with 2.5 vol. of cold redistilled ethanol. Then RNA was layered onto 5—20% sucrose gradients containing 0.10 M Tris-HCl, *p*H 7.5, 0.05 M $MgCl_2$, 5 mM EDTA, 0.5% sodium dodecyl sulphate and was centrifuged for 17—20 hrs at 22,000 rpm in a Beckman SW-25 rotor at 20 °C; 7—12S RNA fractions were collected. This material was put on a poly(U)-sepharose column (0.5 × 1 cm) at 20 °C in a buffer containing 0.1 M NaCl, 0.01 M EDTA, 0.01 M Tris-HCl, *p*H 7.5, 0.2% sodium dodecyl sulphate. The column was washed with the same buffer to 0.02 O.D.$_{260}$. The poly(A)-containing fraction was then eluted with 0.2% sodium dodecyl sulphate at 45 °C. After that, adsorbed and nonadsorbed fractions were passed through a sephadex G-50 column in buffer containing 0.2 M sodium acetate, 0.10 M Tris-HCl, *p*H 7.5. 3—5 µg of each fraction were analysed by electrophoresis in 4.5% polyacrylamide gels [16]. After electrophoresis the gels were stained with 0.2% methylene blue and analysed in a "Chromoscan". For reverse transcription we used poly(A)-containing 9 S RNA.

RNA-dependent DNA polymerase (reverse transcriptase) was purified from avian myeloblastosis virus as described [8]. Samples were incubated for 1 hr at 37 °C in the following mixture; 0.05 M Tris-HCl, *p*H 8.3, 0.04 M KCl, 6 mM $MgCl_2$, 4 mM dithiothreitol, 0.2 mM of each dATP, dTTP, dGTP, 0.2 mM [^{3}H]-dCTP (400 cpm/pmole), 1 µg/ml of mRNA, 5 µg/ml of oligo$(dT)_{10}$, 100 µg/ml of actinomycin D and 40 units of reverse transcriptase (1 unit catalysed the incorporation of 1 mmole dTTP with poly(A)-oligo$(dT)_{10}$ as template at 37 °C for 15 min). Complementary DNA (cDNA) was extracted with phenol-chloroform (1 : 1) and was incubated in 0.3 M KOH for 18 hrs, neutralized and precipitated with 2.5 vol. of cold ethanol in the presence of yeast tRNA as a carrier. Then the pellet was dissolved in 0.2 M sodium acetate, 0.01 M Tris-HCl, *p*H 7.5, and passed through a sephadex G-50 column (0.5 × 50 cm). The size of cDNA was determined by electrophoresis in 5.2% polyacrylamide gels in formamide [17]. tRNA from yeast and 9 S rabbit globin mRNA were used as markers. Gels with nonradioactive material were developed with "Stains all' (Eastman/Kodak), gels with radioactivity were cut into slices of 2 mm and dissolved in "Tissue solubilizer" (Amersham/Searle) and counted in toluene/fluorine mixture.

Hybridization experiments with human globin mRNA and cDNA were carried out in 0.3 M NaCl at 65 °C. DNA was added to mRNA at different concentration but at constant ratio. The mixture was heated for 5 min at 100 °C, cooled to 65 °C and incubated for different times. At the end of the incubation the amount of hybridized DNA resistant to nuclease S_1 was measured by adding 10 µg/ml E. coli denatured DNA, 1/5 vol. of buffer, containing 0.15 M sodium acetate, *p*H 4.5, $6 \cdot 10^{-4}$ M $ZnSO_4$ and 20 units/ml of nuclease S_1 [18] to each reaction mixture. After incubation for 40 min at 55 °C the samples were cooled to 0 °C and acid-insoluble material was precipitated with trichloroacetic acid, collected on membrane filters and counted.

Translation activity of human globin mRNA was determined in wheat germ protein-synthesizing system isolated as described [19]. The reaction mixture contained: 30 mM HEPES (N-2-hydroxyethylpiperazine-N′-2-ethanesulfonic acid), *p*H 7.6, 1 mM dithiothreitol, 1 mM ATP, 80 mM CTP, 16 mM creatine phosphate, 40 µg/ml of creatine phosphokinase, 100 mM KCl, 2.5 mM magnesium acetate, 20 mmoles of each amino acid except leucine and 1.2 mM [^{3}H]-leucine (58 Ci/mmole,) 0.4 vol. of S_{23} fraction and from 5 to 20 µg/ml human globin mRNA. Radioactivity was determined as described [20]. In the case of preparative synthesis 4 mg/ml of carrier human globin was added after incubation and globin was extracted and chromatographed on carboxymethylcellulose [21].

Results and discussion

Fig. 1 shows electrophoretic analysis of mRNA samples. The material adsorbed on poly(U)-sepharose column has one band corresponding to 9 S RNA. Nonadsorbed material comprises a heterogeneous population of molecules, most part of which is represented by 9 S RNA. The amount of these 9 S RNA molecules was much greater than poly(A)-containing mRNAs and is about 80% of 9 S RNA peak. This fact may be explained in the sense that these mRNA molecules have very short or no poly(A)-fragments. It is known that the size of poly(A) varies in human globin mRNA [22]. In most of the papers on the isolation of human globin mRNA that have already been published there are no data on the yield of mRNA from a definite source. For the isolation of human globin mRNA we used normal blood and reticulocyte-rich blood from haemolytic anaemia. From 1000—1500 ml of

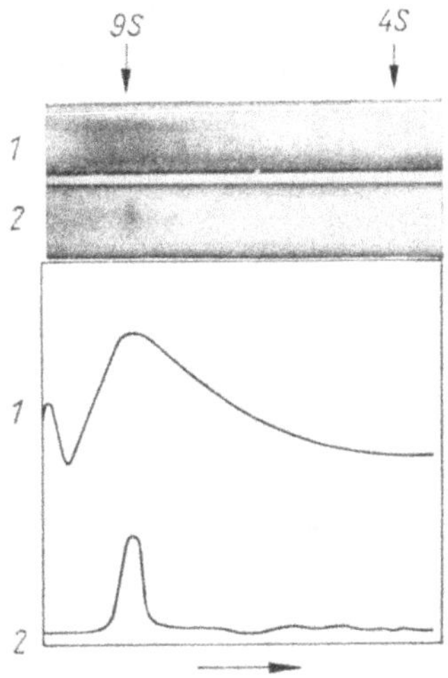

Fig. 1

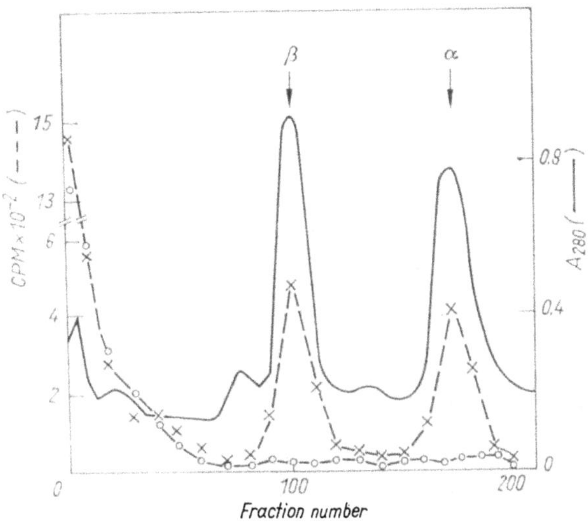

Fig. 2

Fig. 1. Electrophoretic analysis of human reticulocyte 9 S RNA.
1) RNA adsorbed on poly(U)-sepharose; 2) nonadsorbed RNA

Fig. 2. Carboxymethyl cellulose chromatography of proteins, synthesized in vitro. Protein synthesized with human 9 S RNA template (×--------×); without added RNA (o———o); carrier globin protein from normal human erythrocytes (———).

normal donor blood one may obtain 5—10 μg, and from 300 ml of blood with reticulocyte counts of about 7—8%, 15—20 μg of poly(A)-containing 9 S mRNA.

9 S RNA samples were analyzed in a wheat germ protein-synthesizing system. Both adsorbed on poly(U)-sepharose and nonadsorbed material appeared to be able to stimulate protein synthesis in vitro (Table 1). The samples of protein synthesized were studied by chromatography on carboxymethylcellulose. Fig. 2 shows the chromatography of the protein, synthesized in the presence and absence of human 9 S RNA. Both curves nearly coincide in the region of nonglobin proteins that may be synthesized in the system owing to the endogenous templates remaining. The peaks of radioactivity corresponding to α- and β-globin chains are present only in the presence of mRNA, that means that our 9 S RNA from human reticulocytes contains globin mRNA.

In further experiments we studied the template capacity of poly(A) containing human globin RNA for reverse transcription. In the conditions used its template activity was $2 \cdot 10^4$ pmoles cDNA/O.D.$_{260}$. The demonstrated activity is similar to that of the rabbit globin mRNA and is much higher than reported for human globin mRNA [14].

Fig. 3 shows an electrophoretic analysis of cDNA in polyacrylamide gels in formamide. One can see that most part of the cDNA has a mobility corresponding to 9 S chains. These molecules appeared to be full length copies of mRNA.

Table 1
Template activity of human globin mRNA in a wheat germ protein-synthesizing system

Exp. No.	RNA sample	mRNA input [µg/ml]	[^{3}H]-leucine incorporation [cpm]
1	poly(A)+ RNA	10	1000
		20	1200
	poly(A)− RNA	10	1600
2	poly(A)− RNA	5	400
		10	1750
		20	2670

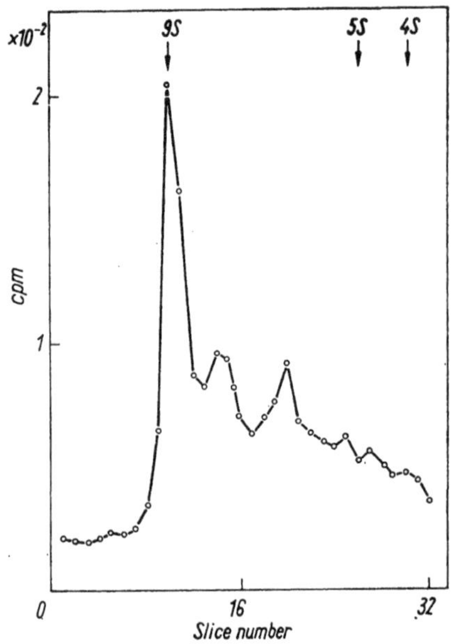

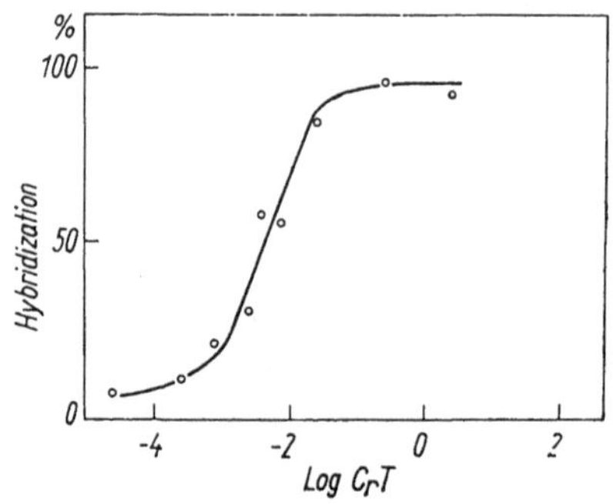

Fig. 4. Hybridization of cDNA in excess of human globin mRNA

Fig. 3. Electrophoretic properties of cDNA synthesized on human globin mRNA template

◄

Fig. 4 shows the hybridization of cDNA in excess of human globin RNA. It is evident that cDNA is complementary to the template used. The hybridization curve is in the CrT regions typical for described curves of human globin sequences [14].

Thus, the present paper demonstrates the isolation of globin mRNA from human reticulocytes and the synthesis of complementary DNA, most part of which is a complete transcript of human globin genes. The obtained samples of cDNA may be used for studies in human molecular genetics; for instance, in research of molecular causes of thalassemia. Moreover these genes may be obtained in preparative amounts by gene engineering methods and used for elucidating the general regularities of organization and function of human globin genes.

We wish to thank Prof. G. P. GEORGIEV for stimulating discussions, Prof. A. E. EVDOKIMOVA, Drs. L. I. IDELSON and S. S. SOBOLEVA for reticulocyte-rich blood samples. We are grateful to Drs. S. N. KHILKO and E. E. MAKAROVSKAYA for their help in translation analysis of mRNA, and Dr. N. E. MALEEVA and Mrs. I. FURYGINA for their assistance.

References

[1] Brown, D. D., and R. Stern: A Rev. Biochem. **43**, 667 (1974)
[2] Brenner, D. J., M. J. Fournier, and B. P. Doctor: Nature, Lond. **227**, 448 (1970)
[3] Shih, T. Y., and M. A. Martin: Proc. natn. Acad. Sci. U.S.A. **70**, 1697 (1973)
[4] Tarantul, V. Z., V. A. Lipasova, and K. G. Gasaryan: Dokl. Akad. Nauk SSSR **226**, 719 (1976)
[5] Kacian, D. L., S. Spiegelman, A. Bank, M. Terada, S. Metafora, L. Dow, and P. A. Marks: Nature New Biol. **235**, 167 (1972)
[6] Verma, I. M., G. F. Temple, H. Fan, and D. Baltimore: Nature New Biol. **235**, 163 (1972)
[7] Harrison, P. A , A. Hell, G. D. Birnie, and J. Paul: Nature, Lond. **239** ,219 (1972)
[8] Kisselev, L. L., L. Yu Frolova, K. G. Gasaryan, V. Z. Tarantul, and V. A. Engelhart: Dokl. Akad. Nauk SSSR **213**, 1203 (1973)
[9] Diggelman, H., C. H Faust, and B Mach: Proc natn. Acad. Sci. U.S.A. **70**, 693 (1973)
[10] Harris, S. E., A R. Means, W. M Mitchell, and B. W. O'Malley: Proc. natn. Acad. Sci. U.S.A. **70**, 3776 (1973)
[11] Ottolenghi, S., W. G. Lanyon, J. Paul, R Williamson, D. J. Weatherall, J. B. Clegg, J. Pritchard, S. Pootrakul, and W. H. Boon: Nature, Lond. **251**, 389 (1974)
[12] Taylor, J. H., A. Dozy, Y. W. Kan, H. E. Varmus, L. E. Lie-Injo, J. Ganezan, and D Todd: Nature, Lond. **251**, 392 (1974)
[13] Ottolenghi, S., W. G. Lanyon, R. Williamson, D. J. Weatherall, J. B. Clegg, and C. S. Pitcher: Proc. natn. Acad Sci. U.S.A. **72**, 2294 (1975)
[14] Velez, R., J. A. Kantor, D. J. Picciano, W. F. Anderson, and A. W. Nienhuis: J. biol. Chem. **250**, 3193 (1975)
[15] Williamson, R., M. Morrison, G. Lanyon, R. Eason, and J. Paul: Biochemistry **10**, 3014 (1971)
[16] Loening, J. E.: Biochem. J. **102**, 251 (1967)
[17] Pinder, J. D., D. L. Stainov, and W B. Gratzer: Biochemistry **13**, 5373 (1974)
[18] Sutton, W. D : Biochim. biophys. Acta **240**, 522 (1971)
[19] Roberts, B. E., and B. M. Patterson: Proc natn Acad Sci. U S A. **70**, 2330 (1973)
[20] Mans, R., and G Novelli: Archs Biochem Biophys. **94**, 48 (1961)
[21] Clegg, J. B., M. A. Naughton, and D J Weatherall: J. molec. Biol. **19**, 91 (1966)
[22] Cann, A , R. Gambino, and J. Bank: J biol. Chem. **249**, 7536 (1974)

Acta biol. med. germ., Band 36, Seite 341—351 (1977)

Institut für Physiologische und Biologische Chemie der Humboldt-Universität Berlin, 104 Berlin, DDR

Stickstoff-Ökonomie und Synthese von Serin und Glyzin in Retikulozyten

M. Müller, S. Rapoport, J. Rathmann und R. Dumdey

Zusammenfassung

Aminosäuren sind das Hauptsubstrat der Retikulozyten. Der Aminosäure-pool der Retikulozyten repräsentiert eine charakteristische Selektion. Die Zusammensetzung des Aminosäurepools ist reifungsabhängig. Die vorrangige Oxydation kurzkettiger Aminosäuren läßt ein Verhältnis von etwa 5:1 zwischen O_2-Verbrauch und NH_3-Bildung erwarten. Nach Korrektur der NH_3-Freisetzung durch Nukleotiddesaminierung übersteigt der Quotient zwischen O_2-Verbrauch und NH_3-Bildung den Erwartungswert um etwa eine Größenordnung. Die geringe Freisetzung von NH_3 bei der Energiegewinnung aus Aminosäuren resultiert aus der Wiederverwertung des Stickstoffs für die Synthese von Serin und Glyzin mit dem C-Skelett der Glukose. Zwischenprodukte der Serinsynthese sind Ⓟ-OH-Pyruvat und OH-Pyruvat. Serin und Glyzin werden bevorzugt für die Hämoglobinsäuresynthese benötigt. Im Retikulozyten existiert eine Kompartmentierung des Glyzins, welche eine unterschiedliche spezifische Aktivität von Serin und Glyzin in Isotopenversuchen bedingt. Aus der zeitlichen Änderung der spezifischen Aktivität des Serins und Glyzins nach Pulsmarkierung des Serins wird der Umfang der Serinsynthese mit 15—30% des Glukoseumsatzes abgeschätzt.

Die meisten Wege des Zwischenstoffwechsels der Retikulozyten sind seit einigen Jahren qualitativ aufgeklärt. Wir wissen heute, daß die Aminosäuren von außerordentlich großer Bedeutung im Stoffwechsel dieser Zellen sind. Das ist zutreffend sowohl für den Energiestoffwechsel als auch für die Syntheseleistungen des Retikulozyten, die im wesentlichen in der Häm- und Globinsynthese bestehen [1]. Diese Kenntnis darf uns nicht darüber hinwegtäuschen, daß wir weder die Zusammensetzung des Substratgemisches noch seine Regulation im Detail kennen. Wir wissen wenig über die Verteilung der Metabolite auf anabole und katabole Stoffwechselwege und noch weniger über die Regulationsprinzipien zwischen mitochondrialen und zytosolischen Systemen.

In bezug auf die Substratverwertung der Retikulozyten verfügen wir über folgende Kenntnisse:

Der Anteil der Glukose an der Gesamt-CO_2-Bildung der Retikulozyten ist verglichen mit anderen Zellen verhältnismäßig klein. Der Anteil der Glukose steigt von weniger als 20% in sehr unreifen Zellen mit der Reifung auf 50% an der Gesamt-CO_2-Bildung an [2].

Aminosäuren haben also den größten Anteil am endogenen und exogenen Substrat der Retikulozyten [3—6].

Es liegt bis jetzt noch keine komplette Analyse der intrazellulären Aminosäurekonzentrationen vor [1, 7].

In Tab. 1 sind unter Einbeziehung von Literaturdaten unsere Kenntnisse über die Zusammensetzung des intrazellulären Aminosäure-pools zusammengestellt. Die

Tabelle 1
Konzentration einiger Aminosäuren in Kaninchenretikulozyten

Aminosäure	Konzentration (mM)	n
Glyzin	1,60 ± 0,60	22
Glutamat	0,80 ± 0,32	26
Aspartat	0,58 ± 0,13	3
Alanin	0,53 ± 0,20	23
Serin	0,36 ± 0,14	10
Ⓟ-Serin	0,29 ± 0,14	6
Threonin	0,28 ± 0,10	3
andere	< 0,05	

Tabelle 2
Konzentration einiger Aminosäuren von Kaninchenretikulozyten in verschiedenen Stadien der Entblutungsanämie

Aminosäure	Konzentration (mM)				P (%)
	4.–6. Tag	n	9.–16. Tag	n	
Glyzin	1,10 ± 0,28	7	1,80 ± 0,45	9	< 1
Glutamat	0,70 ± 0,22	6	1,00 ± 0,35	12	> 5
Alanin	0,40 ± 0,14	7	0,60 ± 0,13	9	< 1

Zusammensetzung des Aminosäure-pools der Retikulozyten repräsentiert eine charakteristische Selektion der Aminosäuren. Glyzin überwiegt, gefolgt von Glutamat, Aspartat, Alanin, Serin und Threonin. Kürzlich konnten wir auch P-Serin mit einer etwa dem Serin entsprechenden Konzentration nachweisen. Die Konzentration aller übrigen Aminosäuren ist etwa eine Größenordnung kleiner. Die relativ großen Schwankungsbreiten lassen sich auf die Tatsache zurückführen, daß die Retikulozyten im Vollblut während einer Entblutungsanämie keine homogene Zellpopulation repräsentieren [1, 8]. Wir finden Differenzen in der Aminosäurekonzentration in verschiedenen Stadien der Anämie.

Tab. 2 zeigt, daß die Aminosäurekonzentrationen im Verlauf einer Entblutungsanämie ansteigen. Der Anstieg ist signifikant für Glyzin und Alanin. In einer frühen Phase der Entblutungsanämie erscheinen sehr unreife Retikulozyten im peripheren Blut. Nach länger andauernder kontinuierlicher Entblutung treten im peripheren Blut reifere Retikulozyten auf. Die Spezifik der Reifung der Retikulozyten besteht im fortschreitenden Abbau der Organellen, Ribosomen und Mitochondrien, im partiellen Abbau der Zellmembran und in der selektiven Aktivitätsabnahme von Enzymen [1]. Aus diesem Reifungsgeschehen resultiert der Konzentrationsanstieg der Aminosäuren. Offenbar erfolgt der aktive Transport der Aminosäuren auch dann noch mit großer Intensität, wenn in reiferen Retikulozyten die Funktionsfähigkeit der Mitochondrien vermindert ist. Der Konzentrationsanstieg wird begünstigt durch den verminderten Verbrauch der Aminosäuren für den Energiestoffwechsel und für Synthesen und durch die Bildung von Aminosäuren durch intrazelluläre Proteolyse.

Die Zusammensetzung des Aminosäure-pools wird insgesamt durch die in Abb. 1 zusammengestellten Prozesse beeinflußt [1]. Die pool-Zusammensetzung resultiert aus der Aminosäureaufnahme aus dem Plasma, dem Katabolismus zelleigener Eiweiße und dem Verbrauch der Aminosäuren für den Energiestoffwechsel und für Synthesen. Neuere Erkenntnisse machen es notwendig, die intrazelluläre Bildung von Aminosäuren beim Studium der Dynamik des Aminosäure-pools zu berücksichtigen. Dieser Aspekt wird Gegenstand der weiteren Diskussion sein.

Welchem Stoffwechsel-Schicksal unterliegen die Aminosäuren? Es soll nicht das Schicksal jeder einzelnen Aminosäure analysiert, sondern es sollen die Probleme dargestellt werden, die sich aus der vorrangigen Verwendung der Aminosäuren für den Energiestoffwechsel ergeben.

Der Anteil einzelner Aminosäuren an der Gesamt-CO_2-Bildung der Zellen ist sehr different und primär eine Funktion der intrazellulären Aminosäurekonzentration [6].

In Übereinstimmung mit dem von WARBURG bestimmten respiratorischen Quotienten von etwa 1 [9], erfolgt die CO_2-Bildung bevorzugt aus aliphatischen Aminosäuren mit weniger als 6 C-Atomen. Dabei dienen Glutamat bzw. Glutamin, Aspartat und Alanin als Metabolite, die in enger Beziehung zum Zitratzyklus stehen, vorrangig der Energiegewinnung [6, 10, 11]. Die hohe CO_2-Bildung aus Serin und Glyzin erfolgt dagegen im Zusammenhang mit der intensiven Porphyrinsynthese der Retikulozyten [1, 6, 12, 13].

Aminosäuren, die in direkter Beziehung zum Zitratzyklus stehen, können bei ihrer Konzentrationserhöhung die unveränderte Gesamt-CO_2-Bildung der Retikulozyten weitgehend monopolisieren [1]. Beispielsweise können Glutamin bzw. Glutamat mit 80% den höchsten Anteil an der Gesamt-CO_2-Bildung erreichen [10].

Da Aminosäuren das Hauptsubstrat der Retikulozyten sind, ist eine beträchtliche Ammoniakbildung zu vermuten. Wie aus Literaturdaten hervorgeht, bleibt aber die Ammoniakfreisetzung weit hinter der nach dem Anteil der Aminosäuren am O_2-Verbrauch erwarteten zurück [14—16]. Die vorrangige Oxydation kurzkettiger Aminosäuren läßt ein Verhältnis von etwa 5:1 zwischen Sauerstoffverbrauch und Ammoniakbildung erwarten.

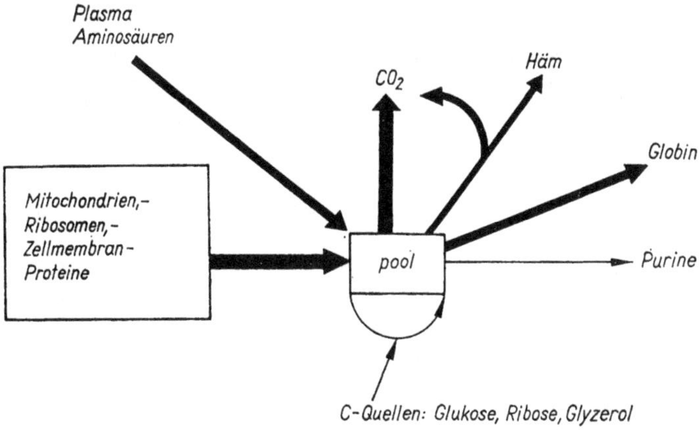

Abb. 1. Dynamik des Aminosäure-pools in Retikulozyten

Selbst wenn die geringe Verwertung von Kohlenhydraten und Fettsäuren für den Energiestoffwechsel in Rechnung gestellt wird, ist das aus Literaturdaten kalkulierte Verhältnis zwischen O_2-Verbrauch und NH_3-Bildung von etwa 10:1 sehr hoch. Dabei muß berücksichtigt werden, daß sich die vorhandene NH_3-Bildung im wesentlichen auf die Desaminierung von Nukleotiden, besonders von Adeninnukleotiden, zurückführen läßt.

Die Nukleotiddesaminierung ist ein schneller Prozeß, der bei kurzzeitigen Inkubationen die NH_3-Freisetzung aus Aminosäuren überwiegt. Wird diese Tatsache berücksichtigt, dann übersteigt der Quotient zwischen O_2-Verbrauch und NH_3-Bildung den Erwartungswert von etwa 5:1 um etwa eine Größenordnung.

Die geringe Freisetzung von NH_3 bei der Energiegewinnung aus Aminosäuren ist ein Ausdruck der hohen Stickstoffökonomie der Retikulozyten. Die Stickstoffökonomie der Retikulozyten stellt seit langem ein ungelöstes Problem dar. Wir hoffen, daß wir durch unsere Untersuchung zu seiner Lösung beitragen können.

Wir gingen davon aus, daß offenbar die Transaminierung die Voraussetzung für den Abbau des C-Skeletts des Glutamats und anderer Aminosäuren im Zitratzyklus ist. Die geringe NH_3-Freisetzung bei der Energiegewinnung aus Aminosäuren spricht weiterhin dafür, daß es in Retikulozyten eine Wiederverwertung des Stickstoffs geben muß.

Tatsächlich konnte der Nachweis erbracht werden, daß die oxydative Desaminierung des Glutamats, das in hohem Maße an der Energiegewinnung der Retikulozyten beteiligt ist, nur eine untergeordnete Rolle spielt. Nur etwa 10% des umgesetzten Glutamats werden oxydativ desaminiert [17].

Im Folgenden versuchten wir die Frage zu klären, wie die Wiederverwertung des Stickstoffs erfolgt. Dabei gingen wir von der Vorstellung aus, daß eine Neusynthese von Aminosäuren erfolgt.

Da der Stoffwechsel der Retikulozyten außerordentlich gut der Hauptfunktion dieser Zellen, der Hämoglobinbildung, angepaßt ist, sind Serin und Glyzin prädestinierte Aminierungsprodukte.

Beide Aminosäuren werden bevorzugt für die Hämoglobinsynthese verwendet. Kaninchenhämoglobin enthält nach BRAUNITZER et al. [18, 19] pro Tetramer 40 Mol Glyzin und 44 Mol Serin.

Zusätzlich sind 32 Mol Glyzin für die Porphyrinsynthese erforderlich. Beide Aminosäuren werden darüber hinaus auch für die Purinsynthese benötigt [1].

Tatsächlich konnte nach Inkubation mit [^{15}N]-Glutamat [^{15}N]-markiertes Glyzin isoliert werden (Tab. 3). Der Umfang der Glyzinneusynthese liegt in dem Bereich,

Tabelle 3
Einbau von ^{15}N in Glyzin aus [^{15}N]-Glutamat (18—30 mM)

Versuch Nr.	O_2-Verbrauch (mMol/l Zellen · Std.)	Konzentration (mM) Glyzin	[^{15}N]-Glyzin
1*	9,4	1,3	0,016
2**	6,7	1,4	0,050
3**	4,2	0,9	0,022

* Ohne Glukose; ** 24 mM Glukose. 60 min Inkubation bei 37 °C

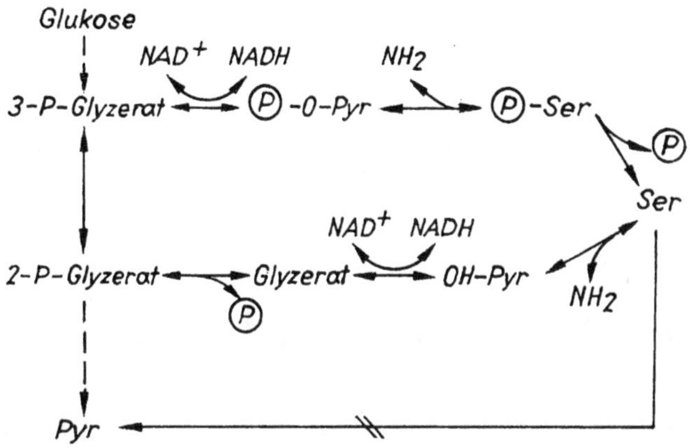

Abb. 2. Alternative Wege der Serinsynthese mit dem C-Skelett der Glukose

der aus der Höhe der CO_2-Bildung aus Glutamat unter Berücksichtigung von pool-Verdünnungen abgeschätzt werden kann.

Sehr wahrscheinlich dienen auch andere Aminosäuren, die an der Transaminierung teilnehmen, direkt oder indirekt als NH_2-Donatoren bei der Glyzinneusynthese.

Aus diesem Ergebnis ergab sich zwanglos die Fragestellung nach dem C-Skelett, nach dem NH_2-Akzeptor für die NH_2-Gruppe der verstoffwechselten Aminosäuren.

Als NH_2-Akzeptoren wurden durch Studien mit U-[^{14}C]-Glukose OH-Pyruvat und P–OH-Pyruvat identifiziert (unveröffentlichte Befunde). Die Neubildung von Serin bzw. Glyzin erfolgt also mit dem C-Skelett der Glukose.

2 Wege werden für die Serinbildung ausgehend vom 2- bzw. 3-P-Glyzerat beschrieben [20, 21] (Abb. 2):

Ausgehend von 2-P-Glyzerat geht einer Dehydrierung eine Dephosphorylierung voraus. Als ein Zwischenprodukt der Serinsynthese wird OH-Pyruvat gebildet. Auf dem 2. Syntheseweg tritt OH-Pyruvat nicht als Zwischenprodukt auf. Die Phosphatabspaltung erfolgt erst nach der Transaminierung, so daß eine Phosphoserinbildung der Serinbildung vorausgeht. In Gewebsextrakten verschiedener Spezies wurde die Existenz beider Synthesewege nachgewiesen [20, 21]. Meistens ist nur einer der Wege vorherrschend. Wir fanden, daß auch in Retikulozyten beide Synthesewege existieren, wobei über den Anteil der verschiedenen Wege an der Serinsynthese noch keine Aussage gemacht werden kann (unveröffentlichte Befunde). Als direkte Transaminierungsprodukte resultieren also Serin bzw. P-Serin. Durch P-Abspaltung wird aus P-Serin Serin gebildet, das durch C_1-Abspaltung in Glyzin übergehen kann [13].

Die Serinspaltung erfolgt in einer Gleichgewichtsreaktion [22, 23]. Nach Inkubation mit U-[^{14}C]-Glukose erwarteten wir daher die gleiche spezifische Aktivität im Serin und im Glyzin. Das ist offensichtlich nicht der Fall (Tab. 4). Nach Zusatz von OH-Pyruvat zum Inkubationsmedium ist erwartungsgemäß eine starke Verminderung der spezifischen Aktivität des Serins nachweisbar. OH-Pyruvat ist, wie erläutert, eine Zwischenprodukt auf dem Syntheseweg des Serins aus dem C-Skelett der Glukose.

Tabelle 4
Beeinflussung der Konzentration und spezifischen Aktivität des Glyzins und Serins durch OH-Pyruvat

Versuche Nr.	Aminosäure	Konzentration (mM) OH-Pyruvat		spez. Akt. (dpm/nMol) OH-Pyruvat	
		–	+	–	+
1*	Glyzin	0,92	1,07	2,38	1,71
2**	Glyzin	1,37	1,13	1,21	1,10
3**	Glyzin	1,31	1,44	1,07	0,84
1*	Serin	0,39	1,25	18,40	2,03
3**	Serin	0,36	0,75	8,10	3,87

* 10 mM OH-Pyruvat; ** 5 mM OH-Pyruvat. 60 min. Inkubation bei 37 °C, 20 mM Glukose

Zunächst bleibt jedoch unverständlich, daß die spezifische Aktivität des Glyzins durch OH-Pyruvatzusatz nur unwesentlich vermindert wird.
Als Nächstens versuchten wir nach Pulsmarkierung mit U-[^{14}C]-Serin aus der zeitlichen Änderung der spezifischen Aktivitäten des intrazellulären Serins und Glyzins eine Abschätzung über den Umfang der Synthese beider Aminosäuren zu gewinnen. Wir erwarteten eine zeitabhängige Annäherung der spezifischen Aktivitäten beider Aminosäuren. Meßbar ist jedoch ein annähernd paralleler Abfall der spezifischen Aktivitäten beider Aminosäuren, der mit einer hohen Anfangsgeschwindigkeit erfolgt. Die spezifische Aktivität (Abb. 3) des Glyzins liegt jedoch entgegen der Erwartung unterhalb der des Serins. Mit steigender intrazellulärer Glyzinkonzentration vergrößert sich die Differenz zwischen den spezifischen Aktivitäten beider Aminosäuren (Abb. 4).

Welche Faktoren könnten die spezifische Aktivität beeinflussen?

Aus den Stoffwechselwegen des Serins und Glyzins in Retikulozyten ergeben sich Anhaltspunkte zur Beantwortung der aufgeworfenen Frage. Die Umwandlung des Serins in Glyzin und Methylen-FH_4 findet ausschließlich in den Mitochondrien der Retikulozyten statt (unveröffentlichte Ergebnisse). In dieser Hinsicht nimmt der Retikulozyt eine Ausnahmestellung ein. In anderen biologischen Systemen ist die Serin-OH-Methyltransferase überwiegend im Zytosol lokalisiert [25]. Die von diesem Enzym katalysierte Reaktion verläuft mit geringer Energietönung reversibel. Demnach kann die gegenseitige Umwandlung von Serin und Glyzin keine Quelle unterschiedlicher spezifischer Aktivitäten sein. Es bleiben noch zwei Denkmöglichkeiten zur Erklärung der unterschiedlichen spezifischen Aktivitäten von Serin und Glyzin. Ihnen ist gemeinsam, daß ein langsamer Austausch einer Aminosäure, Serin bzw. Glyzin, zwischen Zytosol und Mitochondrien angenommen wird.

Die erste Annahme wäre ein langsamer Einstrom von Serin in das Mitochondrium, wodurch alle nachfolgenden Schritte, d. h. die Umwandlung in Glyzin und der Ausstrom des Glyzins aus dem Mitochondrium in das Zytosol, eine geringere spezifische Aktivität aufweisen würden, vorausgesetzt, daß ein inaktiver Pool im Mitochondrium vorhanden ist.

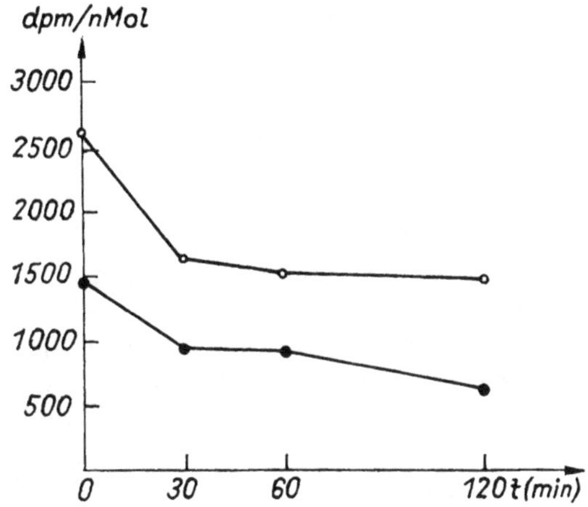

Abb. 3. Zeitabhängige Änderung der spezifischen Aktivität des Serins und Glyzins nach Pulsmarkierung des Serins.
o——o spezifische Aktivität des Serins; •——• spezifische Aktivität des Glyzins. Intrazelluläre Glyzinkonzentration = 0,65 mM; Pulsmarkierung 15 min. Inkubationsbedingungen: 6 mM Glukose, 3 mM $GluNH_2$, *p*H 7,6, 37 °C

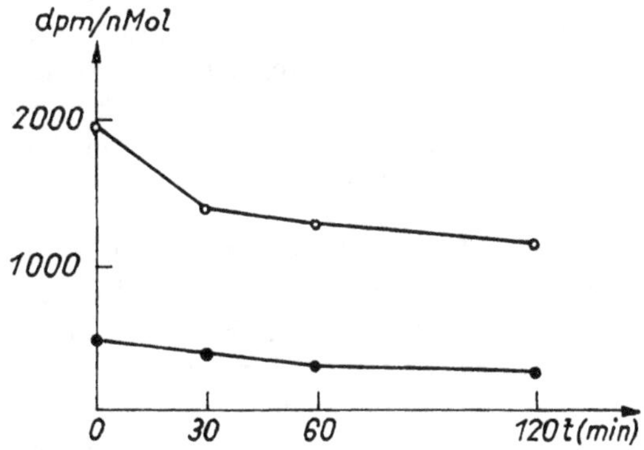

Abb. 4. Zeitabhängige Änderung der spezifischen Aktivität des Serins und Glyzins nach Pulsmarkierung des Serins.
o——o spezifische Aktivität des Serins; •——• spezifische Aktivität des Glyzins. Intrazelluläre Glyzinkonzentration = 1,0 mM; Pulsmarkierung 15 min. Inkubationsbedingungen: 6 mM Glukose, 6 mM $GluNH_2$, *p*H 7,6, 37 °C

Gegen diese Annahme spricht vor allem die Tatsache, daß durch Isotopenexperimente von uns gezeigt werden konnte, daß Serin, und vor allem der aus ihm stammende C_1-Rest, mit hoher Intensität im Stoffwechsel verwertet wird bzw. CO_2 liefert. Da diese Reaktionen im Mitochondrium ablaufen, ergibt sich, daß auch der Einstrom von Serin in das Mitochondrium schnell sein muß. Darüber hinaus gibt es keine Anhaltspunkte für einen beträchtlichen inaktiven Serin-Pool im Mitochondrium.

Die zweite Denkmöglichkeit besteht in der Annahme, daß der Ausstrom des mitochondrial gebildeten markierten Glyzins in das Zytosol ein langsamer Prozeß ist. Dieses Glyzin würde sich mit einem großen Pool inaktiven Glyzins im Zytosol mischen. Das zytosolische Glyzin stammt in erster Linie aus dem Plasma. Höhere Konzentrationen von Glyzin verglichen mit denen von Plasma werden zwar auch in geringem Maße bei den reifen Erythrozyten einiger Spezies beobachtet und sind auf energieunabhängige erleichterte Diffusion zurückzuführen [26]. Demgegenüber haben Retikulozyten eine um vieles stärker ausgeprägte Fähigkeit, Glyzin zu konzentrieren, und diese ist durch Zyanid und Dinitrophenol hemmbar [27, 28]. Es handelt sich offensichtlich um ein aktives Transportsystem, das mit der Reifung der roten Blutzelle verschwindet. Die Aufnahme des Glyzins erfolgt über

glyzinspezifische Wege. Dabei addieren sich zwei Mechanismen, von denen der Hauptweg Na^+-abhängig und der Nebenweg Na^+-unabhängig ist [29, 30].
Eine zweite Quelle für das zytosolische Glyzin stellt der energieabhängige Proteinabbau dar [31, 32]. Wie in früheren Untersuchungen gezeigt wurde, kommt es in Retikulozyten bei Inkubation zu einem beträchtlichen Anstieg des Rest-N, der überwiegend auf die Freisetzung von Aminosäuren zurückzuführen ist. Dieser Proteinabbau ist energieabhängig und wird durch Dinitrophenol gehemmt. Daneben setzt das Dinitrophenol allerdings auch den oxydativen Abbau von Aminosäuren herab. Das Zusammenspiel beider Prozesse ist es wahrscheinlich, was für die Variationen der Glyzinkonzentrationen mit unterschiedlicher Entblutungsdauer verantwortlich ist. Eine völlige Undurchlässigkeit der Mitochondrienmembran für Glyzin ist selbstverständlich ausgeschlossen. Sie würde einen noch wesentlich größeren Unterschied in der spezifischen Aktivität bedingen. Darüber hinaus wäre es unverständlich, wie es zum Einbau von markiertem Glyzin in das Porphyrin kommt.
Von großem Interesse sind erste Versuche, die wir in Gegenwart von 2,4-DNP durchführten. Diese Experimente unterstützen die getroffenen Aussagen.
Unter diesen Bedingungen ist eine starke Erhöhung der spezifischen Aktivität des Serins und des Glyzins nach Inkubation mit U-[^{14}C]-Glukose und eine Annäherung der spezifischen Aktivitäten des Serins und des Glyzins nach Inkubation mit U-[^{14}C]-Glukose und nach Pulsmarkierung des Serins meßbar (Tab. 5).

Wie sind diese Ergebnisse zu deuten?

1) Eine starke Erhöhung der Serinbildung müßte eine Konzentrationserhöhung des Serins und Glyzins zur Folge haben, da die Verwertung der Aminosäuren in Gegenwart von 2,4-DNP vermindert ist. Eine solche Erhöhung der Konzentration von Serin oder Glyzin wurde aber nicht beobachtet. Man muß also eine andere Erklärung für die Erhöhung der spezifischen Aktivitäten der Aminosäuren suchen. Wir führen die Erhöhung der spezifischen Aktivitäten in erster Linie darauf zurück, daß das Serin aus dem C-Skelett der Glukose mit höherer spezifischer Aktivität gebildet wird.
Welcher Mechanismus könnte aber für die Bildung von Serin und erhöhter spezifischer Aktivität verantwortlich sein? Eine Möglichkeit beruht auf der Überlegung, daß die Radioaktivität der Zwischenprodukte der Glykolyse unterhalb von 1,3-Bisphosphoglyzerat, die aus der radioaktiven Glukose stammen, durch den Pool des 2,3-Bisphosphoglyzerats, das ja in hoher Konzentration in roten Blutzellen vorliegt, stark verdünnt wird. Das Dinitrophenol könnte den Grad der Verdünnung dadurch verringern, daß es den Anteil der Glykolyse, der über die Phosphoglyzeratkinase läuft, verglichen mit dem, der über den 2,3-Bisphosphoglyzerat-Nebenweg geht, erhöht. Es kommt nämlich unter dem Einfluß von Dinitrophenol zu einer Erniedrigung der Konzentration von ATP und einer entsprechenden Erhöhung derjenigen von ADP. Dieses ist Substrat der Phosphoglyzeratkinase und würde deren aktuelle Aktivität verstärken. Das aus 2- oder 3-Phosphoglyzerat gebildete Serin widerspiegelt unmittelbar deren spezifische Aktivität.

2) Der Ausgleich der spezifischen Aktivitäten des Serins und des Glyzins muß bedeuten, daß die Barriere, die wir für den Glyzinausstrom aus den Mitochondrien postulieren, aufgehoben ist und beide Aminosäuren im Isotopengleichgewicht

stehen. Diese Veränderung könnte mit dem durch das Dinitrophenol bewirkten ATP-Abfall zusammenhängen, der eine Konformationsänderung der Mitochondrienmembran und eine Veränderung ihrer Permeabilität verursachen könnte. Insgesamt sprechen die Befunde dafür, daß es die Barriere für den Glyzinausstrom aus den Mitochondrien gibt, die durch 2,4-DNP aufgehoben wird.
Es ist von großem Interesse, daß die Aufhebung des Pasteur-Effektes durch Dinitrophenol, die von uns gezeigt wurde, auch mit einer Beseitigung der mitochondrialen Barriere für die Adeninnukleotide einhergeht.
Sicher ist es eine grobe Vereinfachung, wenn bei der Bilanzierung der Serin- bzw. Glyzinneusynthese aus dem C-Skelett der Glukose der sicher vorhandene Austausch zwischen dem Mitochondrien- und Zytosol-pool außer Acht gelassen wird und vereinfachend vorausgesetzt wird, daß die intrazelluläre Proteolyse den

Tabelle 5
Beeinflussung der spezifischen Aktivität des Glyzins und Serins durch 2,4-DNP ($5 \cdot 10^{-5}$ M)

Inkubationszusatz	Aminosäure	spezifische Aktivität (dpm/nMol) −2,4-DNP	spezifische Aktivität (dpm/nMol) +2,4-DNP
U-[^{14}C]-Glukose	Serin	10	24
(20 mM)	Glyzin	< 1	23
U-[^{14}C]-Serin (tracer)	Serin	676	540
Glukose (10 mM)	Glyzin	405	480

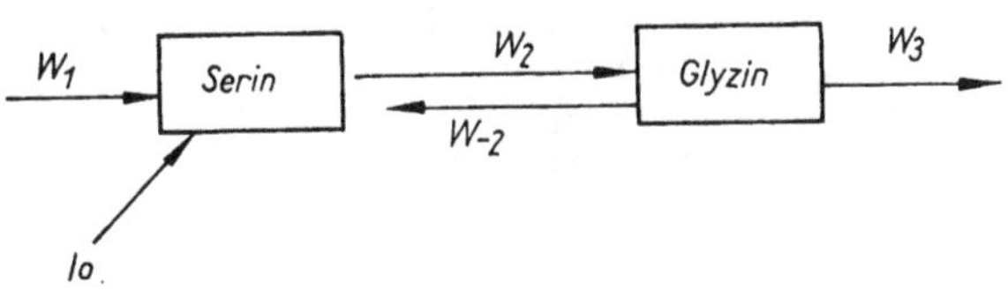

$$\frac{d\alpha}{dt} = -\frac{W_1 \alpha}{[Ser]} + \frac{W_{-2}\ (\beta-\alpha)}{[Ser]}$$

$$\frac{d\beta}{dt} = \frac{W_2\ (\alpha-\beta)}{[Gly]}$$

$$\frac{d[Ser]}{dt} = \frac{d[Gly]}{dt} = 0 \ / \ \alpha = \frac{dpm}{[Ser]}; \quad \beta = \frac{dpm}{[Gly]}$$

$$|W_1| = \frac{\Delta\alpha\,[Ser] + \Delta\beta\,[Gly]}{\beta}$$

Abb. 5. Modell für die Berechnung der Serin- und Glyzinbildung aus dem C-Skelett der Glukose

Serin- und Glyzin-pool in gleichem Maße beeinflußt und deshalb vernachlässigt wird.

Der Versuch einer Bilanzierung der Neusynthese von Serin bzw. Glyzin bezieht sich auf die ersten 30 min der Inkubation. Die offensichtlich verminderte Glyzinbildung aus Serin bei längerer Inkubation führen wir auf eine mitochondriale Glyzinanhäufung zurück, die aus einem verminderten Verbrauch des Glyzins für die Porphyrinsynthese resultiert. Die Ursache für die Verminderung der Porphyrinsynthese während der Inkubation ist nicht bekannt. Eine mögliche Ursache könnte u. a. ein Fe-Mangel sein.

Die Grundlage für unsere Abschätzung des Ausmaßes der Serin- bzw. Glyzinsynthese aus dem C-Skelett der Glukose ist in Abb. 5 dargestellt [24]. Unter diesen Voraussetzungen errechnen wir eine Neusynthese von etwa 0,5—1,0 μMol Serin und Glyzin pro ml Zellen und Stunde aus dem C-Skelett der Glukose. Das bedeutet, daß in Abhängigkeit vom Reifegrad der Zellen etwa 15—30% der Glukose für die Neusynthese von Aminosäuren verwendet wird. Dieses Ausmaß der Aminosäureneusynthese liegt in dem Bereich der notwendigen NH_2-Verwertung und der Hämoglobinsynthese in der entsprechenden Inkubationszeit.

Literatur

[1] Rapoport, S. in: Cellular and Molecular Biology of Erythrocytes. H. Yoshikawa u. S. Rapoport (Hrsg.). University of Tokyo Press, Tokyo 1974, S. 93

[2] Schultze, M., u. S. Rapoport: Acta biol. med. germ. **13**, 310 (1964)

[3] Schweiger, H., S. Rapoport u. E. Schölzel: Hoppe Seyler's Z. physiol. Chem. **306**, 33 (1957)

[4] Schweiger, H., u. S. Rapoport: Hoppe Seyler's Z. physiol. Chem. **313**, 97 (1956)

[5] Brandt, H., u. S. Rapoport: Acta biol. med. germ. **2**, 290 (1959)

[6] Schultze M., I. Scholz, J. Rost, M. Müller u. S. Just: Folia haemat. **89**, 183 (1968)

[7] Buchmann, R., u. S. Rapoport: Acta biol. med. germ. **13**, 819 (1964)

[8] Rosenthal, S.: VI. Internationales Symposium über Struktur und Funktion der Erythrozyten, Berlin 1970 (= Abhandlungen der Deutschen Akademie der Wissenschaften zu Berlin) S. Rapoport u. F. Jung (Hrsg.). Akademie-Verlag, Berlin 1972, S. 513

[9] Warburg, O., F. Kubowitz u. W. Christian: Biochem. Z. **242**, 170 (1931)

[10] Rapoport, S., J. Rost u. M. Schultze: Eur. J. Biochem. **23**, 166 (1971)

[11] Rost, J., u. S. Rapoport: Eur. J. Biochem. **26**, 106 (1972)

[12] Jänig, G., u M. Schultze: Folia haemat. **89**, 212 (1968)

[13] Rapoport, S., M. Müller u. R. Knöfel: Eur. J. Biochem. **10**, 207 (1969)

[14] Schweiger, H., u. S. Rapoport: Acta biol. med. germ. **1**, 422 (1958)

[15] Ababei, L., u. S. Rapoport: Acta biol. med. germ. **5**, 636 (1960)

[16] Ababei, L., S. Sarkar u. S. Rapoport: Acta biol. med. germ. **8**, 266 (1962)

[17] Rost, J., M. Müller, M. Schultze u. S. Rapoport: FEBS Lett. **24**, 15 (1972)

[18] Braunitzer, G., J. Best, U. Flamm u. B. Schrank: Hoppe-Seyler's Z. physiol. Chem. **347**, 207 (1966)

[19] Flamm, U., J. Best u. G. Braunitzer: Hoppe-Seyler's Z. physiol. Chem. **352**, 885 (1971)

[20] Walsh, D., u. H. Sallach: J. biol. Chem. **241**, 4068 (1966)

[21] Fallon, J., E. Hackney u. W. Byrne: J. biol. Chem. **241**, 4157 (1966)

[22] Bertino, J., B. Simmons u. D. Donohue: J. biol. Chem. **237** 1314 (1962)

[23] Wilmanns, W., u. L. Jaenicke: Klin. Wschr. **41**, 1077, 1027 (1963)

[24] Neiman, M., u. D. Gal: The Kinetic Isotope Method and its Application. Akadémiai Kiadó, Budapest 1971, S. 15

[25] Yoshida, T., u. G. Kikuchi: J. Biochem. **73**, 1013 (1973)
[26] Christensen, H. N., T. R. Riggs u. V. E. Ray: J. biol. Chem. **194**, 40 (1952)
[27] Riggs, R., H. N. Christensen u. J. M. Palatine: J. biol. Chem. **194**, 53 (1952)
[28] Winter, C. G., u. H. N. Christensen: J. biol. Chem. **240**, 3594 (1965)
[29] Vidaver, G. A., L. F. Romain u. F. Haurowitz: Archs Biochem. Biophys. **107**, 82 (1964)
[30] Vidaver, G. A.: Biochemistry **3**, 662 (1969)
[31] Schweiger, H. G., u. S. Rapoport: Acta biol. med. germ. **1**, 422 (1958)
[32] Schweiger, H. G., u. S. Rapoport: Hoppe Seylers Z. physiol. Chem. **313**, 97 (1958)

Summary

M. Müller, S. Rapoport, J. Rathmann, and R. Dumdey: N-Economy and synthesis of serine and glycine in reticulocytes

Amino acids constitute the main substrate of the reticulocyte. The amino acid pool of reticulocytes represents a characteristic selection. The composition of the amino acid pool is dependent on maturation. From the preferential oxidation of short-chain amino acids one would expect a ratio about of 5:1 between the oxygen consumption and ammonia formation. If one corrects for NH_3-formation by the deamination of nucleotides the ratio between the oxygen consumption and NH_3-formation is about an order of magnitude higher than the theoretical ratio. The small liberation of NH_3 in energy production from amino acids results from the re-utilization of their α-NH_2-group for the synthesis of serine and glycine, while the C-skeleton stems from glucose. Serine is formed via Ⓟ-OH-pyruvate and OH-pyruvate. Serine and glycine serve preferentially for the synthesis of hemoglobin. In reticulocytes there exists a compartmentation of glycine which accounts for differences between serine and glycine in isotopic experiments. From the time dependent change of the specific activities of pulse-labelled serine and glycine one may calculate that the serine synthesis amounts to 15—30% of the glucose utilization.

Acta biol. med. germ., Band 36, Seite 353—362 (1977)

Department of Pathological Physiology, Faculty of General Medicine, Charles University, 128 53 Prague 2, Czechoslovakia

The onset of hemoglobin synthesis in spleens of irradiated mice after bone marrow transplantation

P. Poňka, O. Fuchs, J. Borová, J. Neuwirt, and E. Nečas

Summary

Messenger RNA (mRNA) for globin was isolated from spleens of irradiated mice in which erythroid differentiation was induced by a bone marrow graft. The globin mRNA was isolated either by means of sucrose gradients of reticulocyte polysomal RNA or by affinity chromatography of total spleen RNA on poly (U)-sepharose. The globin mRNA was tested in a wheat embryo cell-free system. The appearance of mRNA in the spleen erythroid colonies was correlated with other parameters of erythroid differentiation such as globin synthesis, activity of δ-aminolevulinic acid synthetase and iron uptake. Poly(A) containing mRNA did appear already on the 3rd day after grafting. However, significant translational activity of globin mRNA could be demonstrated only one day later together with the increase in globin synthesis and δ-aminolevulinic acid synthetase and enhanced iron uptake. In the second part of this study mouse spleen cells rich in erythroid elements were incubated with a specific heme synthesis inhibitor (isonicotinic acid hydrazide, INH) and the synthesis of 9 S RNA was estimated. It was found that a 40-minute incubation with INH reduced uridine incorporation into 9 S RNA fraction by about 40%.

Introduction

Erythroid cell development in spleens of heavily irradiated mice grafted with bone marrow cells represents a suitable in vivo model for the study of erythroid differentiation. It has been known for about 15 years that the transplantation of normal bone marrow cells into lethally irradiated mice leads to seeding of pluripotent stem cells in the spleens of recipient mice [1]. After some delay the transplanted pluripotent stem cells, called CFU-S, enter in active cell cycle and proliferate until a critical compartment size is achieved at which time they return to a resting or G_0 state [2]. For both the CFU-S and the compartment of committed erythroid precursors it is believed that a critical size must be achieved before differentiation may occur [2]. In other words, in spleens of mice after irradiation and bone marrow transplantation, erythroid cells differentiate and mature as a cohort and so provide relatively homogenous populations. It should be emphasized that in irradiated animals, the level of erythropoietin — which is essential for erythroid cell differentiation is significantly increased [3, 4].

The aim of this study was to develop the techniques for the isolation of globin mRNA from the spleen erythroid cells and to follow the time-course of this mRNA appearance in spleens of irradiated mice after bone marrow transplantation. Moreover, some other parameters of hemoglobin synthesis were estimated during erythroid cell development in this system.

Materials and methods

Treatment of animals

NMRI mice were used both as donors and recipients of bone marrow cells. Mouse recipients were irradiated 2 hrs before injection of bone marrow cells with a dose of irradiation of 850 R. 5–7 million cells in a final volume of 0.2 ml of media were injected to irradiated recipients [5]. On the third, fourth and fifth day after transplantation the spleens were removed and examined further. In experiments concerning RNA isolation, 50 mouse spleens were usually used. In some experiments 6 hrs ^{59}Fe incorporation into spleens was determined [6].

Isolation of poly(A) containing RNA

Poly(A) containing RNA was isolated from mouse spleens using the method of Nokin et al. [7]. This method consists in the extraction of total RNA from spleen tissue without previous polysome isolation. Spleens were removed, frozen in liquid nitrogen and ground until a powder was obtained at −80 °C. This powder was then rapidly added to 0.1 M Tris-HCl buffer, *p*H 0.9, 0.1 M NaCl, 1 mM EDTA, 1% SDS (20 ml per gram of tissue) and an equal volume of phenol-chloroform (1:1). The whole mixture was then homogenized using a waring blendor for 5 min. The individual phases were separated by centrigfuation at 20,000 × *g* for 10 min and the phenol-chloroform phase was repeatedly extracted with the above buffer. The pooled aqueous phases were made 0.24 M in ammonium acetate, the RNA was precipitated by addition of 2 volumes of cold ethanol and the mixture was kept at −20 °C overnight. The pellet of RNA was obtained by low speed centrifugation and resuspended in 2 M LiCl using a magnetic stirrer (1 hr at 4 °C). This procedure was repeated twice. Under these conditions the final RNA preparation does not contain DNA, residual proteins and light molecular weight RNA species (i.e. below 5 S).

Poly(A) containing RNA was purified by affinity chromatography of ethanol-precipitated total RNA on the column of poly(U)-Sepharose [8].

Hybridization technique

The amount of globin mRNA in poly(A) containing preparation and in total spleen RNA was estimated using the hybridization technique with [^{3}H] cDNA for mouse globin in excess of RNA according to Harrison et al. [9]. [^{3}H] cDNA was mixed with various amounts of RNA's and with a constant excess of yeast tRNA and incubated at 45 °C in hybridization buffer (0.5 M NaCl, 25 mM Tris-HCl, *p*H 6.8, 0.5 mM EDTA, 50% formamide). Before hybridization, the mixture was heated for 10 min at 100 °C. After 36 hrs of incubation at 45 °C each sample was put on a hydroxyapatite column (2 ml packed hydroxyapatite) equilibrated with 0.03 M phosphate buffer, *p*H 6.8, at 60 °C. The single-stranded material was eluted with 0.12 M phosphate buffer, *p*H 6.8, and finally the hybrid was eluted with 0.4 M phosphate buffer, *p*H 6.8, at 60 °C. Each fraction was precipitated with trichloroacetic acid and after collecting precipitates on nitrocellulose filters, the radioactivity was measured in 10 ml of scintillation liquid SLT 41 (Spolana, Neratovice, Czechoslovakia) in a liquid scintillation counter Mark II.

Fractionation of RNA in sucrose gradient

Total RNA isolated from spleen cells (see above) after incubation was layered over 12 ml of a 5 to 20% (w/v) linear sucrose gradient, in 5 mM Tris-HCl, *p*H 7.4, and centrifuged in a IEC B 60 ultracentrifuge, rotor SB 283 (40,000 r.p.m. for 11 hrs at 4 °C). Individual samples were collected, precipitated with trichloroacetic acid, collected on nitrocellulose filtres and counted in SLT 41 as described above.

Incubation of spleen cells

The suspension of spleen cells was prepared by a gentle homogenization of small pieces of spleen in a medium (MEM, ref. [10]) with 20% fetal calf serum using a syringe. After passing through the nylon net and washing with the incubation medium, the cells ($4 \cdot 10^6$/ml) were incubated with L-leucine-[4.5−^{3}H] in the medium containing 10% (v/v) fetal calf serum and 15% mouse serum at 37 °C for various time intervals. The cells were collected by centrifugation at 1000 × *g* for 15 min and lysed with 1.5 ml (per $4 \cdot 10^6$ cells) of 5 mM Tris-HCl buffer,

*p*H 8.0, in an ice bath for 10 min. The carrier hemoglobin (80 mg) was added and the lysate was centrifuged at 20,000 × *g* for 20 min. The mouse globin was prepared using the method of Rossi-Fanelli et al. [11] and washed twice with acetone-HCl, 4times with acetone and dried under vacuum.
Radioactivity in the trichloroacetic acid-precipitable protein was determined according to the method of Mans and Novelli [12].

Wheat embryo cell-free system

The wheat embryo cell-free system was prepared according to Marcus et al. [13] and Roberts and Paterson [14]. The optimal translation of globin mRNA proceeded at 50 mM KCl, 2 mM magnesium acetate, 1 mM ATP, 0.8 mM GTP and 18 mM creatine phosphate.
The preincubation of the postmitochondrial supernatant (S-23) with amino acids and the source of energy was not found to be necessary and was, therefore, omitted.

Separation of globin chains

The separation of α and β chains of mouse globin was performed on CM-cellulose column according to the chromatographic method of Clegg et al. [15]. The first peak represents the unadsorbed material.

Estimation of δ-aminolevulinic acid synthetase activity

The activity of δ-aminolevulinic acid synthetase was determined in the spleen homogenate according to the method of Urata and Granick [16], modified by Marver et al. [17].

Results and discussion

Table 1 shows the relative proportion of various types of erythroid cells on the 3rd, 4th and 5th day after transplantation. On the 4th day there is a considerable increase in the percentage of erythroid elements in the spleen which is caused mainly by the significant elevation of basophilic and polychromatophilic erythroblasts.
In these experiments the weight of spleens was determined and also the amounts of total RNA and poly(A) containing RNA isolated per g of spleen tissue were estimated. We observed practically a linear increase in all these parameters from the third to the fifth day after irradiation and bone marrow transplantation (Fig. 1).
In the same experiment (Fig. 2) the amount of poly(A) containing RNA isolated per g of spleen tissue was compared with other parameters that characterize the cell as erythroid, i.e. iron uptake and the activity of δ-aminolevulinic acid synthetase. It can be seen that both the level of iron incorporation into spleen and the activity of δ-aminolevulinic acid synthetase are very low on the third day after

Table 1

The distribution of various types of erythroid cells in spleens of mice after irradiation and bone marrow transplantation

	Days after irradiation and bone marrow transplantation		
	3	4	5
Proerybl. (%)	0.9	1.4	1.4
Baso. erybl. (%)	0.9	4.1	5.6
Poly. erybl. (%)	4.6	28.0	51.6
Oxy. erybl. (%)	0	0.6	0.6
Total erythroid line (%)	6.4	34.1	59.2

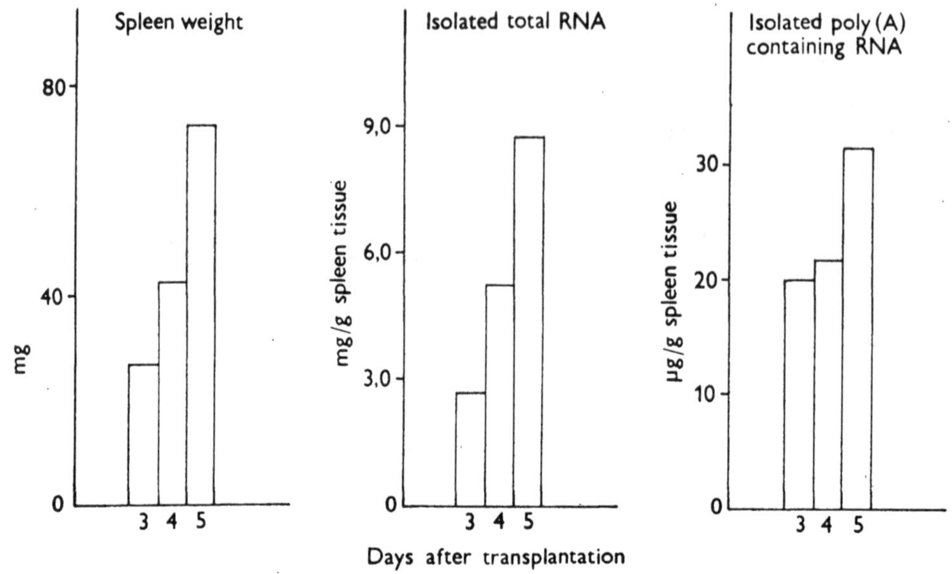

Fig. 1. Changes in some spleen parameters at different times after transplantation

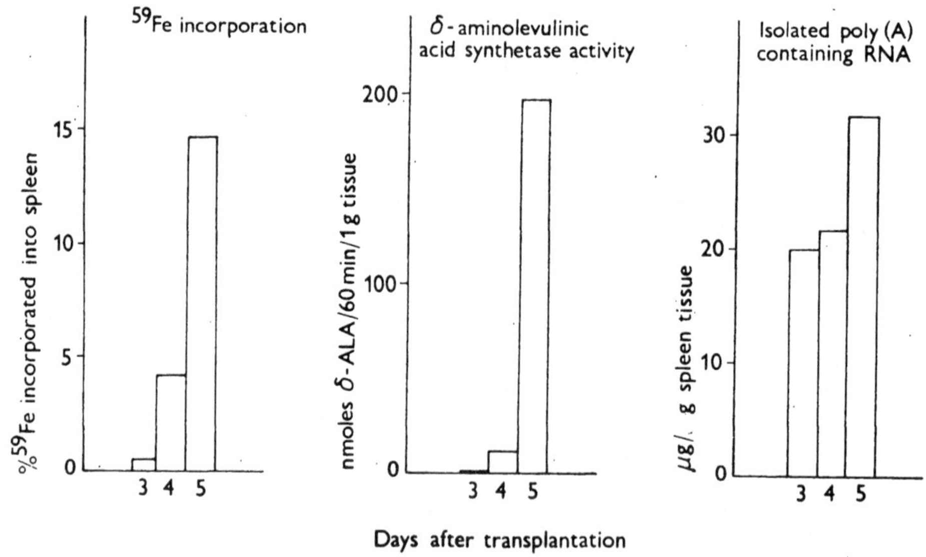

Fig. 2. See Fig. 1

bone marrow transplantation when compared with the significant amount of poly(A) containing RNA that was isolated from the spleen tissue. However, poly(A) containing RNA fraction is a mixture of various messengers and by no means can be considered to be globin mRNA.

Therefore, in the next experiment the biological activity of this poly(A) containing RNA isolated from the spleen tissue was tested. Fig. 3 shows that a significant ability of poly(A) containing RNA to direct globin synthesis in wheat embryo cell-free system can be observed only on the fourth and fifth day after transplantation. The very low translational activity of poly(A) containing RNA isolated from spleens on the third day after transplantation suggests that in this time interval the poly(A) containing fraction is contaminated by various other messengers in addition to globin mRNA.

In this connection we would like to present our preliminary results concerning hybridization technique with labeled cDNA for estimation of globin mRNA sequences in mouse spleen RNA. Table 2 shows that on the 4th day after irradiation and transplantation of bone marrow, the poly(A) containing RNA isolated from the spleen is very rich in mouse globin mRNA sequences. This conclusion ensues from the finding of a high percentage of labeled cDNA in RNA $\times$ cDNA hybrid. This value is almost as high as in a control experiment performed with globin mRNA isolated from mouse reticulocytes. On the other hand, there is a very low extent of hybridization of mouse globin cDNA with the yeast tRNA. At present we do not know whether poly(A) containing RNA isolated on the 3rd day after irradiation and transplantation contains already globin mRNA sequences. The translational activity of poly(A) containing RNA isolated in this time interval is, however, very low (see Fig. 3).

The aim of further experiments was to determine whether the appearance of biologically active globin mRNA, that occurs on the 4th day after irradiation and transplantation, is associated with the synthesis of hemoglobin in spleen cells. The level of total protein synthesis estimated by leucine incorporation is very low on the third day but increases gradually on the 4th and 5th day after bone marrow graft (Fig. 4). The following experiment demonstrates that already on the 4th day after transplantation, there is a significant incorporation of labeled leucine into both α- and β-chains of globin (Fig. 5). These results, therefore, indicate that the increase in the amount of biologically active globin mRNA is associated with the increasing ability of cells to synthesize globin.

There are control mechanisms which closely coordinate the synthesis of heme and globin during erythroid development. Various reports indicate that in erythroid cells heme plays a central role in ensuring that heme and globin are produced in equal amounts. Heme regulates its own synthesis by the repression of the production of δ-aminolevulinic acid synthetase [19] and by inhibition of iron release from transferrin [20, 21]. On the other hand, heme is required for the translation of globin mRNA [22—25]. However, the mechanisms which insure that both heme and globin synthesis are parallelly induced during erythroid cell differentiation are obscure. Since heme seems to be typical of a class of general regulatory molecules, it is tempting to speculate that heme might play a part not only in globin mRNA translation but also in the transcription of this molecule. The involvement of heme in this process could explain the coordination of heme and globin synthesis during early erythropoiesis.

In order to test this possibility, mouse spleen cells rich in erythroid elements (after administration of phenylhydrazine to mice) were incubated with a specific heme synthesis inhibitor and the synthesis of 9S RNA was estimated. Isonicotinic acid hydrazide (INH) was used as the inhibitor of heme synthesis [20, 23]. The following experiment presents evidence that this agent inhibits significantly the

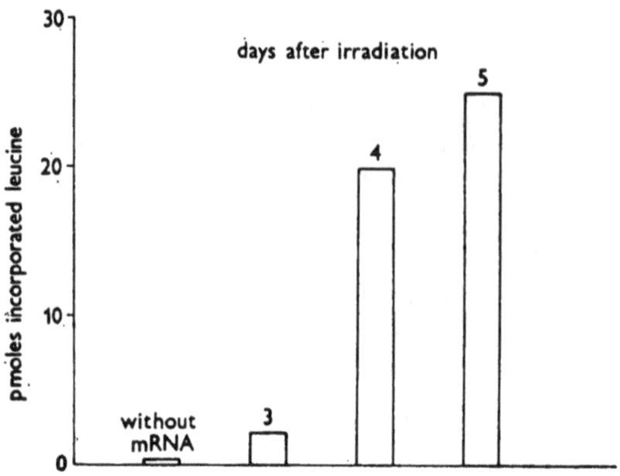

Fig. 3

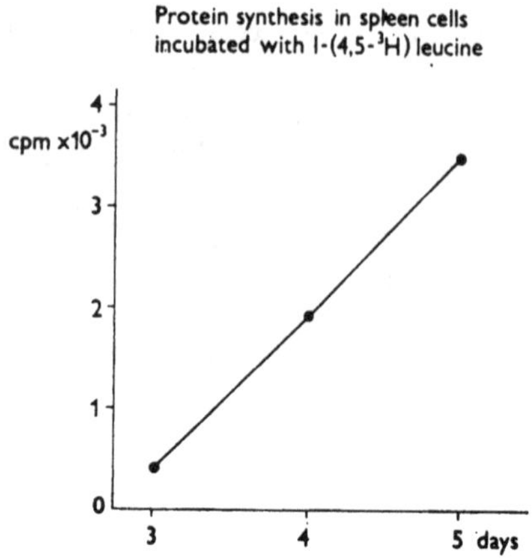

Fig. 4

Fig. 3. Biological activity of poly(A) containing RNA from spleens of irradiated mice in various time intervals after bone marrow transplantation. The cell-free system contained 25 mM Tris-acetate buffer, *p*H 7.6, 50 mM KCl, 2 mM magnesium acetate, 2 mM dithiotreitol, 20 amino acids (concentration 30 μM), 1 mM ATP, 0.8 mM GTP, 15 mM creatine phosphate, creatine phosphokinase (60 μg/ml cell-free system), dialyzed S-23 (ref. [13, 14]) (1 ml per 3 ml cell-free system), 200 μCi [^{3}H]leucine per ml cell-free system and mRNA (32 μg/1 ml cell-free system).

The incubation was carried out at 25 °C for 90 min. The procedure of Mans and Novelli [12] was used for the preparation of the protein synthesis product of cell-free system.

Ordinate: pmoles leucine incorporated per 0.1 ml cell-free system at 25 °C for 90 min

Fig. 4. Incorporation of [^{3}H]leucine into trichloroacetic acid precipitable product formed during incubation of spleen cells in vitro (60 min).

Abscissa: days after irradiation and bone marrow transplantation. Ordinate: cpm per mg of protein precipitated by trichloroacetic acid

Table 2

Comparison of the extent of hybridization of spleen RNA and reticulocyte mRNA with globin [^{3}H] cDNA

RNA fraction	$\left[\frac{\text{cDNA}}{\text{RNA}}\right]$	% [^{3}H] cDNA in hybrid
Control (only yeast tRNA)		6.1
Mouse globin mRNA	10	75.3
from reticulocytes	100	83.1
Spleen poly(A)-containing RNA	10	54.3
	100	77.2
Total spleen RNA	100	77.0

mRNA was isolated from polyribosomes of mouse reticulocytes. Reticulocytosis was induced by phenylhydrazine treatment. 0.4 ng (15,200 cpm) of mouse globin [^{3}H] cDNA and excess yeast tRNA (2 μg) were mixed with various amounts of RNA tested. Each solution was prepared in bidistilled water. After lyophylization, the samples were diluted in 4 μl of hybridization buffer, transferred to glass microcapillaries and then preincubated, incubated and further analyzed as described in Materials and methods.

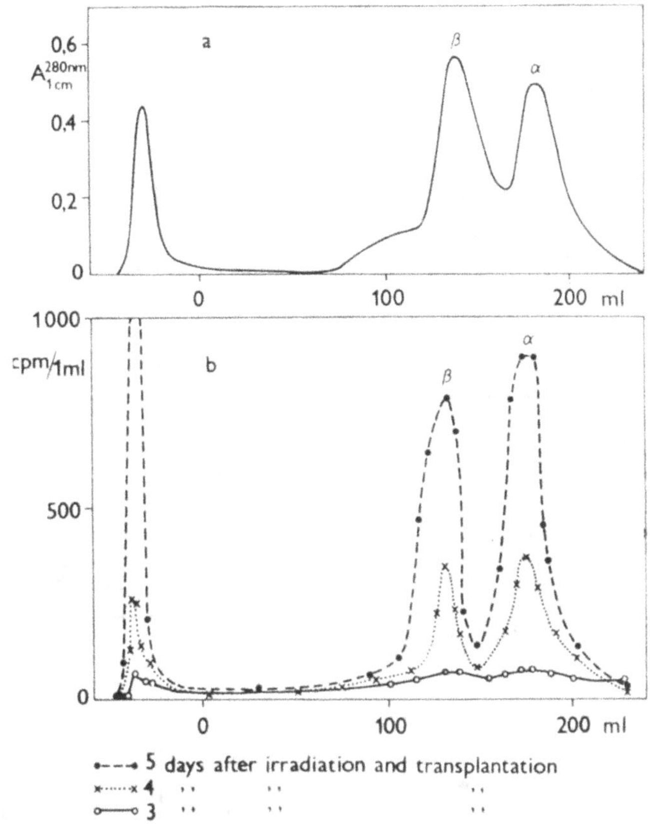

Fig. 5. Synthesis of α- and β-chains of mouse globin in mouse spleen cells incubated in vitro. Spleens were removed and analyzed at various time intervals after irradiation and bone marrow transplantation. 50 mg of globin were put on the column of CM-cellulose (1.2 × 20 cm) and the elution was performed with a linear gradient of 1 to 30 mM phosphate buffer, 8 M urea, 50 mM 2-mercaptoethanol, *p*H 6.7. 1 ml samples were taken from 5 ml fractions, diluted by water to decrease urea concentration and precipitated with trichloroacetic acid (final concentration 10%). Radioactivity of precipitate was determined as described in Materials and methods. a) Chromatography of purified mouse globin on the column of CM-cellulose; b) chromatography of products of protein synthesis in spleen cells incubated in vitro

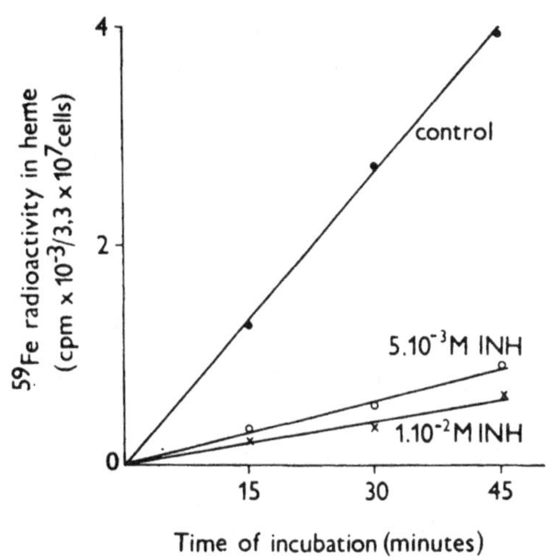

Fig. 6. The effect of INH on ^{59}Fe incorporation into heme of spleen cells incubated in vitro. Spleens were removed from mice after phenylhydrazine treatment. 10^8 spleen cells were incubated in a total volume of 10 ml. The incubation medium (see Materials and methods) contained 15% mouse serum with bound ^{59}Fe (10 μCi per ml of serum). In incubated time intervals samples of 3.3 ml were removed and the incorporation of ^{59}Fe into heme was determined according to Borová et al. [18]

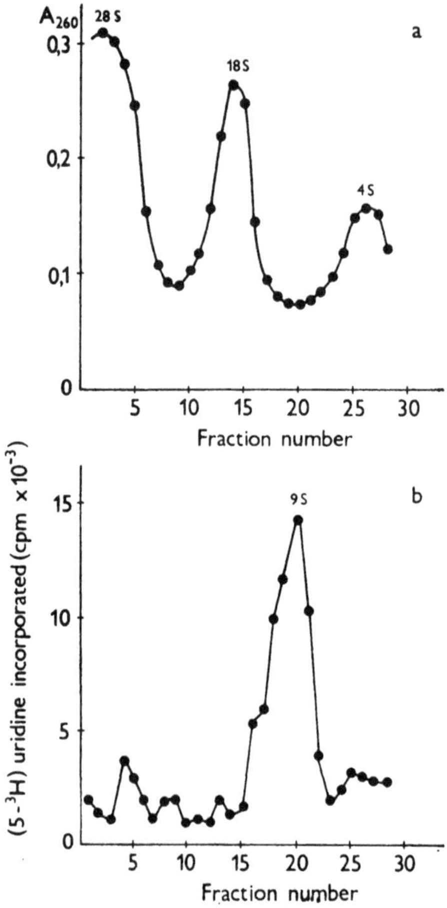

Fig. 7

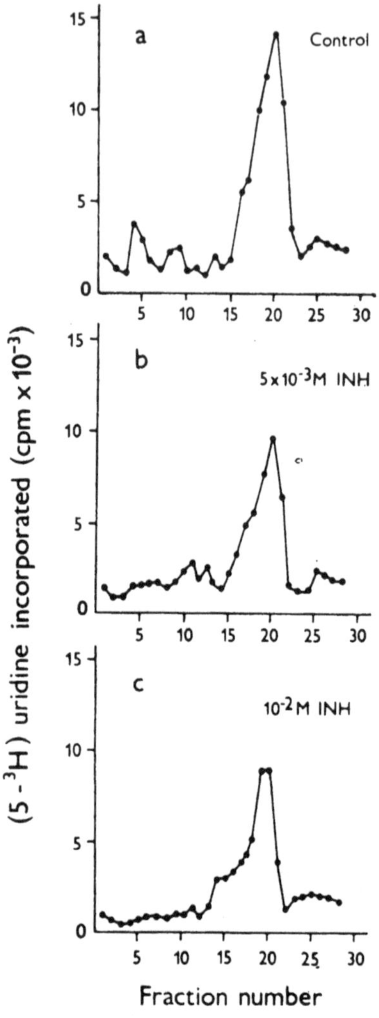

Fig. 8 ▶

Fig. 7. a) Sedimentation profile of total RNA isolated from mouse reticulocytes in 5 to 20% (w/v) linear succrose gradient in 5 mM Tris-HCl, *p*H 7.4.
b) Sedimentation profile of total spleen RNA isolated from mouse spleen cells after incubation. The cells were preincubated for 20 min, then [^{3}H]uridine (150 μCi/40 ml of cell suspension) was added and incubation continued for further 20 min. 180 μg of total spleen RNA were layered over 12 ml of 5 to 20% (w/v) sucrose gradient in 5 mM Tris-HCl, *p*H 7.4. Centrifugation conditions are described in Materials and methods

Fig. 8. Sedimentation profiles of total spleen RNA isolated after a) 20-minute preincubation and 20-min incubation with [^{3}H]uridine (see Fig. 7b); b) 20-min preincubation with INH ($5 \cdot 10^{-3}$ M) which was also present during further 20-min incubation with [^{3}H]uridine; c) the same as in b) but the concentration of INH was 10^{-2} M. In each experiment, the same amount (180 μg) of total spleen RNA was analyzed on gradients (centrifugation conditions see Fig. 7)

synthesis of heme in mouse spleen cells incubated in vitro (Fig. 6). Fig. 7 represents a control experiment showing [^{3}H]uridine incorporation into 9S RNA fraction of spleen cells after in vitro incubation. The effect of 20-minute preincubation with

INH on the incorporation of labeled uridine into 9 S RNA is shown in Fig. 8. It is evident that a 40 min incubation with the inhibitor of heme synthesis reduces uridine incorporation into 9 S RNA fraction by about 40 per cent. This result suggests that heme might be involved in the synthesis of 9 S RNA, probably globin mRNA. After addition of INH the synthesis of heme was apparently more depressed than the synthesis of 9 S RNA (Fig. 6 and Fig. 8). However, nucleated erythroid cells contain preformed heme [26] that may remain available even though new heme synthesis is markedly inhibited by INH.

On the basis of these results we wish to present the following tentative hypothesis concerning biochemical events during erythroid differentiation. Erythropoietin, probably indirectly [27], triggers the appearance of globin mRNA [28—31] and the first globin chains that are synthesized combine with preformed heme. It should be pointed out that a low level of heme synthesis must occur already in erythroid precursors. The decreased level of intracellular heme induces the synthesis of δ-aminolevulinic acid synthetase and facilitates the release of iron from transferrin (see above). This is accompanied by a significant increase in the rate of heme synthesis which is followed by further stimulation of globin mRNA transcription.

Acknowledgement

We are grateful to Dr. Gerard Marbaix for mouse globin [^{3}H]cDNA which was used in hybridization experiments.

References

[1] Till, J. E., and E. A. McCulloch: Radiat. Res. **14**, 213 (1961)
[2] Stohlman, F. Jr., P. Quesenberry, and W. S. Tyler: Prog. Hemat. **8**, 259 (1973)
[3] Stohlman, F. Jr.: Ann. N.Y. Acad. Sci. **77**, 710 (1959)
[4] Stohlman, F. Jr., and G. Brecher: Proc. Soc. exp. Biol. Med. **100**, 40 (1959)
[5] Nečas, E., P. Poňka, and J. Neuwirt: Cell Tissue Kinet. **9**, 223 (1976)
[6] Trávníček, T., and J. Neuwirt: Acta Univ. Carol. Med. Monogr. **34**, 219 (1968)
[7] Nokin, P., A. Burny, Y. Cleuter, G. Huez, G. Marbaix, and H. Chantrenne: Eur. J. Biochem. **53**, 83 (1975)
[8] Lindberg, V., and T. Persson: Eur. J. Biochem. **32**, 420 (1972)
[9] Harrison, P. R., G. D. Birnie, A. Hell, S. Humphries, B. D. Young, and J. Paul: J. molec. Biol. **84**, 539 (1974)
[10] Eagle, H.: Science, N.Y. **130**, 432 (1959)
[11] Rossi-Fanelli, A., E. Antonini, and A. Caputo: Biochim. biophys. Acta **30**, 608 (1958)
[12] Mans, R. J., and G. D. Novelli: Archs Biochem. Biophys. **94**, 48 (1961)
[13] Marcus, A., D. Efron, and D. P. Weeks: Meth. Enzymol. **30**, 749 (1974)
[14] Roberts, B. E., and B. M. Paterson: Proc. natn. Acad. Sci. U.S.A. **70**, 2330 (1973)
[15] Clegg, J. B., M. A. Naughton, and D. J. Weatherall: Nature, Lond. **207**, 945 (1965)
[16] Urata, G., and S. Granick: J. biol. Chem. **238**, 811 (1963)
[17] Marver, H. S., D. P. Tschudy, M. G. Perlroth, A. Collins and G. Hunter: Analyt. Biochem. **14**, 53 (1966)
[18] Borová, J., P. Poňka, and J. Neuwirt: Biochim. biophys. Acta **320**, 143 (1973)
[19] Sassa, S., and S. Granick: Proc. natn. Acad. Sci. U.S.A. **67**, 517 (1970)
[20] Poňka, P., and J. Neuwirt: Blood **33**, 690 (1969)
[21] Poňka, P., J. Neuwirt, and J. Borová: Enzyme **17**, 91 (1974)
[22] Bruns, G. P., and I. M. London: Biochem. biophys. Res. Commun. **18**, 236 (1965)
[23] Poňka, P., J. Neuwirt, M. Šperl, and Z. Březik: Biochem. biophys. Res. Commun. **38**, 817 (1970)

[24] Rabinovitz, M., M. L. Freedman, J. M. Fisher, and C. R. Maxwell: Cold Spring Harb. Symp. quant. Biol. **34**, 567 (1969)
[25] Mathews, M. B., T. Hunt, and A. Brayley: Nature New Biol. **243**, 230 (1973)
[26] Glass, J., L. M. Lavidor, and S. H. Robinson: 18th Annual Meeting of the American Society of Hematology, Dallas, Texas, 1975, Abstract No. 395
[27] Chang, C. S., and E. Goldwasser: Develop. Biol. **34**, 246, (1973)
[28] Gross, M., and E. Goldwasser: Biochemistry **8**, 1795 (1969)
[29] Gross, M., and E. Goldwasser: J. biol. Chem. **246**, 2480 (1971)
[30] Terada, M., L. Cantor, S. Metafora, R. A. Rifkind, A. Bank, and P. A. Marks: Proc. natn. Acad. Sci. U.S.A. **69**, 3575 (1972)
[31] Ramirez, F., R. Gambino, G. M. Maniatias, R. A. Rifkind, P. A. Marks, and A. Bank: J. biol. Chem. **250**, 6054 (1975)

Acta biol. med. germ., Band 36, Seite 363—372 (1977)

Institut für Physiologische und Biologische Chemie der Humboldt-Universität Berlin, 104 Berlin, DDR

Degradation of mitochondria by cytosolic factors in reticulocytes

T. SCHEWE, W. HALANGK, CH. HIEBSCH, and S. RAPOPORT

Summary

The degradation process of mitochondria in rabbit reticulocytes proceeds predominately directly in the cytosol rather than in secondary lysosomes as judged by electronmicroscopy. At least five cytosolic protein factors are present in reticulocytes, which could be related to the degradation of mitochondria: the two inhibitory proteins of the respiratory chain RF and RC and three enzymes which cause a lysis of mitochondria in vitro (lipoxygenase, proteinase, phospholipase A). The properties of these factors are the subject of this paper. A hypothetic scheme of the degradation of mitochondria in reticulocytes is proposed. The degradation of mitochondria in reticulocytes is viewed as a complex interplay of various cytosolic factors and the functional state of the mitochondrial membranes. The lipoxygenase damages the membranes and triggers the penetration of the respiratory inhibitors. In this manner, a catastrophic cycle is initiated which leads to the complete breakdown of the mitochondria.

Introduction

The reticulocyte represents an intermediate stage of maturation of the red cell, which is characterized by the elimination or inactivation of the nucleus on the one hand, and by the presence of functional ribosomes and mitochondria, on the other. The last stage of maturation; i.e. the transition from the reticulocyte to the normocyte, is characterized among other things by the largely complete disappearance of mitochondria, which leads to a basic switch-over in the energy metabolism of the cell to exclusive glycolysis. The breakdown of mitochondria goes hand in hand with a strong decline or complete loss of mitochondrial enzymes. Some maturation-dependent enzymes are listed in Table 1.

Our research has been focused on the following question: Which are the mechanisms of the selective inactivation and breakdown of mitochondria as well as of the corresponding enzymes during the maturation of the red cell? The speed of these events argues in favour of active processes rather than of stochastic ones. The nature of these processes will be the subject of the present contribution.

The reticulocyte is an excellent model for the investigation of the basic principles of the degradation of mitochondria. It may be hoped that the events

Table 1

Strongly maturation-dependent mitochondrial enzymes in reticulocytes

Inner membrane
Cytochrome oxidase
Succinate dehydrogenase
Matrix
δ-Aminolaevulinic acid synthetase
NADP-dependent isocitrate dehydrogenase
Aspartate aminotransferase
α-Oxoglutarate dehydrogenase
Outer membrane
Glutaminase
Desoxyribonuclease

in the reticulocyte are applicable to other kinds of cells, despite their specialization. The choice of an appropriate cellular model object is a prerequisite for the study of the principles of the degradation of mitochondria. The reticulocyte possesses the following advantages in comparison with other cells:

1) The absence of new-formation of organelles due to the absence of an active nucleus and, consequently, of RNA synthesis,

2) the practical absence of the endoplasmic reticulum — except remnants [1] — and, therefore, the absence of the interplay with this organelle,

3) the nearly synchronized degradation of mitochondria during the maturation process,

4) the convenient availability of large amounts of reticulocytes.

Morphological studies

Some conclusions as to the mechanism of degradation may be drawn from the morphological appearance of the mitochondria in intact reticulocytes during the maturation process; these have been studied by KRAUSE et al. [2] as well as by GASKO and DANON [3]. The mitochondria of immature reticulocytes exhibit mostly orthodox structures (Fig. 1 a). Both condensed (Fig. 1 b) and vacuolized forms occur (Fig. 2) in the same cell. The partly degraded, vacuolized mitochondria predominate in the more mature cells. A few of the degradation forms of the mitochondria are located in autophagic vacuoles (secondary lysosomes, (Fig. 3)). In these vacuoles activity of β-glycerophosphatase as a marker enzyme of lysosomes was detected by Dr. W. KRAUSE by means of a histochemical method (Fig. 4). The digestion of mitochondria in phagosomes seems to play an important role in a variety of cells such as Kupffer cells [4] but appears to be of minor importance for the reticulocytes. Since most of the partly degraded forms of mitochondria

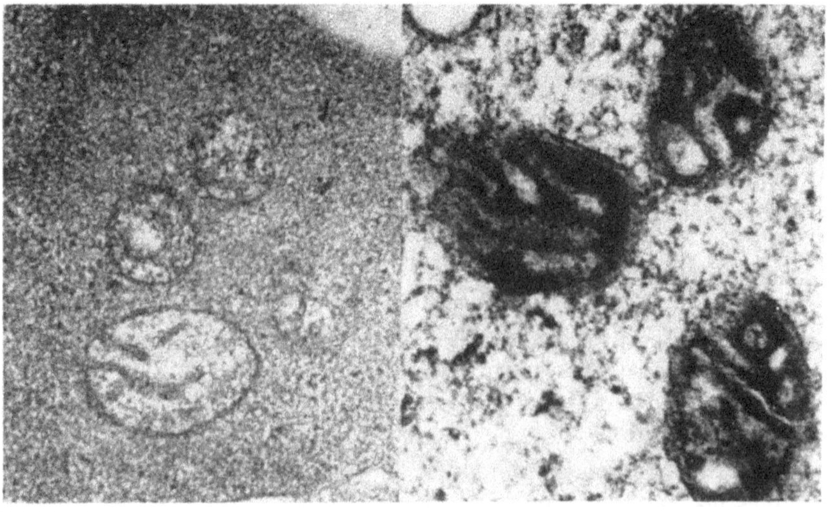

Fig. 1. Mitochondria of rabbit reticulocytes in situ.
1 a: orthodox forms; 1 b: condensed forms; magnification: 50,000:1

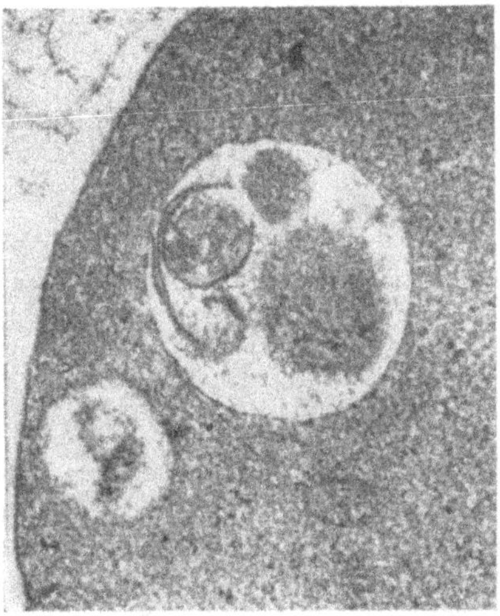

Fig. 2

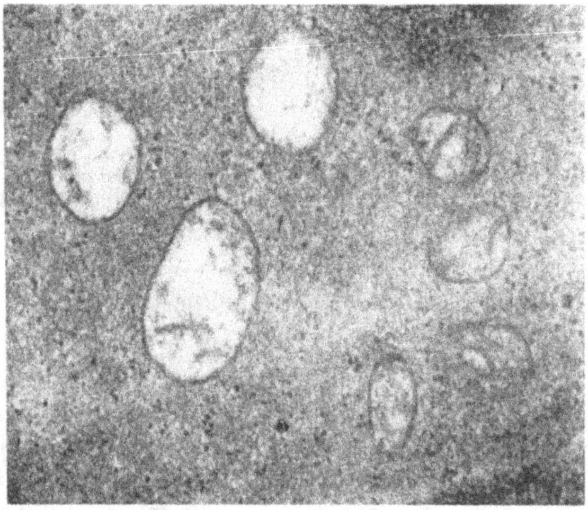

Fig. 3

Fig. 2. Phagocytic vacuoles in a rabbit reticulocyte.
Magnification: 50,000:1

Fig. 3. Partially degraded swollen and vacuolized mitochondria on the left and normal orthodox ones (on the right) in the same reticulocyte.
Magnification: 40,000:1

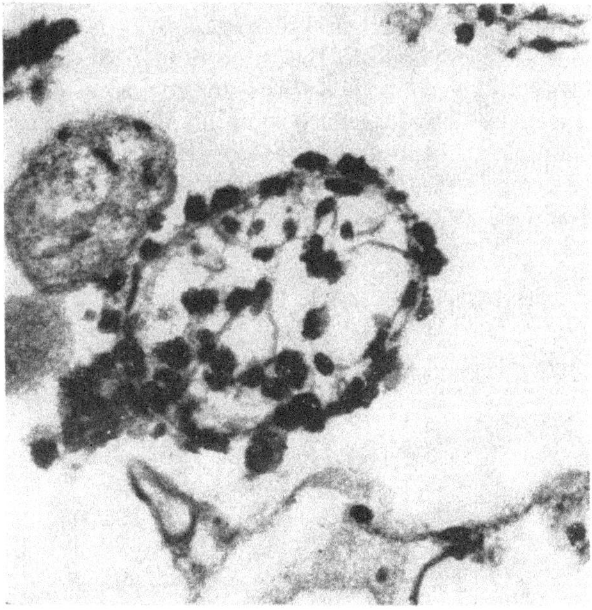

Fig. 4. Acid phosphatase in autophagic vacuoles of the reticulocyte. Reticulocytes from rabbits were cautiously haemolyzed by freezing and thawing to remove the bulk amount of haemoglobin which disturbs the histochemical reaction. The detection of acid phsophatase with β-glycerophosphate as substrate was carried out by the method of MATANO and ISHII [19]. The dark spots represent precipitations of lead phosphate.
Magnification: 28,000:1

appear directly in the cytosol rather than in autophagic vacuoles, the degradation process has to be brought about by means of cytosolic factors. Therefore, we have searched for factors in the cytosol of reticulocytes which are related to the degradation of mitochondria. We could detect five protein factors which were identified as:

— an inhibitor of the iron-sulphur region of the respiratory chain (RF) [5,6],
— an inhibitor of the cytochrome oxidase (RC) [7,8],
— lipoxygenase [9],
— a proteolytic enzyme [10],
— a Ca^{2+}-stimulated phospholipase A [11].

The respiratory inhibitors

Fig. 5 shows the action sites of the inhibitors.

RF suppresses both the NADH and the succinate oxidase system of submitochondrial particles on the level of the oxidized forms of the iron-sulphur proteins of complex I and complex II. The other inhibitor, RC, acts on the oxidized species of the cytochrome oxidase. These action sites were established by measuring the enzymatic activity of partial systems of the respiratory chain, by a comparative analysis with other respiratory inhibitors and by investigation of the influence of the redox state of the submitochondrial particles on the inhibitory action [6,8]. The effect of RF is additive in the succinate oxidase system with that of the iron-chelating agent TTFA, and with rotenone and amytal in the NADH-system [6]. The action of both inhibitors is prevented by succinate under anaerobic conditions, whereas subsequent aeration leads to the inhibition [6, 8]. The points of action of the two inhibitors could be confirmed with isolated respiratory enzymes. As Fig. 6 shows, soluble succinate dehydrogenase is inhibited by RF. Fig. 7 demonstrates the inhibition of soluble cytochrome oxidase by RC. In Table 2 some molecular properties of RF are compiled. The molecular properties of RF including the amino acid composition as well as the purification procedure in detail are the subject of the contribution by WIESNER et al. [12]. RF seems to be present only in reti-

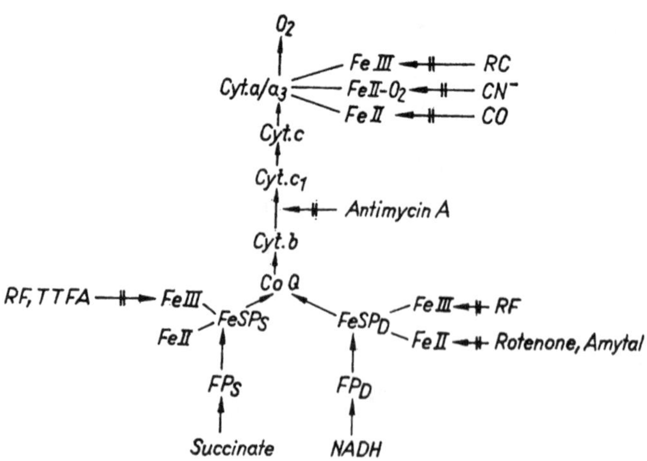

Fig. 5. Scheme for the action sites of RF and RC in the respiratory chain

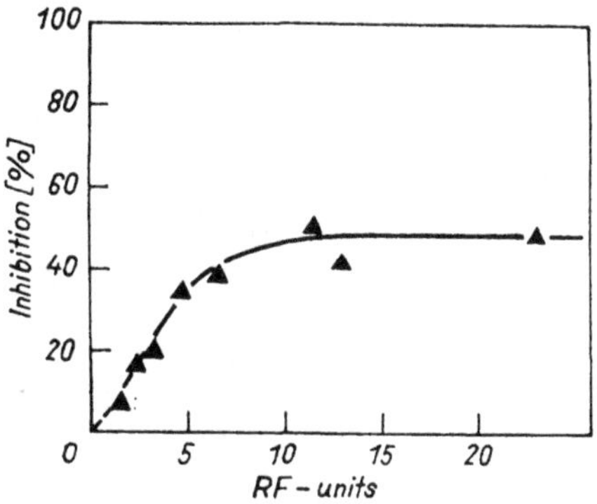

Fig. 6

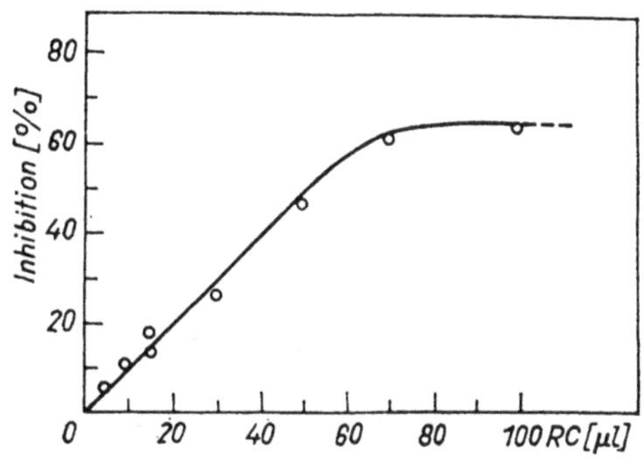

Fig. 7

Fig. 6. Inhibition of soluble succinate dehydrogenase by RF

Fig. 7. Inhibition of soluble cytochrome oxidase by RC

culocytes. TANNERT failed to detect the protein by immunodiffusion both in bone marrow cells and in mature red cells (unpublished results). A screening for RF protein in other tissues of the rabbit by means of immunodiffusion did not supply any hint of the occurrence in cells other than reticulocytes. The various reticulocyte populations appearing in the course of bleeding anaemia differ considerably with respect to their RF content [13]. The RF activity in separated cell fractions revealed a marked biological dynamics as shown in Fig. 8. Here the cytochrome oxidase as a mitochondrial marker enzyme and the RF activity are plotted versus the decreasing RNA content which is a measure of the maturational stage of the reticulocyte. In very young, ribosome-rich cells no inhibitor is detectable. The maximal activity is reached at the very time of the steepest decline of mitochondrial constituents. Subsequently the inhibitor disappears. This may be due to the interaction with phospholipids which are liberated during the degradation process of mitochondria and which inactivate the inhibitor. From this biological dynamics it is evident that 1) the degradation of mitochondria and that of ribosomes have different kinetics and 2) RF is very likely responsible for the relatively rapid inactivation of the respiratory system of the mitochondria.

Table 2
Some molecular properties of RF

Property	Evidence
Single polypeptide chain with MW of 80,000	SDS electrophoresis; ultracentrifugation
N-terminal glycine	dansylation method
IP: 5.55	isoelectric focusing
Carbohydrate moiety	staining on electropherograms; anthrone method
Functional Fe^{II}	chelating agents
Essential SH-group(s)	SH-reagents; amino acid analysis
Affinity to phospholipids	electrophoresis; inactivation by phospholipids
Tendency to aggregation and inactivation → counteracted by divalent metals	storage and disc electrophoresis

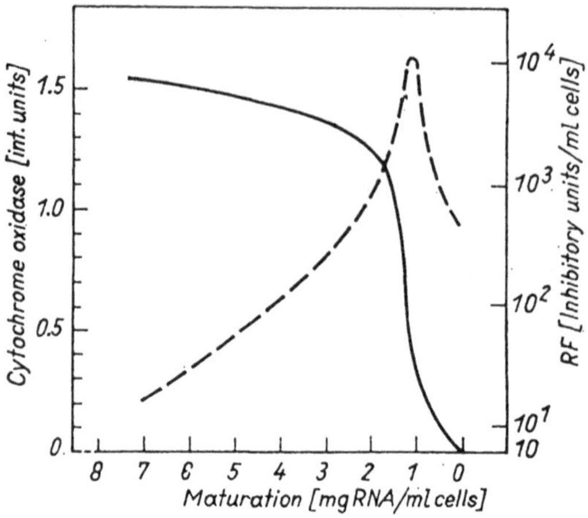

Fig. 8. Biological dynamics of cytochrome oxidase (———) and RF (— — —) during the maturation process of the reticulocyte. The age of the cell populations increases with decreasing RNA content.

Factors causing lysis of mitochondria

RF as a macromolecule should not be able to penetrate through the mitochondrial membranes. Nevertheless, we have repeatedly observed that crude preparations of RF inhibit the respiratory chain of intact mitochondria in a relatively short time. Therefore, we have searched for factors in the reticulocyte supernatant, which induce lysis of mitochondria. We used the release of enzymes of the matrix and of the intermembrane space as a test system for such lysis factors. Three lysis factors could be detected and partially characterized (Fig. 9). One factor turned out to be a low-molecular calcium-stimulated phospholipase A [11]. This enzyme seems to be only of secondary importance for the breakdown of mitochondria; it was not further characterized by us. Some hints for the presence of a phospholipase D were obtained by the group of AUGUSTIN [14]. Another lysis factor has been identified by us as a proteolytic enzyme. It liberates free amino groups from both haemoglobin and intact rat liver mitochondria. In all attempts to purify this factor, such as salt precipitation, cation exchange chromatography, membrane filtration etc. it behaved as a concomitant of haemoglobin. This proteinase contributes a considerable part of the total lytic activity of the stroma-free supernatant fluid of reticulocytes, but the extent of release of malate dehydrogenase never exceeded values of 20—30%, showing that this factor alone is not able to cause a complete rupture of the mitochondrial membranes.

By contrast, the main lysis factor which later turned out to be a lipoxygenase causes under proper conditions a total release of matrix enzymes. The action of the lipoxygenase leads to drastic damages of the mitochondrial membranes as may be seen from electronmicrographs. Fig. 10 shows the appearance of mitochondria before and after action of lipoxygenase for 5 and 15 minutes respectively. One can observe disappearance of the outer membranes and the cristae, loss of matrix constituents as well as deformation of the remaining inner membrane. These alterations are accompanied by a formation of malonyl dialdehyde, which is a secondary product of the peroxidation of membrane phospholipids. The formation of malonyl dialdehyde and the release of malate dehydrogenase coincide with each other largely in their extent with respect to the dependence on the amount of lipoxygenase, on time and on temperature. From the temperature dependence an energy of activation of about 11 kcal/moles and a Q_{10} value of 1.8 were calculated [9]. The action of the lipoxygenase on mitochondria shows little dependence on *p*H. Some properties of the reticulocyte lipoxygenase are listed in Table 3. The properties of the enzyme differ from the arachidonate lipoxygenase recently described in blood platelets [15]. Furthermore, there are both common features and differences as compared with the known lipoxygenases of beans and cereals.

A striking phenomenon is the dependence of the action of the lipoxygenase on the functional state of the mitochondria. The action on mitochondria is prevented by either ATP or succinate plus ADP (Table 4). The protective effect by substrates

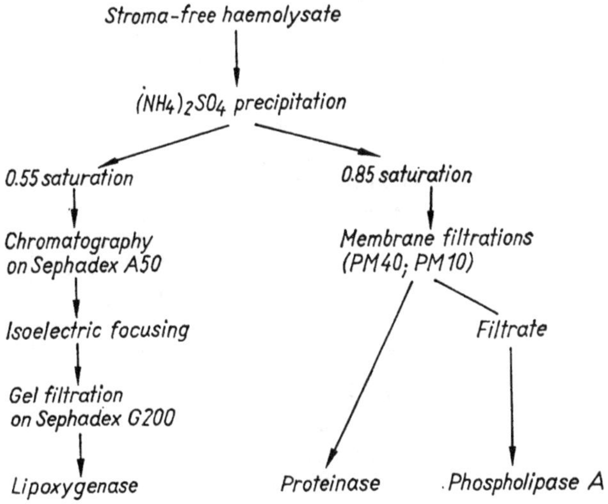

Fig. 9. Scheme of the separation of the mitochondrialysis factors from rabbit reticulocytes

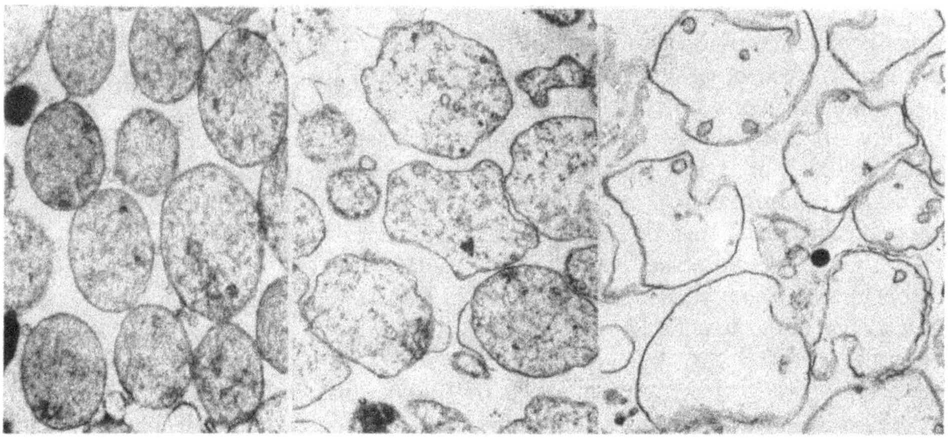

Fig. 10. Action of the reticulocyte lipoxygenase on isolated rat liver mitochondria. From the left to the right: mitochondria after 0, 5 and 15 minutes incubation with MLF at 37° in 0.3 M buffered sucrose.
Magnification: 50,000:1

Table 3
Some properties of the reticulocyte lipoxygenase

M.W.:	120,000	*p*H-optimum:	7.4–7.8 (linoleic acid)
I.P.:	5.55	K_m:	$1 \cdot 10^{-4}$ M (linoleic acid)
Substrates:	mitochondria	Inhibitors:	antioxidants
	phospholipids		– propyl gallate
	– phosphatidyl ethanolamine		– α-naphthol (not tocopherol!)
	– lecithin		chelating agents
	– cardiolipin		– KCN
	polyunsaturated fatty acids		– tiron
	– linoleic acid (18:2)		serum albumin (if fatty acids are substrate), product(s) of linoleic acid peroxidation, haemoglobin
	– linolenic acid (18:3)		
	– eicosatrienoic acid (20:3)		
	– arachidonic acid (20:4)		

Table 4
Suppression by 10 mM succinate plus 1 mM ADP of the lipoxygenase action on rat liver mitochondria

Incubation	Method	Action [% MDH release or E_{532})	Effect [%]
Aerobic	MDH-release	90	
Aerobic + substrate	MDH-release	2	−98
Anaerobic	MDH-release	56	−38
Anaerobic + substrate	MDH-release	10	−89
Aerobic	MDA-formation	0.182	
Aerobic + substrate	MDA-formation	0.078	−57

Incubation 30 min 37 °C; MDH = malate dehydrogenase, MDA = malonyl dialdehyde (determined by the thiobarbituric acid method)

is partly counteracted by respiratory inhibitors or uncouplers. One may assume that the protection against lipoxygenase is caused by energization of the inner mitochondrial membrane, but further experiments are necessary to elucidate the relationship. It should be mentioned that the susceptibility against lipoxygenase behaves very similar as certain fluorescence probes used to detect the energy state of the mitochondria [16]. By contrast, no correlation could be found between lipoxygenase attack and the conformational state of mitochondria [17].

A possible objection to the assumption of energization of the mitochondria as the reason for the lipoxygenase resistance is the fact that the first point of attack of the lipoxygenase must be the outer membrane, whereas energization takes place in the inner membrane.

The basis for the varying susceptibility to lipoxygenase may consist in the state of protein-lipid interactions in the membranes. The degree of hydrophobic interactions between the fatty acid moieties of the phospholipids and helical segments of the membrane proteins may lead to different restrictions of the fluidity of the lipids and, therefore, to varying structural constraints, so that the isolene groupings as the site of action of lipoxygenase are masked more or less. The importance of the protein-lipid interactions is furthermore indicated by the opposite temperature dependence of the formation of malonyl dialdehyde from mitochondria and from pure phospholipids. Whereas only little malonyl dialdehyde was formed from mitochondria below 15°, the amounts from phospholipids at low temperatures were 20 times as high as at 37°.

Hypothesis

One may assume that a complex interplay exists between the various protein factors described here and the functional state of the mitochondria, which leads to the complete breakdown of mitochondria. A hypothetic scheme is presented in Fig. 11.

The numbers indicate the possible order of events. The lipoxygenase may induce holes in both membranes by peroxidation of their lipids (event 1). The attack on the inner membrane is dependent on the functional state which could be influenced by the age of the mitochondria. The lipoxygenase action permits the penetration of the respiratory inhibitors which suppress the respiration irreversibly (event 2). The interruption of the energy production via the respiratory chain leads to an

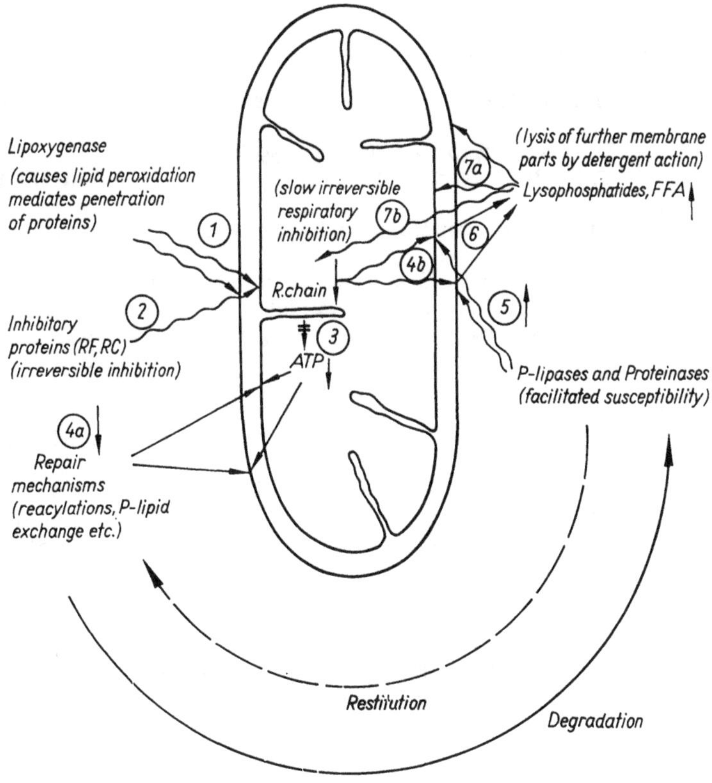

Fig. 11. Proposed scheme for the degradation of mitochondria in reticulocytes. The numbers indicate the possible order of events (see text).

ATP deficiency (event 3). Therefore, repair mechanisms such as reacylation of lysophosphatides, phospholipid exchange and mitochondrial protein synthesis are diminished (event 4a). On the other hand, the absence of an electron flux is known to cause a greater susceptibility of the inner membrane to proteinases and phospholipases (event 4b) [18]. Both effects lead to a preponderance of the action of degrading enzymes over the repair and renewal reactions (arrows below, event 5) Furthermore, phospholipases produce lysophosphatides and free fatty acids (event 6) which are multi-site damaging agents of the mitochondrial metabolism (events 7a and 7b). In this manner, the degradation process — once initiated — is apt to have a self-accelerating character leading to a catastrophic cascade with a complete degradation of mitochondria. Further efforts have to be made to elucidate the complex interplay of the degradation process of mitochondria in reticulocytes.

Acknowledgement

The authors are indebted to Dr. W. KRAUSE, Humboldt-University, Pathological Institute, Division of Electronmicroscopy, Berlin, GDR, for the kind performance of the electronmicrographs.

References

[1] AUGUSTIN, W., F. GELLERICH, I. WISWEDEL, and I. STEINBRECHT: persönliche Mitteilung
[2] KRAUSE, W., H. DAVID, I. UERLINGS, and S. ROSENTHAL: Acta biol. med. germ. **28**, 779 (1972)
[3] GASKO, O., and D. DANON: Expl Cell Res. **75**, 159 (1972)
[4] GLAUMANN, H., I. K. BEREZESKY, J. L. E. ERICSSON, and B. F. TRUMP: Lab. Invest. **33**, 239 (1975)
[5] RAPOPORT, S., and W. GRISCHER-MOTHES: Hoppe-Seyler's Z. physiol. Chem. **302**, 167 (1955)
[6] SCHEWE, T., CH. HIEBSCH, and S. RAPOPORT: Acta biol. med. germ. **29**, 189 (1972)
[7] ALTENBRUNN, H.-J., and S. RAPOPORT: Acta biol. med. germ. **2**, 599 (1959)
[8] WIESNER, R., and S. RAPOPORT: Acta biol. med. germ. **31**, 289 (1973)
[9] SCHEWE, T., W. HALANGK, CH. HIEBSCH, and S. M. RAPOPORT: FEBS Lett. **60**, 149 (1975)
[10] HALANGK, W., and CH. HIEBSCH: Unpublishing results (1974)
[11] GARCIA PARRA, M., T. SCHEWE, and S. M. RAPOPORT: Acta biol. med. germ. **34**, 1075 (1975)
[12] WIESNER, R., CH. TANNERT, and E. BARZ: In Vorbereitung
[13] WIESNER, R., S. ROSENTHAL, and CH. HIEBSCH: Acta biol. med. germ. **30**, 631 (1973)
[14] WISWEDEL, I., G. LUTZE, J. BARANSKA, J. ZBOROWSKI, and W. AUGUSTIN: Acta biol. med. germ. **36**, 567 (1977)
[15] NUGTEREN, D. H.: Biochim. biophys. Acta **380**, 299 (1975)
[16] BEREITER-HAHN, J.: Biochim. biophys. Acta **423**, 1 (1976)
[17] KRAUSE, W., and W. HALANGK: Acta biol. med. germ. **36**, 381 (1977)
[18] LUZIKOV, V. N., and L. V. ROMASHINA: Biochim. biophys. Acta **267**, 37 (1972)
[19] MATANO, S., and S. ISHII: Okajimas Folia anat. jap. **44**, 1 (1967)

Acta biol. med. germ., Band 36, Seite 373–374 (1977)

Laboratory of Biological Chemistry, Department of Molecular Biology, University of Brussels, Brussels, Belgium

Some properties of globin messenger RNA from a free cytoplasmic nucleoprotein of rabbit reticulocytes

G. Huez, G. Marbaix[1], P. Nokin, and Y. Cleuter

Summary

Free cytoplasmic globin messenger RNA of rabbit reticulocytes is less labelled than polysomal mRNA if anaemic rabbits are injected with [^{32}P]-orthophosphate 15 hrs before their sacrifice. It also possesses a poly(A) segment shorter than that of polysomal mRNA. We thus conclude that free cytoplasmic globin mRNA appears during the maturation of rabbit reticulocytes.

A free nucleoprotein (mRNP) containing mainly α globin messenger RNA (mRNA) is found in the cytoplasm of rabbit reticulocytes [1].

This entity might represent the form under which mRNA travels from the nucleus to the translation sites in the cytoplasm: in this case, the mRNA it contains should be younger than polysomal mRNA. Free mRNP might also be the last form under which one finds mRNA in maturing reticulocytes: in this case, its mRNA should be older than polysomal mRNA.

In order to see which one of these two possibilities is true, we decided to compare polysomal and free cytoplasmic mRNA isolated from reticulocytes of anaemic rabbits injected with [^{32}P]-orthophosphate 15 hrs before their sacrifice. It is known that, in these labelling conditions, mRNA from young reticulocytes which just appeared in the peripheral blood will be the most radioactive [2]. Polysomes and a *p*H 5 precipitate from the post-ribosomal supernatant of these labelled rabbit reticulocytes have been prepared according to Gianni et al. [1]. Globin mRNA from both sources has been isolated by proteinase K digestion and poly(U)-sepharose affinity chromatography.

In our experiment, 19% of globin mRNA was found free in the cytoplasm and its specific radioactivity was 75% of that of polysomal mRNA. Free mRNA thus looks older than polysomal mRNA.

55% of the free cytoplasmic mRNA was eluted from the poly(U)-sepharose column at 25 °C with low ionic-strength buffer whilst only 25% of the polysomal mRNA was eluted in the same conditions. This means that free mRNA has a shorter poly(A) stretch and is thus older [3] than polysomal mRNA.

It is thus proven that, in mammalian reticulocytes, free mRNA represents mainly the last stage of the message life. This is in perfect agreement with our previous observations [2] showing that rabbit reticulocytes ageing is associated with a decrease in size of polysomes and thus with an increasing inefficiency of initiation of translation. In the oldest reticulocytes, lack of initiation leads to the appearance of free message in the cytoplasm.

[1] Chercheur qualifié du FNRS.

Most of the free message is coding for α globin and this mRNA requires haemin to be translated [4]. What becomes limiting in the initiation of translation when reticulocytes age is thus perhaps haemin.

References

[1] GIANNI, A. M., B. GIGLIONI, S. OTTOLENGHI, P. COMI, and G. G. GUIDOTTI: Nature New Biol. **240**, 183 (1972)
[2] MARBAIX, G., A. BURNY, G. HUEZ, B. LEBLEU, and J. TEMMERMAN: Eur. J. Biochem. **13**, 322 (1970)
[3] NOKIN, P., G. HUEZ, G. MARBAIX, A. BURNY, and H. CHANTRENNE: Eur. J. Biochem. **62**, 509 (1976)
[4] GIGLIONI, B., A. M. GIANNI, P. COMI, S. OTTOLENGHI, and D. RUNGGER: Nature New Biol. **246**, 99 (1973)

Address of the authors: Laboratory of Biological Chemistry, Department of Molecular Biology, University of Brussels, 67, rue des Chevaux, B-1640 Rhode St-Genèse, Belgium.

Acta biol. med. germ., Band 36, Seite 375—379 (1977)

Forschungsabteilung der Kinderklinik des Bereichs Medizin (Charité) der Humboldt-Universität zu Berlin, DDR

Veränderungen der 2,3-DPG-Konzentration in roten Blutzellen und der Plasmaerythropoetinaktivität bei hypoxischen Neugeborenen

A. Michel, R. Pohle, B. Adrian, A. Daniel und J. Gross

Zusammenfassung

Es wurden die 2,3-DPG-Konzentration in roten Blutzellen und die Plasmaerythropoetinaktivität bei normalen und hypoxischen Neugeborenen und Säuglingen auf ihren diagnostischen Wert zur Beurteilung der Hypoxie überprüft.

Erhöhungen der Erythropoetinaktivität des Plasma wurden nur bei klinisch schweren hypoxischen Zuständen gefunden (idiopathisches Atemnotsyndrom (ANS), Transposition der großen Arterien (TGA) und Hydrops universalis bei Rh-Inkompatibilität). Patienten mit TGA im Lebensalter vom 14.—56. Tag hatten trotz deutlicher Hypoxie (O_2-Partialdruck im Kapillarblut unter 22 Torr) normale Plasmaerythropoetinaktivitäten. Hierfür könnte das Auftreten des Erythropoesehemmfaktors verantwortlich sein. Erhöhungen der 2,3-DPG-Konzentration der roten Blutzellen wurden bei fast allen von uns untersuchten hypoxischen Kindern gefunden (Rh-Inkompatibilität, TGA, transitorisches ANS). Neugeborene mit idiopathischem ANS zeigen normale 2,3-DPG-Konzentrationen bei Vorliegen einer Azidose (*p*H 7,1—7,3).

Erhöhungen der Plasmaerythropoetinaktivitäten sprechen für das Vorliegen einer schweren Hypoxie. Die Bestimmung der 2,3-DPG-Konzentration scheint ein empfindlicheres Kriterium zur Beurteilung der Hypoxie des Neugeborenen zu sein als die Erythropoetinaktivität.

Einleitung

Die fetale und neonatale Hypoxie hat wesentlichen Anteil an der Genese verschiedener Erkrankungen des Neu- und Frühgeborenen. Darüber hinaus ist die Hypoxie Ursache für viele unter dem Sammelbegriff „frühkindlicher Hirnschaden" zusammengefaßter Krankheitsbilder, deren Erkennung und genaue Beurteilung in der Perinatalperiode für die rechtzeitige Prophylaxe und Therapie von Bedeutung sind. Ziel der vorliegenden Untersuchungen ist es, die Veränderung der Erythropoetinaktivität im Plasma und der 2,3-DPG-Konzentration in roten Blutzellen bei hypoxischen Zuständen des Neugeborenen zu prüfen. Es ist bekannt, daß es unter hypoxischen Bedingungen einerseits zur erhöhten 2,3-DPG-Bildung mit den bekannten Auswirkungen auf die Sauerstoffdissoziationskurve und andererseits zur Erythropoetinerhöhung im Plasma und Stimulierung der Erythropoese kommt [1—4]. Uns schien ein Vergleich des Plasmaerythropoetins und des 2,3-DPG von besonderem Interesse, da beide Kriterien über verschiedene Mechanismen vom pO_2 abhängig sind.

Abkürzungen: 2,3-DPG = 2,3-Diphosphoglyzerat; ANS = Atemnotsyndrom; TGA = Transposition der großen Arterien; Pcap O_2 = Sauerstoffpartialdruck im Kapillarblut

Material und Methoden

Es wurden folgende Patientengruppen untersucht:

- gesunde Neugeborene als Kontrollgruppe,
- Neugeborene mit Rh-Inkompatibilität,
- Neugeborene mit Atemnotsyndrom,
- Neugeborene und Säuglinge mit zyanotischen Herzmißbildungen (komplette Transposition der großen Arterien).

Die Untersuchungen erfolgten innerhalb der ersten 48 Lebensstunden. Neugeborene mit kompletter Transposition der großen Gefäße wurden darüber hinaus longitudinal untersucht (innerhalb der ersten 14 Lebenstage, vom 14.—56. Lebenstag und vom 56. Lebenstag bis zum 1. Lebensjahr).

Außer der Erythropoetinaktivität im Plasma und der 2,3-DPG-Konzentration in roten Blutzellen wurden der Hämatokrit, die Retikulozytenzahl und der O_2-Partialdruck im Kapillarblut (Pcap O_2) bestimmt.

Die 2,3-*DPG-Bestimmung* der roten Blutzellen erfolgte nach der von Luisada-Opper [5] angegebenen Methode.

Die *Erythropoetinaktivität* des Plasmas wurde mittels eines biologischen Tests an hypertransfundierten jungfräulichen weißen Mäusen, Gewicht 20—25 g, in Anlehnung an die von Nečas und Neuwirt [6] angegebene Methode gemessen.

Die *Signifikanzberechnungen* wurden nach dem Wilcoxon-Test [7] durchgeführt.

Ergebnisse

Tab. 1 gibt einen Überblick über die Erythropoetinaktivität im Plasma und die 2,3-DPG-Konzentration in roten Blutzellen gesunder und hypoxischer Neugeborener.

Das Plasma von gesunden Neugeborenen zeigt Eiseneinbauraten von durchschnittlich <1%; die 2,3-DPG-Konzentration von roten Blutzellen beträgt 4,35 µMol/ml

Tabelle 1

Plasmaerythropoetinaktivität (% ^{59}Fe-Einbau) und 2,3-DPG-Konzentration (µMol/ml Zellen) der roten Blutzellen bei gesunden und hypoxischen Neugeborenen.
a gegenüber der Norm mit $p < 0{,}01$ signifikant; b gegenüber der Norm mit $p < 0{,}01$ signifikant

Diagnose	^{59}Fe-Einbau (%)	Hämatokrit (Vol. %)	Retikulozytenzahl (%)	2,3-DPG (µMol/ml Zellen)
Gesunde Neugeborene (Nabelschnurblut, $n = 17$, >2500 g)	0,65 ± 0,61	49,7 ± 9,2	40 ± 18	4,35 ± 0,27
Rh-Inkompatibilität (Venenblut)				
Anaemia neonatorum ($n = 3$)	0,30 ± 0,10	36,3 ± 0,4	112 ± 18	7,66^b ± 1,06
Hydrops universalis ($n = 1$)	31,0	16,5	319	5,25^b
Atemnotsyndrom (ANS)				
transitorisches ANS ($n = 4$)	0,30 ± 0,10	50,0 ± 2,0	—	5,47^b ± 0,24
idiopathisches ANS	3,00 ± 1,10^a ($n = 6$)	43,1 ± 10,0 ($n = 4$)	40 ± 17 ($n = 4$)	4,32 ± 0,70 ($n = 4$)
idiopathisches ANS ($n = 1$)	16,0	58,0	—	6,93^b

Zellen. Neugeborene mit klinisch nachweisbarer Hypoxie wiesen unterschiedlich hohe Erythropoetinaktivitäten im Plasma und 2,3-DPG-Konzentrationen in roten Blutzellen auf.

Deutlich erhöhte Erythropoetinaktivitäten fanden wir bei einem Patienten mit Hydrops universalis bei Rh-Inkompatibilität, einem Patienten mit idiopathischem Atemnotsyndrom (dieser Patient verstarb am 4. Lebenstag; es wurden pathologisch-anatomisch pulmonale hyaline Membranen nachgewiesen) und bei 4 Patienten mit einem idiopathischen ANS. Bei Neugeborenen mit einer Anaemia neonatorum bei Rh-Inkompatibilität oder transitorischem ANS finden wir normale Erythropoetinaktivitäten.

Die 2,3-DPG-Konzentration in roten Blutzellen wurde bei fast allen hypoxischen Neugeborenen erhöht gefunden. Besonders hohe Werte fanden wir bei Patienten mit Anaemia neonatorum bei Rh-Inkompatibilität und bei einem verstorbenen Neugeborenen mit idiopathischem ANS.

In Tab. 2 sind die Plasmaerythropoetinaktivität und die 2,3-DPG-Konzentration in roten Blutzellen von Neugeborenen und Säuglingen mit kompletter Transposition der großen Arterien enthalten.

Die Erythropoetinaktivität im Plasma ist in der Zeit vom 1.—14. Lebenstag deutlich erhöht. Vom 14.—56. Lebenstag fällt sie trotz weiter bestehender Hypoxie (Pcap O_2 22 Torr) ab. Danach kommt es zu einem erneuten Anstieg des Erythropoetins im Plasma.

Die 2,3-DPG-Konzentration ist ebenfalls bereits in den ersten 14 Lebenstagen deutlich erhöht. Mit zunehmendem Lebensalter steigt sie weiter an.

In Abb. 1 ist die Plasmaerythropoetinaktivität in ihrer Beziehung zum Pcap O_2 dargestellt. Nabelschnurblute gesunder Neugeborener zeigen Eiseneinbauraten von 0—4%. Erhöhungen der Erythropoetinaktivität fanden wir bei Kindern mit ANS oder TGA bei pO_2-Werten unter 50 Torr, und zwar in den ersten 14 Lebenstagen und nach dem 56. Lebenstag. Jenseits des 14. bis zum 56. Lebenstag wurden bei pO_2-Werten unter 50 Torr keine erhöhten Erythropoetinaktivitäten gefunden. Bei pO_2-Werten über 50 Torr waren normale Eiseneinbauraten feststellbar.

Tabelle 2

Plasmaerythropoetinaktivität (% ^{59}Fe-Einbau) und 2,3-DPG-Konzentration (μMol/ml Zellen) der roten Blutzellen von Neugeborenen und Säuglingen mit zyanotischen Herzmißbildungen

Diagnose Lebensalter	^{59}Fe-Einbau (%)	Hämatokrit (Vol. %)	Retikulozytenzahl (%)	2,3-DPG (μMol/ml Zellen)	pO_2 (Torr)
Transposition der großen Gefäße					
1.—14. Lebenstag ($n = 8$)	4,40 ± 1,90	48,4 ± 6,3	17 ± 13	5,22 ± 0,78	28 ± 6
14.—56. Lebenstag ($n = 12$)	0,60 ± 0,40	49,1 ± 3,4	17 ± 5	5,49 ± 0,64	22 ± 5
56. Lebenstag ($n = 6$)	5,70 ± 3,60	58,0 ± 3,6	20 ± 7	5,83 ± 1,57	32 ± 7

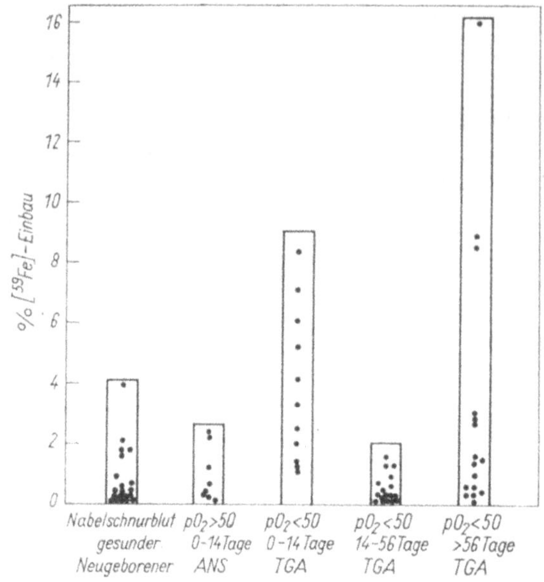

Abb. 1. Erythropoetinaktivität (% ^{59}Fe-Einbau) im Nabelschnurblut normaler Neugeborener und im Venenblut von Patienten mit ANS und TGA in Abhängigkeit vom PO_2

Diskussion

Die von uns ermittelten Durchschnittswerte der Erythropoetinaktivität im Plasma und der 2,3-DPG-Konzentration in roten Blutzellen gesunder Neugeborener stimmen gut mit den in der Literatur angegebenen Werten überein [2, 3, 8, 9]. Die deutlichen Erhöhungen der Erythropoetinaktivität bei einem Patienten mit Hydrops universalis, einem idiopathischen ANS mit postnatal nachgewiesenen pulmonalen hyalinen Membranen und von Patienten mit idiopathischem ANS sprechen für das Vorliegen langanhaltender schwerer hypoxischer Zustände. Bei Patienten mit transitorischem ANS im Sinne von postnatalen Adaptationsstörungen und den Patienten mit Rh-Inkompatibilität ist die Hypoxie nicht so hochgradig und sie führt daher nicht zu Erhöhungen der Erythropoetinaktivität im Plasma. Hierfür sprechen auch unsere Befunde, daß bei pO_2-Drücken unter 50 Torr erhöhte Erythropoetinaktivitäten beobachtet werden (Abb. 1). Es ist bemerkenswert, daß Patienten mit TGA im Lebensalter vom 14. bis zum 56. Tag trotz anhaltender Hypoxie (Pcap O_2 22 Torr) keine Erhöhungen der Erythropoetinaktivität aufweisen. Wahrscheinlich ist das ein Ausdruck für das Auftreten des Erythropoesehemmfaktors. Bisher hatte man angenommen, daß er nur bei gesunden Neugeborenen im Laufe des 1. Trimenons auftritt [3, 4, 10—13]. Damit scheint der Erythropoesehemmfaktor zu einem bestimmten Zeitpunkt postnatal aufzutreten, um im Laufe des 1. Trimenons wieder zu verschwinden, unabhängig davon, ob das Kind gesund ist oder ob eine Form der Hypoxie vorliegt. Das Auftreten des Hemmfaktors könnte auch die Ursache für den relativ langsamen Anstieg des Hämatokrits bei Patienten mit TGA während des 1. Trimenons sein.

Die 2,3-DPG-Konzentration in roten Blutzellen scheint ein empfindlicheres Kriterium als die Erythropoetinaktivität im Plasma zu sein, da bei Patienten ohne Erythropoetinveränderungen im Plasma deutliche Erhöhungen nachweisbar sind, mit Ausnahme der Neugeborenen mit idiopathischem ANS (gering erhöhte

Erythropoetinaktivität im Plasma). Bei diesen Kindern fanden wir keine 2,3-DPG-Erhöhungen. Die Patienten hatten eine Azidose (*p*H 7,1—7,3), die bekanntlich zur Verminderung der 2,3-DPG-Konzentration führt [8, 9, 13, 14]. Als Resultate von 2,3-DPG-Anstieg durch die Hypoxie und einem 2,3-DPG-Abfall durch die Azidose ergibt sich bei diesen Patienten eine normale 2,3-DPG-Konzentration. Sowohl die Erythropoetinaktivität im Plasma als auch die 2,3-DPG-Konzentration in roten Blutzellen sind bei den von uns untersuchten Patientengruppen als Kriterien zur Beurteilung einer perinatalen Hypoxie geeignet, sofern man die Untersuchungen in den ersten 48 Lebensstunden durchführt. Dabei sprechen Erhöhungen der Erythropoetinaktivität im Plasma für das Vorliegen schwerer hypoxischer Zustände des Neugeborenen. Die Veränderungen der 2,3-DPG-Konzentration sind nur bei gleichzeitiger Bestimmung des *p*H-Wertes von diagnostischem Wert.

Für die klinische Praxis, wo es auf einfache und schnelle Methoden zur Diagnostik ankommt, ist die Erythropoetinaktivität als Kriterium wegen des hohen Aufwandes kaum einsetzbar. Es ist daher notwendig, einfache Kriterien im Sinne eines Siebtestes zu finden, die es gestatten, die Hypoxie und ihre Auswirkung auf das Erythropoetin und somit auf die Erythropoese widerzuspiegeln. Möglicherweise ist die Dichte der roten Blutzellen ein brauchbarer Parameter [15,16].

Literatur

[1] Benesch, R., u. R. E. Benesch: Biochem. biophys. Res. Commun. **26**, 162 (1967)
[2] Halvorson, S., u. P. H. Finne: Br. med. J. **1963** 1132
[3] Finne, P. H.: Acta pediat. scand. **53**, 269 (1964)
[4] Halvorson, S., u. P. H. Finne: Ann. N.Y. Acad. Sci. **149**, 579 (1968)
[5] Luisada-Opper, A. V.: Clin. Chem. **19**, 113 (1973)
[6] Nečas, E., u. J. Neuwirt: Life Sci. **8**, 1221 (1969)
[7] Wilcoxon-Test, in: Documenta Geigy **7**, 193 (1969)
[8] Gross, J., B. Göldner, G. Göldner, R. Wauer, u. A. Daniel: Clinica chim. Acta (im Druck)
[9] Versmold, H., J. Wenner, u. K. Riegel: Z. Kinderheilk. **113**, 1 (1972)
[10] Krantz, S. B.: Med. Clins N. Am. **54**, 173 (1970)
[11] Lindemann, R.: Israel J. med. Sci. **7**, 1007 (1971)
[12] Lindemann, R.: Acta pediat. scand. **63**, 764 (1974)
[13] Skjelaaen, P., u. S. Halvorson: J. Lab. clin. Med. **84**, 625 (1974)
[14] Suemcintosh, M. D.: J. Pediat. **86**, 202 (1975)
[15] Syllm-Rapoport, I., E. L. Grauel, M. Böttcher, u. J. Gross: Density of Red Blood Cells as a Possible Index of Previous Fetal Hypoxic. 4th Congress of Perinatal Medicine, Prag 28.—31. 8. 1974
[16] Grauel, E. L., M. Böttcher, A. Michel, R. Staak, u. J. Gross: Density of Red Blood Cells as a Possible Index of Previous Hypoxia. 5th European Congress for Perinatal Medicine, Uppsala 9.—12. 6. 1976

Acta biol. med. germ., Band 36, Seite 381—387 (1977)

Abteilung für Elektronenmikroskopie der Charité am Pathologischen Institut der Humboldt-Universität Berlin, 104 Berlin, DDR und
Institut für Physiologische und Biologische Chemie der Humboldt-Universität Berlin, 104 Berlin, DDR

Beziehungen zwischen der Konformation isolierter Rattenleber-mitochondrien und ihrer Angreifbarkeit durch Lipoxygenase aus Kaninchenretikulozyten

W. Krause und W. Halangk

Zusammenfassung

Lipoxygenase aus Kaninchenretikulozyten bewirkt an isolierten Rattenlebermitochondrien drastische Schädigungen der Membranen. Der Nachweis dieser Schädigung erfolgte elektronenmikroskopisch und durch Bestimmung der Freisetzung von Malatdehydrogenase aus der Matrix. Die stärkste Schädigung tritt bei den orthodoxen Strukturformen der Mitochondrien in Abwesenheit von Substrat auf. ATP oder ADP plus Sukzinat üben einen starken Schutzeffekt aus; die Mitochondrien bleiben hierbei kondensiert bzw. teilweise kondensiert. Der Schutzeffekt durch Substrat wurde auch in Gegenwart des Entkopplers 2,4-Dinitrophenol beobachtet, obwohl die Mitochondrien in die orthodoxe Form umgewandelt wurden. In hypotoner Saccharose ist die Mitochondrienschädigung ausgeprägter als in hypertoner Saccharose; auch kondensierte Mitochondrien werden durch Lipoxygenase geschädigt. Daher besteht keine Beziehung zwischen der Lipoxygenasewirkung und dem Konformationszustand der Mitochondrien. Die Zugänglichkeit der Fettsäurekomponente der Phospholipide in der Membran für Lipoxygenase hängt vermutlich sowohl von metabolischen als auch strukturellen Einflüssen ab, die möglicherweise durch Veränderungen in den Protein-Lipid-Wechselwirkungen bedingt sind.

Einleitung

Die von Schewe et al. [1] nachgewiesene Lipoxygenase aus Kaninchenretikulozyten schädigt die Membranen der Rattenlebermitochondrien in drastischer Weise. Der Nachweis dieser Schädigungen erfolgte elektronenmikroskopisch [2] durch die Freisetzung von MDH [3] sowie durch die Bildung von Malondialdehyd [4]. Isolierte Mitochondrien unterliegen einem ultrastrukturellen Strukturwechsel in Abhängigkeit vom metabolischen Status [5, 6]. Zur Klärung der Frage, ob der Konformationszustand der Mitochondrien entscheidend für die Angreifbarkeit durch Lipoxygenase ist, wurde der Einfluß von Substraten, Adeninnukleotiden und der Osmolarität untersucht.

Material und Methodik

Rattenlebermitochondrien wurden in 0,3 M Saccharose nach der Methode von de Duve et al. isoliert [7]. Die Inkubation der Mitochondrien erfolgte bei 37 °C für 15 min unter verschiedenen Bedingungen:
1. AS; 2. AS, Sukzinat plus ADP; 3. AS plus ATP; 4. AS, Sukzinat plus 2,4-DNP; 5. AS in 0,15 M Saccharose; 6. AS in 0,45 M Saccharose (1.—4. in 0,3 M Saccharose). Zur Kontrolle

Abkürzungen: AS = Proteinfällung bei 0,55 Sättigung Ammoniumsulfat aus dem stromafreien Hämolysat von Retikulozyten; MDH = Malatdehydrogenase; 2,4-DNP = 2,4-Dinitrophenol wurde nur mit Saccharose der entsprechenden Osmolarität inkubiert. Jeder Versuchsansatz enthielt 0,5 ml Mitochondriensuspension (3 mg Protein/ml) in einem Endvolumen von 0,6 ml. Der Austritt von MDH aus der Mitochondrienmatrix während der Inkubation wurde als prozentualer Anteil der Gesamtfreisetzung nach Behandlung mit 0,2% Triton X 100 nach SIEGEL und ENGLARD [8] bestimmt.

Zur elektronenmikroskopischen Untersuchung wurden die Proben nach der Inkubation mit 3%igem Glutaraldehyd vorfixiert und mit 1%igem Osmiumtetroxid nachfixiert. Die Einbettung erfolgte in Vestopal. Die Ultradünnschnitte wurden mit dem LKB-Ultrotom und die elektronenmikroskopischen Aufnahmen mit dem Elmiskop Ia, Siemens AG (Berlin-West) angefertigt.

Ergebnisse

Isolierte Rattenlebermitochondrien sind in 0,15 M, 0,30 M und 0,45 M Saccharose kondensiert (Abb. 1). Während der Kontrollinkubation mit 0,15 M und 0,30 M Saccharose gehen die meisten Mitochondrien in die orthodoxe Strukturform über (Abb. 2). In 0,45 M Saccharose bleiben die Mitochondrien kondensiert. Durch Lipoxygenase aus Retikulozyten werden orthodoxe Mitochondrien zu Vesikeln umgewandelt (Abb. 3). ATP oder ADP plus Sukzinat verhindern die Mitochondrienschädigung; die Mitochondrien liegen dann in der kondensierten bzw. in der teilkondensierten Konformation vor (Abb. 4 und 5). In Gegenwart von 2,4-DNP wurde auch der Schutzeffekt des Substrates beobachtet, obwohl die Mitochondrien in der orthodoxen Strukturform vorlagen (Abb. 6).

Die Schädigung der Mitochondrien ist in 0,15 M Saccharose stärker als in 0,45 M Saccharose ausgeprägt (Abb. 7 und 8). Die Freisetzung der MDH während der Inkubation stimmt mit dem Grad der Mitochondrienschädigung überein (Tab. 1).

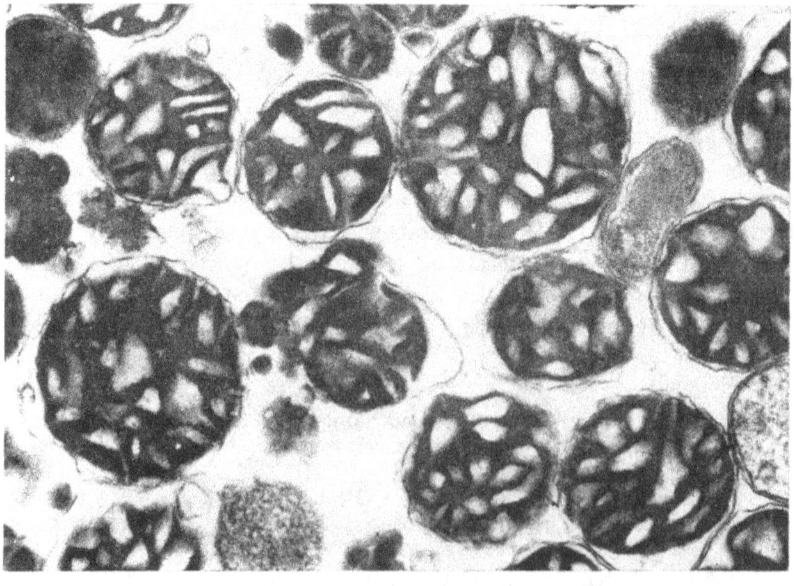

Abb. 1. Isolierte Rattenlebermitochondrien in 0,15 M, 0,3 M oder 0,45 M Saccharose in der kondensierten Strukturform. × 46000

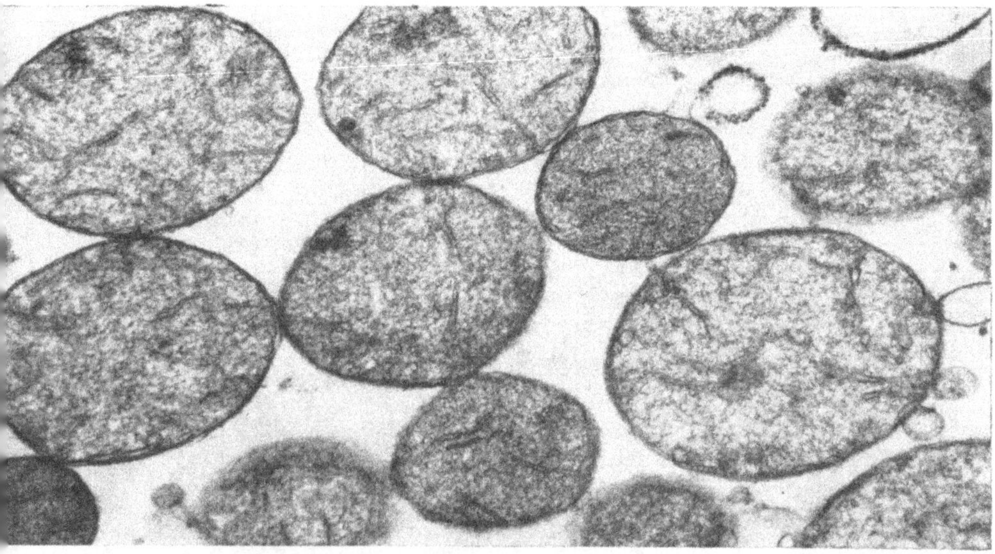

Abb. 2. Kontrollinkubation in 0,15 M und 0,3 M Saccharose bei 37 °C für 15 min. Übergang von der kondensierten in die orthodoxe Strukturform. In 0,45 M Saccharose bleiben die Mitochondrien kondensiert. × 46000

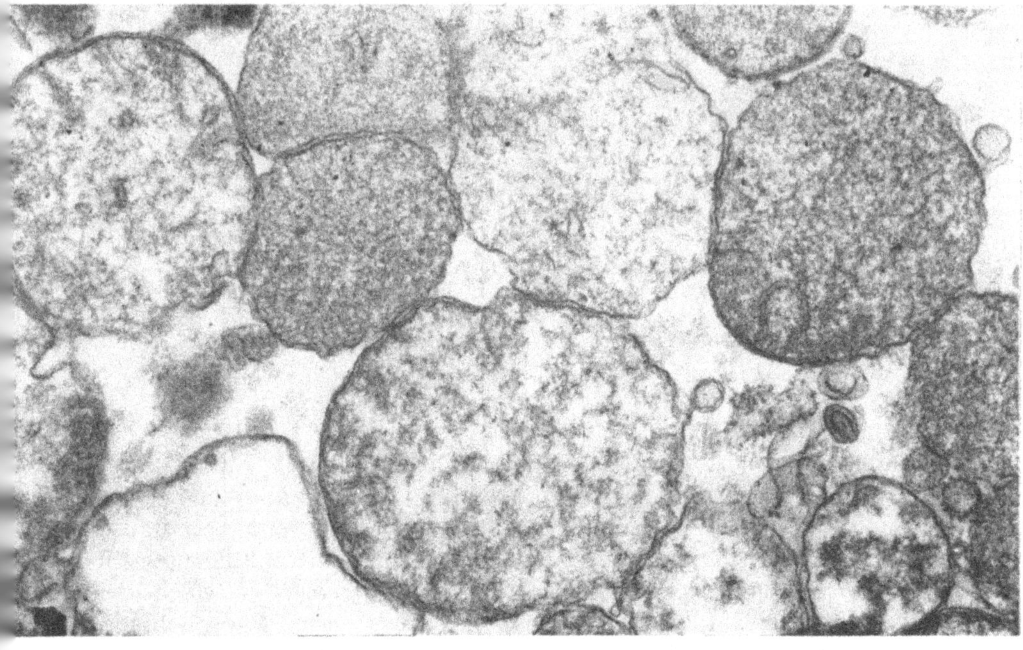

Abb. 3. Einwirkung von Lipoxygenase ohne Substrate. × 46000

Tabelle 1
Freisetzung von MDH aus intakten Rattenlebermitochondrien durch Lipoxygenase

Versuchsansatz	MDH-Freisetzung [%]	ultrastrukturelle Schädigung
AS in 0,3 M Saccharose	25	++ (Abb. 3)
AS, ADP, Sukzinat	3	– (Abb. 4)
AS, ATP	5,5	– (Abb. 5)
AS, 2,4-DNP, Sukzinat	9	+ (Abb. 6)
AS in 0,15 M Saccharose	41	+++ (Abb. 7)
AS in 0,45 M Saccharose	23	(++) (Abb. 8)

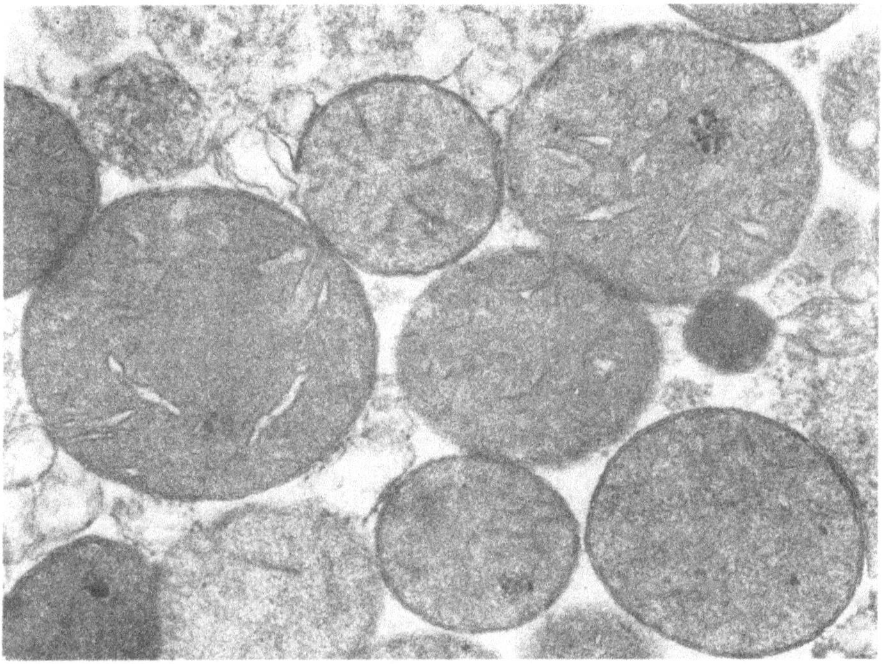

Abb. 4. Einwirkung von Lipoxygenase, 1 mM ADP und 10 mM Sukzinat. × 46000

Diskussion

Die Schädigung isolierter Rattenlebermitochondrien durch Lipoxygenase aus Retikulozyten erfolgt durch Lipidperoxydation der Membranlipide [1]. Isolierte Phospholipide und freie mehrfach ungesättigte Fettsäuren werden ebenfalls durch Lipoxygenase angegriffen [9]. Bestimmte Parameter der Retikulozyten-Lipoxygenase, wie z. B. das Temperaturverhalten und der *p*H-Wert, unterscheiden sich bei Verwendung von intakten Mitochondrien oder isolierten Phospholipiden [9]. Ohne Substratzusatz werden isolierte Mitochondrien durch Lipoxygenase zu Vesikeln umgewandelt (vgl. Abb. 3). Die Schädigung wird jedoch verhindert, wenn mit ATP oder ADP plus Sukzinat der ultrastrukturelle Übergang der Mitochondrien von der kondensierten in die orthodoxe Strukturform unterbunden wird (vergl. Abb. 4 und 5). Diese Befunde führten zunächst zu der Annahme, daß nur orthodoxe Mitochondrien durch Lipoxygenase geschädigt werden können. Durch 2,4-DNP wird die Schutzfunktion des Substrates bei der Mitochondrienschädigung durch Lipoxygenase nur teilweise aufgehoben. Die ultrastrukturelle Transformation von der kondensierten in die orthodoxe Strukturform findet in diesem Experiment statt, jedoch war die Schädigung der Mitochondrien nur schwach ausgeprägt (vgl. Abb. 6). Der Angriff auf kondensierte Mitochondrien im hypertonen Medium (0,45 M Saccharose ohne Substrat) durch Lipoxygenase sowie die Versuche mit 2,4-DNP zeigen somit, daß keine Beziehung zwischen der

Abb. 5. Einwirkung von Lipoxygenase und 1 mM ATP. × 46000

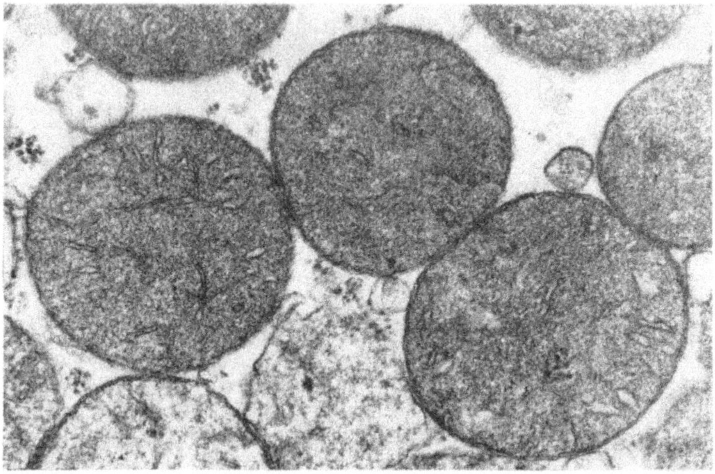

Abb. 6. Einwirkung von Lipoxygenase, 1 mM ADP, 10 mM Sukzinat und 10 µM 2,4-DNP. × 46000

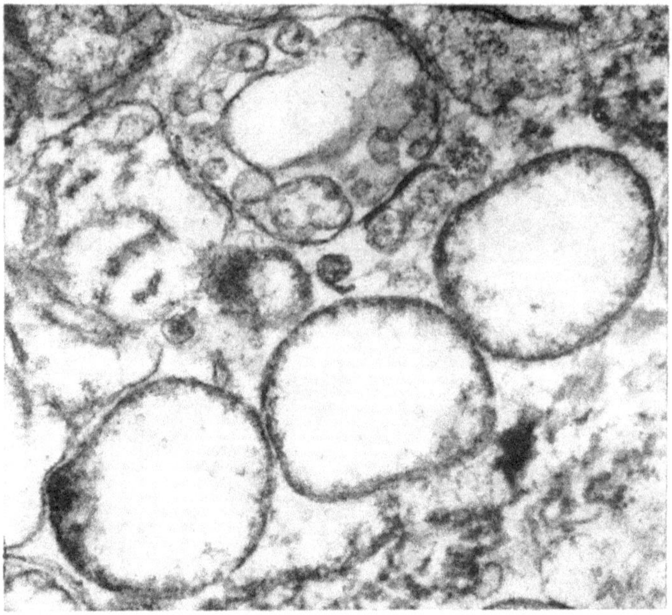

Abb. 7. Einwirkung von Lipoxygenase in 0,15 M Saccharose. × 46000

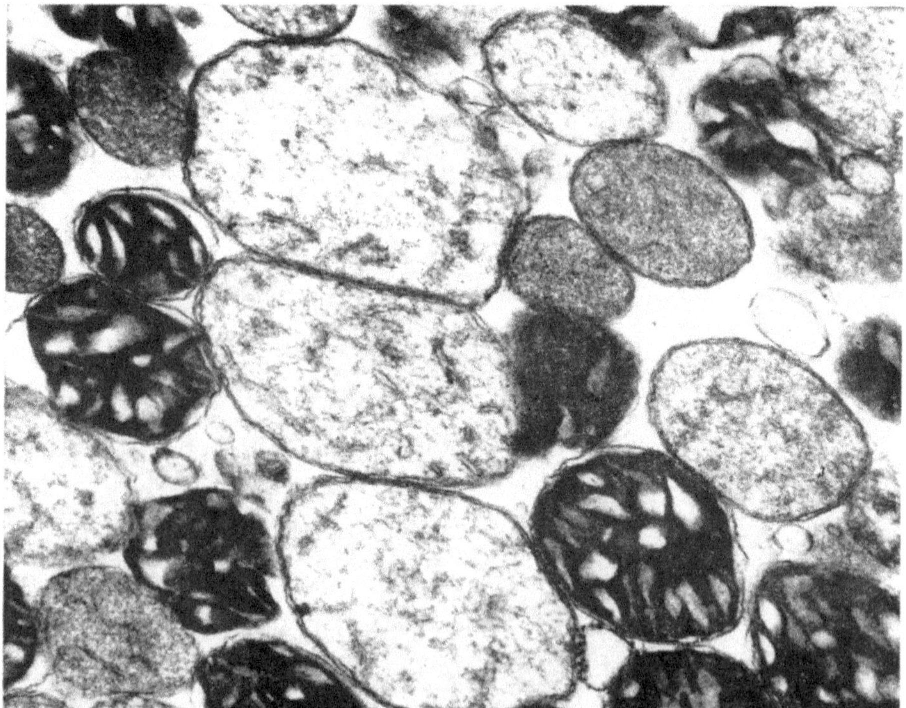

Abb. 8. Einwirkung von Lipoxygenase in 0,45 M Saccharose. × 46000

Lipoxygenase-Wirkung und dem Konformationszustand der Mitochondrien zu bestehen scheint.
Die Zugänglichkeit der Fettsäurekomponente der Phospholipide für Lipoxygenase ist möglicherweise von metabolischen und strukturellen Einflüssen abhängig, die durch Veränderungen der spezifischen Protein-Lipid-Wechselwirkungen in der Membran bedingt werden.

Literatur

[1] SCHEWE, T., W. HALANGK, CH. HIEBSCH, u. S. RAPOPORT: FEBS Lett. **60**, 149 (1975)
[2] KRAUSE, W., T. SCHEWE u. D. BEHRISCH: Acta biol. med. germ. **34**, 1609 (1975)
[3] HALANGK, W.: Diplomarbeit, Humboldt-Universität Berlin, Bereich Medizin, 1973
[4] SCHEWE, T., W. HALANGK, CH. HIEBSCH, u. S. RAPOPORT: Acta biol. med. germ. **36**, 363 (1977)
[5] HACKENBROCK, C. R.: J. Cell Biol. **30**, 269 (1966)
[6] HACKENBROCK, C. R.: J. Cell Biol. **37**, 345 (1968)
[7] DE DUVE, C., B. C. PRESSMAN, R. GIONETTO, R. WATTIAUX, u. F. APPELMANS: Biochem. J. **60**, 604 (1955)
[8] SIEGEL, J., u. E. ENGLARD: Biochim. biophys. Acta **54**, 67 (1961)
[9] HALANGK, W., T. SCHEWE, CH. HIEBSCH, u. S. RAPOPORT: Acta biol. med. germ. **36**, 405 (1977)

Summary

W. KRAUSE and W. HALANGK: Relationship between conformational state and susceptibility to reticulocyte lipoxygenase of rat liver mitochondria

Lipoxygenase from rabbit reticulocytes causes disruption of mitochondrial membranes and peroxidation of their lipids as judged by electronmicroscopy, release of matrix enzymes and formation of malonyldialdehyde. Without substrate mitochondria become orthodox and strong lysis by lipoxygenase appears. The lysis is prevented by ATP or ADP plus succinate; in this case mitochondria remain condensed or partly condensed. The protection by substrate was even observed in the presence of 2,4-DNP, although the mitochondria were transformed to the condensed state. Lysis was more pronounced in hypotonic than in hypertonic sucrose, condensed mitochondria are also attacked. No relation seems to exist between lipoxygenase attack and the conformational state of mitochondria. Lysis of mitochondria is dependent on the susceptibility of the fatty acid moiety of phospholipids, which may be influenced by both metabolic and structural events via alteration of protein-lipid interactions.

Acta biol. med. germ., Band 36, Seite 389—391 (1977)

Institut für Physiologische und Biologische Chemie der Humboldt-Universität Berlin, 104 Berlin, DDR

Activity of the respiratory inhibitor RF in human red cells in anaemia

K. ZIEM and T. SCHEWE

Summary

Stroma-free haemolysates of red cells from patients with acute or latent anaemia inhibit the NADH oxidase activity of submitochondrial particles from beef heart. The inhibitory activities appeared significantly more frequently in anaemic patients than in normal persons and are likely due to the action of a factor which is identical or similar to the inhibitory protein RF present in rabbit reticulocytes. The possible use of the inhibitor assay for clinical purpose is under discussion.

Introduction

The inhibitory protein RF (formerly designated as RÜ-inhibitor) was first detected in 1955 by RAPOPORT and GERISCHER-MOTHES [1] in rabbit reticulocytes. It suppresses irreversibly the mitochondrial electron transfer system at two points which are located in the two iron-sulphur regions of the NADH-ubiquinone reductase and of the succinate-ubiquinone reductase complex [2]. Until now there have been no studies on the occurrence of this inhibitor in reticulocyte-rich blood of other species with the exception of a report on inhibitory activity in chicken red cells [3]. An occurrence in human red cells should be of special interest for clinical biochemistry. Studies on rabbits have revealed that the various reticulocyte populations occurring in the course of a bleeding anaemia differ strongly with respect to their RF activity [4]. In as much as different kinds of reticulocytes may be poured out in anaemic patients dependent on the type of anaemia, variation of the inhibitory activites is conceivable which could serve as potential parameters for their characterization.

The present communication deals with the following questions:

1. Is inhibitory activity detectable in human red cells?
2. Are there any relations between inhibitory activity and anaemia?

Materials and methods

The 44 blood specimens investigated originated from the following individuals: 23 normal persons, 4 anaemias due to G6PD deficiency, 8 haemolytic anaemias of unknown cause, 1 spherocytosis, 3 pancytopenias, 1 tumour anaemia, 1 osteomyelofibrosis, 1 toxic depression of bone marrow after cobalt irradiation, 1 disorder of bone marrow maturation, 1 presumable disorder of iron incorporation.

The specimens were treated according to the following scheme:

Washed blood cells
↓
Osmotic haemolysis
(1 volume of 30% suspension plus 2 volumes of distilled water)
↓
High-speed centrifugation
(6 min, 18,000 × g)
↓
Stroma-free haemolysate (SFH)

Electron transfer particles (ETP) from beef heart mitochondria prepared according to CRANE et al. [5] served as test object, ETP (50—100 μg protein-Lowry-method) and SFH (10 or 20 μl) were incubated for 15 min at 37 °C. Thereafter the NADH oxidase activity was measured spectrophotometrically at 340 nm (0.05 M Tris-H_2SO_4 buffer, *p*H 8.0, containing 1 mg/ml human serum albumin, 37 °C). As a criterion of RF activity the diminution of the NADH oxidase activity with elevation of the amount of SFH was used. With low amounts of SFH the inhibitory effect has often escaped detection due to the presence in erythrocytes of an unidentified respiration-enhancing factor [6], which interferes with the inhibitor, but which is overcome with increasing amounts of SFH. For this reason, the inhibitory effect does not show proportionality with the amount of SFH.

Results and discussion

Table 1 shows a summary of the results. From the χ^2 test it is evident that inhibitory activity appears significantly more frequently in specimens from anaemic patients than in those from normal persons (1% probability of error). No correlation was observed between the magnitude of inhibitory activity and the reticulocyte count. In some cases inhibitory activity appeared even in specimens of anaemic patients during phases of nearly normal haematological data (Table 2). In Table 3 the distribution of the inhibitor-positive specimens with respect to the kind of anemia is listed. No prevalences are apparent.

Even though the number of specimens tested is still too low, there is some reason for the assumption that inhibitory activity in the SFH indicates an acute or latent anaemia. Therefore this assay may be useful for clinical purposes. The expense of the assay is low, however it has to be further improved in order to reduce the percentage of questionable results (Table 1). Furthermore, it has to be clarified why inhibitory activites were not detected in all anaemic bloods.

The reason for the appearance of inhibitory activity with acute and latent anaemias may consist in the occurrence of predominantly juvenile red cells (reticulocytes

Table 1
Screening for RF activity in blood specimens

	n	positive	questionable	negative
Normal persons	23	1	7	15
Patients with latent or acute anaemia	21	10	6	5

Table 2
Inhibitory activity in latent anaemias

	Tumour-anaemia		G6PD-deficiency	
Hb (g/100 ml)	11.1		13.8	
Hk (%)	37.0		48.4	
Reticulocytes (%)	2.1		0.4	
Addition of SFH (μl)	activity ΔE_{340}/min	inhibition (%)	activity ΔE_{340}/min	inhibition (%)
0	190		190	
10	184	3	190	0
20	66	65	135	29

Table 3
Cases with inhibitory activity

2 anaemias due to G6PD deficiency
2 haemolytic anaemias of unknown cause
1 spherocytosis
2 pancytopenias
1 tumour anaemia
1 osteomyelofibrosis
1 toxic depression of bone marrow after cobalt irradiation

and young normocytes) in these cases. In older cells the inhibitor could be inactivated or degraded as shown for rabbit reticulocytes. Another possibility is an unusual maturation process of the reticulocytes by which the inhibitor persists. The present results do not unequivocally prove that the inhibitory activites are due to a factor identical or similar to the inhibitory protein RF from rabbit reticulocytes. However the assumption of the role of RF as universal participant in the maturation process of the red cells would argue in favour of the presence of RF in human blood. The possible identity with rabbit RF has to be checked in further experiments.

References

[1] Rapoport, S., and W. Gerischer-Mothes: Hoppe-Seyler's Z. physiol. Chem. **302**, 167 (1955)
[2] Schewe, T., Ch. Hiebsch, and S. Rapoport: Acta biol. med. germ. **29**, 189 (1972)
[3] Augustin, H. W., and S. Rapoport: Acta biol. med. germ. **3**, 433 (1959)
[4] Wiesner, R., S. Rosenthal, and Ch. Hiebsch: Acta biol. med. germ. **30**, 631 (1973)
[5] Crane, F. L., J. L. Glenn, and D. E. Green: Biochim. biophys. Acta **22**, 475 (1956)
[6] Rapoport, S., W. Gerischer-Mothes, and Ch. Nieradt: Naturwissenschaften **42**, 128 (1955)

Acta biol. med. germ., Band 36, Seite 393—403 (1977)

Institut für Physiologische und Biologische Chemie der Humboldt-Universität, 104 Berlin, DDR

Reinigung und Charakterisierung des Atmungshemmstoffes RF aus Kaninchenretikulozyten

R. Wiesner, Ch. Tannert, G. Hausdorf, T. Schewe und S. Rapoport

Zusammenfassung

Im stromafreien Hämolysat von Kaninchenretikulozyten kommt ein Atmungshemmstoff (RF) mit Proteinnatur vor. Er konnte folgendermaßen gereinigt werden: $(NH_4)_2SO_4$-Fällung → DEAE-Sephadex-Ionenaustauschchromatographie → Isoelektrische Fokussierung → Gelfiltration an Sephadex G-200.

Neben der Aktivitätsbestimmung wurde die Identität des Hemmstoffes immunologisch auf allen Reinigungsstufen nachgewiesen. Mit der SDS-Gelelektrophorese wurde ein Molekulargewicht von 80000 bestimmt. Das Protein hat einen isoelektrischen Punkt von 5,55—5,65 und besteht aus einer Polypeptidkette, wie aus SDS-Merkaptoäthanol-Gelelektrophorese und N-terminaler Aminosäurebestimmung hervorgeht. Als N-terminale Aminosäure konnte nach Dansylierung Glyzin bestimmt werden. Die Aminosäureanalyse liefert nur nach vorheriger Oxydation mit Perameisensäure reproduzierbare Werte. Der Anteil polarer Aminosäuren mit 41 Mol-% entspricht dem für lösliche globuläre Proteine. Auffallend hoch ist der Leu- und Try-Gehalt. Wird die saure Hydrolyse ohne vorherige Perameisensäureoxydation durchgeführt, tritt Huminbildung auf.

Durch Reaktion von SDS-Gelen mit Schiff's Reagens und Kohlenhydratbestimmung mit Anthron-Reagens konnte die Glykoproteinnatur des Hemmstoffes nachgewiesen werden.

Einleitung

Seit einigen Jahren ist bekannt, daß es sich bei dem Hemmstoff RF, der den steilen Atmungsabfall bei der Reifung der roten Blutzelle bewirkt [1—5], um ein Protein handelt. Der Hemmstoff greift die oxydierten Eisenschwefelproteine von Komplex I und II der Atmungskette an [6]. Neben diesem Hemmstoff gelang es Rapoport und Mitarbeitern [7, 8] einen zweiten Hemmstoff (RC) mit Proteinnatur nachzuweisen, der auf die Zytochromoxydase wirkt. Viele Eigenschaften des Hemmstoffes RF wurden bisher beschrieben. So enthält er essentielle SH-Gruppen und zweiwertiges Eisen [9], wird durch Serumalbumin [10] und Phospholipide [11, 12] inaktiviert und zeigt eine ausgeprägte biologische Dynamik [11, 13]. Darüber hinaus wurden Kenntnisse über spezifische und unspezifische Hemmstoffbindungen an ETP und deren Beeinflussung durch chaotrope Agenzien gewonnen und eine mögliche Mikroheterogenität des Hemmstoffproteins diskutiert [14].

Die vorliegende Arbeit befaßt sich mit der systematischen Reinigung dieses Proteins und stellt einige physikochemische und chemische Eigenschaften vor.

Abkürzungen: RÜ = Retikulozytenhämolysat-Überstand nach Zentrifugation; AS = Hemmstoff, der durch 0,55 $(NH_4)_2SO_4$-Sättigung aus RÜ erhalten wird; ETP = Elektronentransportpartikel; IF = isoelektrische Fokussierung; IP = isoelektrischer Punkt; ÄDTA = Äthylendiamintetraazetat; PAGE = Polyakrylamidgelelektrophorese; LM = Laufmittel; Dans = dansylierte Aminosäuren bzw. Proteine

Material und Methoden

AS

Der Hemmstoff RF wurde bis zur Stufe der 0,55 Ammoniumsulfatsättigung (= AS) nach einem schon früher beschriebenen Verfahren [6] aus Kaninchenretikulozyten hergestellt. Alle weiteren Präparationsschritte erfolgten unter weitgehend anaeroben Bedingungen, da sich der Hemmstoff mit zunehmender Reinheit als hoch autoxydabel erwies.

Anionenaustauschchromatographie

Die Ionenaustauschchromatographie wurde an DEAE-Sephadex A-50 (Pharmacia, Schweden) durchgeführt. Das Gel wurde in 0,1 mM ÄDTA gequollen und vor der Äquilibrierung mit 0,03 M Tris-HCl (Me^{2+}-frei), *p*H 7,2, mit Lauge und Säure behandelt und unter Stickstoff geschützt. Das Packen der Säule geschah ebenfalls unter Stickstoff mit einer Flußgeschwindigkeit von 12 ml/Std. Zur Äquilibrierung wurde mit 200 ml Puffer, der 1 mM $MgCl_2$ enthielt, gespült. Alle verwendeten Lösungen wurden unter Stickstoff gehalten. Zur Entfernung von Schwermetallverunreinigungen aus den verwendeten Lösungen wurde „Dowex-Chelating-Resin A-1" (Serva, Heidelberg) im Säulenverfahren angewendet. Eluiert wurde mit einem linearen NaCl-Gradienten von 0,01 – 0,40 M. Die Fraktionierung erfolgte in 3,8 ml-Fraktionen und im Bereich der Hemmstoffelution unter anaeroben Bedingungen (N_2-Zelt).

Isoelektrische Fokussierung

Die IF wurde in einer LKB 110 ml-Elektrofokussierungssäule durch eine Modifizierung der bei VESTERBERG [15] beschriebenen Methode durchgeführt. Ein Saccharosegradient zwischen 0 und 50% wurde in der Säule erzeugt. Der Gradient enthielt 2% Ampholine, *p*H 5–8 (LKB-Instruments). Zwischen die Anodenlösung (am Ausfluß der Säule) und den Saccharosegradienten wurde eine 2%ige Glutaminsäurelösung, die 50% Saccharose enthielt, geschichtet. Die Probe wurde etwa in der Mitte des Gradienten eingetragen. Die Fokussierungszeit betrug 60 Std. bei 600 V/24 Std. nach Einstellung einer konstanten Stromstärke wurde in 1,3 ml-Proben fraktioniert. Die Extinktionsbestimmung der Proben bei 280 nm erfolgte am Spektralphotometer VSU-II (Carl Zeiss Jena), die *p*H-Messungen am MV 85 (Präcitronic).

Gelfiltration

Die Gelfiltration erfolgte mit Sephadex G-200 (Pharmacia, Schweden) in einer K 15/90-Säule. Das Gel wurde in 1 mM ÄDTA-Lösung gequollen; Gel und Elutionspuffer wurden wieder unter Stickstoff geschützt. Das Gießen der Säule geschah unter einem Stickstoffzelt. Eluiert wurde mit 0,05 M Kaliumphosphat-Puffer, *p*H 6,0, Me^{2+}-frei und 1 mM $MgCl_2$. Vor dem Auftragen der Probe wurde die Säule mit dem 3fachen Totalvolumen Puffer gespült. Die Flußgeschwindigkeit betrug 7,0 ml/Std.; Proben zu 1,8 ml wurden gesammelt.

SDS-Polyakrylamidgelelektrophorese

Die PAGE wurde nach WEBER und OSBORN [16] bei *p*H 7,0 in Gegenwart von 0,1% SDS durchgeführt. Es wurde mit einem 10%igen Gel gearbeitet. Die Proteinbehandlung erfolgte in 1% SDS und 1% Merkaptoäthanol für 5 min in siedendem Wasserbad. Zur Frontmarkierung diente Bromphenolblau und zur Proteinanfärbung Coomassie-Blau BB R-250 in Methanol-Eisessig. Zum Glykoproteinnachweis wurden die Gele nach dem Lauf mit Perjodsäure-Schiff's-Reagens behandelt [17]. Zur Entfernung von Saccharoseresten aus der IF wurden die Proteinpräparate einer Langzeitdialyse (130 Std.) unterzogen.

Immunodiffusionstest

Die Identität des Hemmstoffs wurde mit Hilfe eines Anti-RF-Serums mit der radialen Immunodoppeldiffusionsmethode nach OUCHTERLONY [18] untersucht. Das Antiserum wurde durch Immunisierung von Meerschweinchen mit gereinigten RF-Präparaten (IF-Stufe) hergestellt. Die Immunisierung erfolgte durch mehrmalige subkutane bzw. intrakutane Injektion von 100–170 µg RF-Eiweiß in physiologischer Kochsalzlösung mit komplettem Freund'schen Adjuvans (1. Immunisierung: Fußballen), inkomplettem Freund'schen Adjuvans (2. Immunisierung: Fußballen, Unterschenkel) bzw. ohne Adjuvans (3. und 4. Immunisierung: Bauchhaut) im Abstand von 8 Wochen (2. Immunisierung) bzw. einer Woche (3. und 4. Immunisierung).

N-terminale Endgruppenbestimmung

Die N-terminale Endgruppenbestimmung erfolgte mit der Danysl-Methode nach GROS und LABONESSE [19] am perameisensäureoxydierten Protein. Etwa 10 nMol Protein, TCE-gefällt und mit Azeton gewaschen, wurden 2 Std. bei 0 °C mit Perameisensäure oxydiert. Das Protein wurde nach zweimaligem Lyophilisieren in 350 µl 0,1 M Phosphatpuffer, *p*H 8,75, unter Zusatz von 250 mg entionisiertem Harnstoff gelöst. Nach Zusatz von 250 µl Dimethylformamid und 100 µl azetonischer 0,1%iger Dansylchloridlösung (Fa. Sigma) folgten 2 Std. Reaktionszeit bei Zimmertemperatur. Nach TCE-Fällung, Waschen mit 0,1 M HCl/Azeton und Trocknen im Wasserstrahlvakuum, wurde mit 250 µl azeotroper Salzsäure 4 Std. bei 110 °C hydrolysiert. Das Hydrolysat wurde im Vakuum getrocknet, mit 50 µl 0,1 M Zitratpuffer, *p*H 3,6, aufgenommen und mit wassergesättigtem Äther extrahiert.

Identifizierung der Dansyl-Aminosäure:

1. Zweidimensionale Dünnschichtchromatographie auf Kieselgel G nach Stahl (Fa. Merck), Glasplatten 20 × 20 cm, 0,25 mm Schichtdicke. Das extrahierte Dansylierungsprodukt wurde zur Auftragung in Azeton — 1 M HCl (19:1 V/V) gelöst. LM 1: Toluol/Pyridin/Eisessig (75/25/1,75 ml); LM 2: Toluol/Äthylenchlorhydrin/25% wäßriges NH_3 (50/40/3,35 ml).
2. Eindimensionale Mikro-Dünnschichtchromatographie auf Polyamidfertigfolien, 8 × 4 cm, Typ K 5 41 V (Fa. Eastman). Das extrahierte Dansylierungsprodukt wurde zur Auftragung in Äther aufgenommen. LM 1: Wasser/85% Ameisensäure (200:3,2 V/V); LM 2: Benzol/Eisessig (9:1 V/V).

Alle Laufmittel wurden aus handelsüblichen, frisch destillierten Lösungen hergestellt.

Aminosäurenanalyse

10 nMol Protein wurden zur Aminosäurenanalyse unter Zusatz von Norleuzin als inneren Standard 3mal mit Perameisensäure oxydiert [20]. Die saure Hydrolyse wurde in 6 N HCl für 24, 48 und 72 Std. bei 110 °C ± 1 °C in Ampullen durchgeführt. Die Hydrolysate wurden am Rotationsverdampfer getrocknet und im Startpuffer gelöst. Die Aminosäurenanalyse erfolgte am automatischen Aminosäuren-Analysator BC 200 (Fa. Biocal) nach einem Einsäulen-Dreipuffer-Verfahren. Tryptophan wurde spektrophotometrisch am Spektralphotometer „Cary 14" nach EDELHOCH [22] bestimmt.

Hemmstoffaktivität wurde im $NADH_2$-Oxydase-Testsystem von ETP am Spektralphotometer „Unicam SP 800" durchgeführt [6].

Eiweißbestimmungen erfolgten mit Biuret-Reagens. Ein qualitativer Test auf Kohlenhydrate im Protein wurde nach ROE [23] und SPIRO [24] mit Anthron-Reagens gemacht.

Ergebnisse

Als Ausgang für die Reinigung des Hemmstoffproteins diente ein aus Kaninchenretikulozyten hergestelltes Hämolysat. Als Reinigungsverfahren wählten wir die Kombination von Ammoniumsulfatfällung, Ionenaustauschchromatographie, isoelektrischer Fokussierung und Gelfiltration. Die genannten Verfahren allein führten auch durch Variation der Bedingungen oder Rechromatographie nicht zu einem reinen Protein. Während der Reinigung verfolgten wir die Hemmstoffaktivität. Wie Tab. 1 zeigt, ist auch unter Anwendung anaerober und schwermetallfreier Bedingungen der Hemmstoff nur im AS stabil. Inaktivierungen gehen mit der Reinigung einher, die aber normalerweise nicht zu einer Verschiebung des Proteinprofils im Chromatogramm führen. Die Aktivitätsbestimmung wurde demzufolge nur zur Identifizierung des Hemmstoffs im Elutionsverlauf herangezogen. Die Disk-Elektrophorese im SDS-System und der immunchemische Nachweis im Ouchterlony-Immundiffusionstest waren für uns die Kriterien für die Beurteilung der Reinigungsschritte.

Tabelle 1
Stabilität von Hemmstoff RF auf verschiedenen Reinigungsstufen

Probe	Stabilität[1]
AS	stabil
AS → DEAE-Sephadex	unstabil
AS → CM-Zellulose	unstabil
AS → Isoelektrische Fokussierung	unstabil
AS → Sephadex-G 200	unstabil

[1] Stabilität bedeutet Erhalt der Hemmstoffaktivität, gemessen im $NADH_2$-Oxydase-Testsystem

DEAE-Sephadex-Ionenaustauschchromatographie

Ein typischer Elutionsverlauf für die Reinigung an DEAE-Sephadex A-50 ist in Abb. 1 dargestellt. Unter den angegebenen Bedingungen passiert Hämoglobin ungebunden die Säule. Im zweiten Hauptproteinpeak (Fr. 50—70) wird der Hemmstoff eluiert (Fr. 58—66), der quantitativ am Austauscher gebunden war. Die Elution des Hemmstoffes beginnt bei 0,07 M NaCl und ist in Peak-Mitte 0,1 M an NaCl. Obwohl ein linearer Gradient vorgelegt wurde, entsteht unter den Chromatographiebedingungen (Schrumpfen der Gelpartikel durch Erhöhung der Ionenstärke) der in Abb. 1 eingezeichnete Gradient. Der Hemmstoffpeak enthält

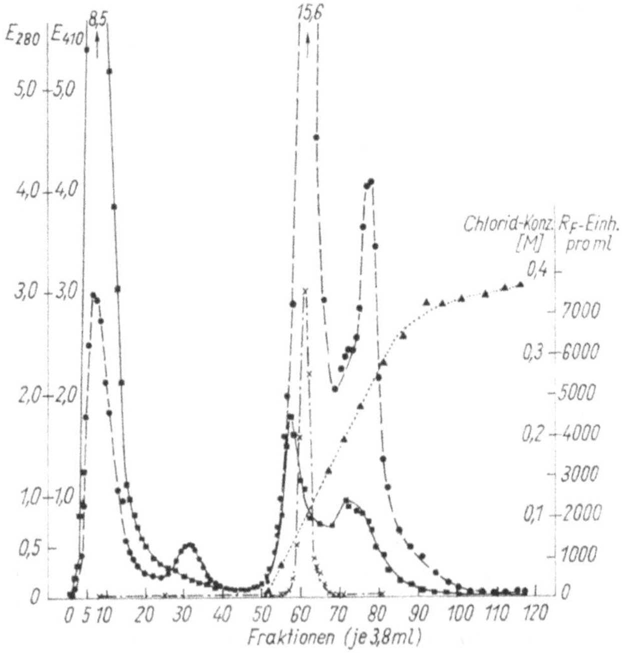

Abb. 1. Ionenaustauschchromatographie an DEAE-Sephadex A-50 von AS. Probenvolumen: 15 ml AS mit 900 mg Protein; Elution mit NaCl-Gradienten von 0,01 bis 0,40 M in 0,03 M Tris-HCl, *p*H 7,2, 1 mM $MgCl_2$ (Me^{2+}-frei).
▲------▲ NaCl-Gradient; ●-------● E_{280}; ■———■ E_{410}; × — · — × RF-Einheiten/ml

~ 70 mg Eiweiß, etwa 7% der aufgetragenen Proteinmenge (unter Zugrundelegung von E_{280}-Werten).

Die spezifische Aktivität (RF-Einheiten/mg Protein) steigt von 200 im Ausgang auf 500 nach der DEAE-Chromatographie an.

Diskelektrophoretisch sind 4 Haupt- und 5 Nebenbanden nachweisbar. In Abb. 2 ist das Diskelektropherogramm von AS abgebildet. Durch die DEAE-Chromatographie werden alle Proteine abgetrennt, die im AS unterhalb der Gelmitte erkennbar sind.

Isoelektrische Fokussierung und Gelfiltration

Der Elutionsverlauf mit einem an DEAE-Sephadex vorgereinigten Hemmstoffpräparat ist in Abb. 3 dargestellt. Der IP des Proteins ist in einer Reihe von Versuchen zu 5,55—5,65 bestimmt worden. Bei der IF im *p*H-Bereich von 5—8 stellt der Hemmstoff den Hauptanteil des bei E_{280} absorbierenden Materials dar. Die Gipfelfraktionen (Fr. 34—36) zeigen in der Diskelektrophorese nur eine Haupt- und zwei Nebenbanden, deren Anteil auf maximal 1% vom Gesamtprotein geschätzt wird, und die durch Gelfiltration (Abb. 4) an Sephadex G-200 abgetrennt wurden (s. Abb. 2).

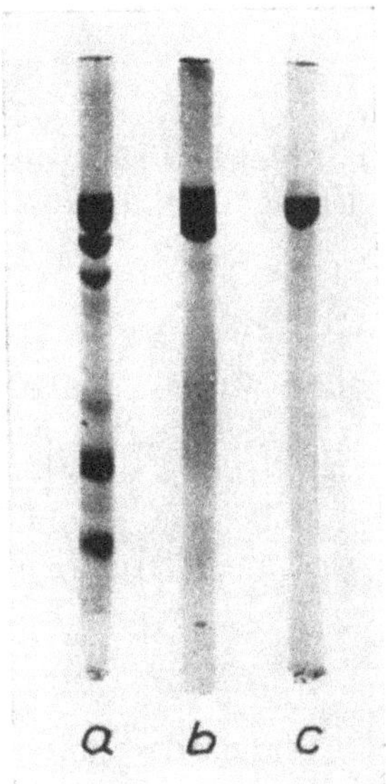

Abb. 2

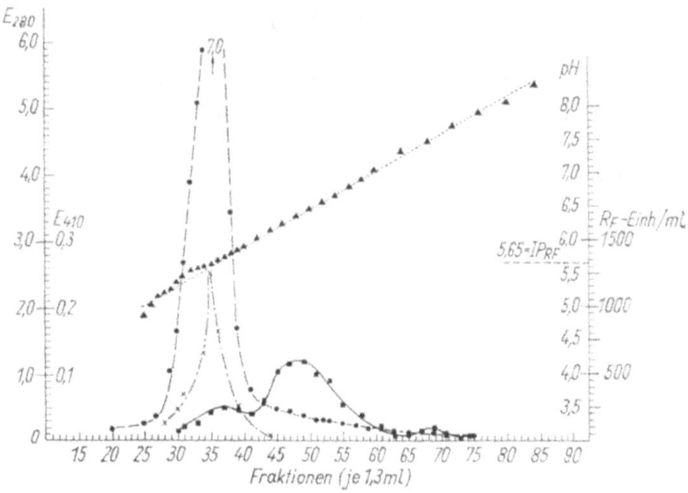

Abb. 3

Abb. 2. Polyakrylamidgelelektrophorese bei *p*H 7,0 im SDS-System.
a) Ausgangsmaterial AS; b) AS, gereinigt durch DEAE-Sephadex-Chromatographie und isoelektrische Fokussierung; c) Proteine von b) weiter gereinigt an Sephadex G-200. Die Auftragsstellen sind oben in der Abbildung

Abb. 3. Isoelektrische Fokussierung im *p*H-Bereich 5—8. Probe: Hemmstoffprotein der DEAE-Sephadex-Chromatographie (4 ml Fraktion G 1 u. G 2).
● — — ● E_{280}; ■———■ E_{410}; × —·— × RF-Einheiten/ml; ▲-------▲ *p*H-Gradient

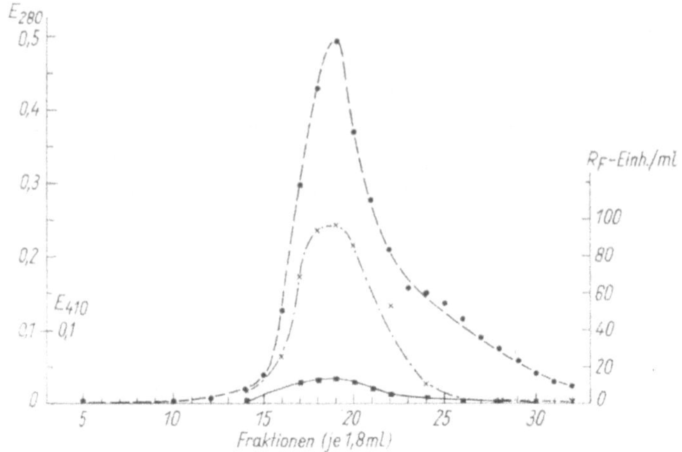

Abb. 4. Gelfiltration an Sephadex G-200.
Probe: 0,8 ml Fraktion 36 der isoelektrischen Fokussierung; eluiert wurde mit 0,05 M Kaliumphosphatpuffer, pH 6,0, 1 mM $MgCl_2$, Me^{2+}-frei.
●— — —● E_{280}; ■———■ E_{410}; × — · — × RF-Einheiten/ml.

Immunodiffusionstest

Im radialen Immunodoppeldiffusionstest (Abb. 5) kann in allen Präparationsstufen (RÜ, AS, DEAE-Chromatographie, IF) ein Protein mit gleichen Determinanten gegen das verwendete Antiserum nachgewiesen werden. Es ist eine Präzipitationslinie ohne Sporenbildung erkennbar.

Molekulargewichtsbestimmung durch PAGE

Die PAGE wurde bei neutralem pH im SDS-System durchgeführt. Zur Molekulargewichtsbestimmung wurden Hemmstoffpräparate eingesetzt, die im zuvor beschriebenen Reinigungsgang nach der IF bzw. nach der Gelfiltration erhalten wurden. In Abb. 6 ist die relative Mobilität [16] von Markerproteinen und Hemmstoff gegen log MW aufgetragen. In allen Fällen fanden wir ein Molekulargewicht für den Hemmstoff von 78000—80000.

N-terminale Aminosäurebestimmung

Die N-terminale Aminosäurebestimmung wurde nach einem Standardverfahren mittels Dansylchlorid durchgeführt (siehe Material und Methoden).

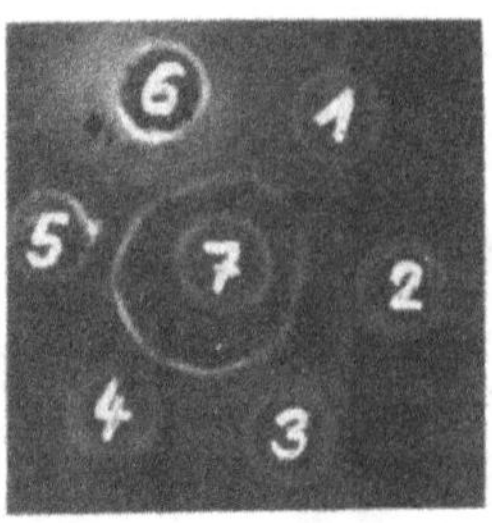

Abb. 5. Radiale Immundoppeldiffusion nach Ouchterlony zum Identitätsnachweis von RF während der Reinigung.
Hemmstoff auf den Reinigungsstufen: 1 = IF; 2 = AS; 3 = IF; 4 = DEAE-Chromatographie; 5 = IF; 6 = RÜ; 7 = Meerschweinchen-Anti-IF-Serum

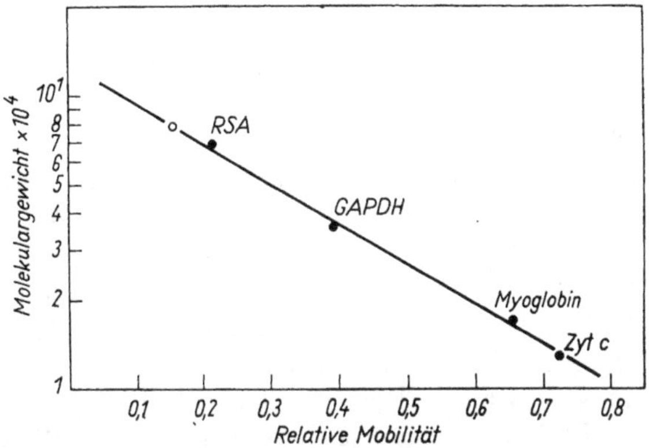

Abb. 6. Molekulargewichtsbestimmung von Hemmstoff RF mittels SDS-Diskelektrophorese. Aufgetragen ist die relative Beweglichkeit gegen das MG × 10^4.
RSA = Rinderserumalbumin, MG = 68000; GAPDH = Glyzerinaldehyd-3-phosphat-Dehydrogenase, MG = 36000; Myoglobin, MG = 17200; Zytochrom c, MG = 13000; o = Hemmstoff RF

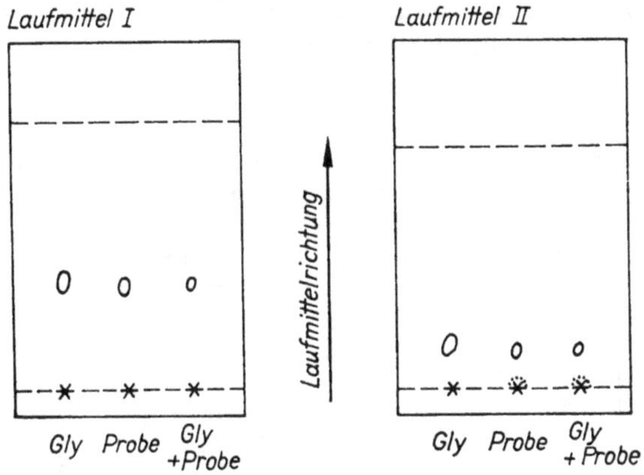

Abb. 7. Endgruppenbestimmung. Dünnschichtchromatographie auf Polyamid-Platten. In der Reihenfolge der Auftragung: Dans-Gly; Extrakt des dansylierten Hemmstoffs; Mischung aus Referenz Dans-Gly und Extrakt

Der Extrakt des hydrolysierten Dansylproteins wurde für einen Übersichtslauf auf Dünnschicht-Kieselgel G-Glasplatten aufgetragen. Dabei wurde ein Hauptfleck und nahe der Auftragsstelle ein Nebenfleck erhalten. Aus der Lage im zweidimensionalen Dünnschichtchromatogramm wurden die Referenz-Dansylaminosäuren ausgesucht, mit denen Gly als N-terminale Aminosäure bestimmt wurde. Im Rückstand des hydrolysierten und extrahierten Dansylproteins war keine Dansylaminosäure nachweisbar. Zur Identifizierung der N-ständigen Aminosäure wurden der Extrakt, Gly und Extrakt + Gly in einer Mikro-Dünnschichtchromatographie auf Polyamidfolien (s. Material und Methoden) aufgetragen und getrennt eindimensional in zwei Laufmitteln chromatographiert. Das Ergebnis zeigt (Abb. 7), daß die Dansylaminosäure des Hemmstoffproteins Glyzin ist. Die

N-terminale Aminosäurebestimmung erfolgte mit Proteinpräparaten, die durch verschiedene IF erhalten wurden und die sich als N-terminal identisch erwiesen.

Aminosäurenzusammensetzung

Zur Aminosäurenanalyse wurde Hemmstoffprotein verschiedener IF eingesetzt. In Langzeitdialysen bis zu 130 Std. unter 6—8maligem Wechsel des Dialysepuffers (0,01 M Phosphatpuffer, *p*H 7,0) wurde die Saccharose entfernt. Fällen des Proteins mit TCE, Waschen mit TCE und 5mal mit Azeton sollten Spuren von Saccharose ausschließen. Das Protein löst sich nicht in 85%iger Ameisensäure und bei der Hydrolyse in 6 N HCl tritt Verfärbung von rot-violett nach schwarz auf unter Abscheidung schwarzer Flocken. Diese Huminbildung wird für Glykoproteine beschrieben [29]. Die Aminosäureanalyse lieferte Wiederfindungswerte von 35—45 Gew.-%. Erst eine dreifache Perameisensäureoxydation führte dazu, daß das Protein nach Aufnahme in 85%iger Ameisensäure und Erhitzen in siedendem Wasserbad farblos gelöst vorlag. Während der Hydrolyse bei 110 °C in 6 N HCl trat keine Huminbildung auf. Tab. 2 zeigt ein vorläufiges Ergebnis der Aminosäurenzusammensetzung des Hemmstoffes. Während Tryptophan spektrophotometrisch nach EDELHOCH [22] bestimmt werden konnte, wurden für Tyrosin mit dieser Methode keine reproduzierbaren Werte erhalten. Auffallend in der Aminosäurenzusammensetzung ist der hohe Gehalt an Leuzin und Tryptophan (s. Tab. 2).

Tabelle 2

Aminosäurezusammensetzung des Hemmstoff-RF-Proteins nach Perameisensäureoxydation und 72 Std. Hydrolyse in 6 N HCl

Aminosäure	Gew.-%	Aminosäure	Gew.-%
Asp	7,0	Met[3]	2,0
Thr[1]	2,9	Ileu	3,8
Ser[1]	3,2	Leu	11,2
Glu	12,2	Tyr	?
Pro	4,9	Phe	5,2
Gly	3,1	Try[4]	5,0
Ala	4,2	Lys	5,8
Cys/2[2]	1,6	His	2,1
Val	5,0	Arg	6,3
Summe			85,6

[1] Unkorrigierte Werte; [2] analysiert als Zysteinsäure; [3] analysiert als Methioninsulfon; [4] Try wurde spektrophotometrisch nach EDELHOCH [22] bestimmt.

Durch die bei der Proteinvorbereitung zur Aminosäureanalyse erhaltenen Hinweise auf die Glykoproteinnatur des Hemmstoffs wurde ein qualitativer Test auf Kohlenhydrate mit Anthron-Reagens und Glykoproteinfärbung von SDS-Gelen durchgeführt. Das nach der IF erhaltene Protein wurde dialysiert (s. Material und Methoden). Sowohl der qualitative Test mit Anthron-Reagens als auch die Glykoproteingelfärbung waren positiv. Die quantitative Kohlenhydratbestimmung und die Bestimmung der Kohlenhydratkomponenten werden Gegenstand weiterer Arbeiten sein.

Diskussion

In dem vorliegenden Ergebnisteil wird der Reinigungsgang des Hemmstoffproteins dargestellt. Bei der Entwicklung eines systematischen Reinigungsverfahrens benutzten wir mehrere Eigenschaften des Hemmstoffproteins, wie Ladungszustand, isoelektrischer Punkt und Molekülgröße, über die in Vorversuchen Kenntnisse gewonnen waren. Techniken, die nur eines dieser Charakteristika ausnutzen, führen auch bei Variation der Bedingungen nicht zu einem reinen Protein. Es sei darauf hingewiesen, daß die Kenntnisse über die Proteineigenschaften bereits am Rohextrakt gesammelt wurden. Für die Reinigung war erschwerend, daß der Hemmstoff höchst sauerstoff- und schwermetallempfindlich und nur auf der Stufe der AS-Fraktion stabil ist; die Reinigung mußte unter anaeroben und schwermetallfreien Bedingungen durchgeführt werden.

Folgende Schritte wurden angewendet:

Retikulozytenhämolysat → 0,55 $(NH_4)_2SO_4$-Sättigung → DEAE-Ionenaustauschchromatographie → isoelektrische Fokussierung → Gelfiltration an Sephadex G-200.

Von ROBINSON und LEABACK [25] wird die Gelfiltration als erster Reinigungsschritt empfohlen, der Trennung nach Molekulargewichten folgen Ionenaustauschchromatographie und isoelektrische Fokussierung. Mit dem von uns gewählten Reinigungsgang konnte von größeren Eiweißmengen ausgegangen werden, da die Kapazität der Austauscher sehr hoch ist. Quantitativ gesehen konnten die Hauptfremdproteine, wie Hämoglobin und Katalase, weitestgehend entfernt werden. Unter den angegebenen Bedingungen passiert Hämoglobin ungebunden den Austauscher, während Katalase und andere Fremdproteine nur z. T. vom Hemmstoff getrennt werden (s. Abb. 1), bzw. am Austauscher verbleiben. Die unvollständige Abtrennung der Katalase vom Hemmstoff und das Auftreten mehrerer Katalasepeaks ist auf eine Heterogenität der Katalase, wie sie bei der Chromatographie an DEAE-Zellulose beschrieben wurde [26], zurückzuführen. Die IF im *p*H-Bereich von 5—8 trennt den Hemmstoff mit einem IP von 5,55—5,65 mit hoher Reinheit ab. Mit der Zwischenschichtung von Glutaminsäure zwischen die saure Anodenlösung und den Ampholin-*p*H-Gradienten werden Unstetigkeiten um $pH = 5$, die durch Diffusion der Phosphorsäure (Anodenlösung) in den Gradienten zustande kommen, vermieden.

Aus SDS-Diskelektropherogrammen (s. Abb. 2) der Gipfelfraktionen der IF kann abgeschätzt werden, daß die Verunreinigungen nur etwa 1% vom Gesamtprotein ausmachen. Das führte uns zu der Berechtigung, derartige IF-Präparate zur Immunisierung und eiweißchemischen Charakterisierung heranzuziehen. Nur in einigen Fällen wurden die Präparate über Sephadex G-200 weiter gereinigt.

Die Immunodiffusion gegen Anti-RF-Serum (Abb. 5) zeigt, daß das zur Immunisierung eingesetzte Präparat eine hohe Reinheit hatte, da offensichtlich nur Antikörper gegen ein Antigen gebildet wurden, wie aus dem Auftreten nur einer Präzipitationslinie hervorgeht.

Eine weitgehende oder vollständige Identität des RF-Proteins auf den verschiedenen Präparationsstufen (RÜ — AS — DEAE — Sephadex-IF) ist vorhanden. Ein Verlust potentieller Determinanten während der Präparation ist mit der verwendeten Methode (Antiserum gegen gereinigten RF der IF-Stufe) nicht auszuschließen.

Die Bestimmung nur einer N-terminalen Aminosäure (Glyzin) ist ein weiteres empfindliches Kriterium für die hohe Einheitlichkeit des verwendeten Präparats und läßt den Schluß zu, daß das relativ hochmolekulare Protein mit einem Molekulargewicht von 80000 (s. Abb. 6) nur aus einer Polypeptidkette besteht. Mit der N-terminalen Einheitlichkeit und einem konstanten Molekulargewicht des Proteins kann die am Kationenaustauscher und bei der präparativen alkalischen Diskelektrophorese erhaltene Heterogenität des Hemmstoffs [14] nur auf eine Abspaltung weniger C-terminaler Aminosäuren bzw. auf Kohlenhydratverlust zurückzuführen sein.

Vergleicht man die Aminosäurezusammensetzung des Hemmstoffes mit einer von REECK und FISHER [27] durchgeführten statistischen Analyse der Aminosäurezusammensetzung von 200 Proteinen, so fällt der ungewöhnlich hohe Anteil an Leuzin und Tryptophan auf (Tyrosin noch nicht bestimmt), wodurch sich der Hemmstoff von anderen Proteinen unterscheidet. Der Anteil polarer Aminosäuren mit 41 Mol-% liegt innerhalb der angegebenen Grenzen von 47 ± 6 Mol-% für lösliche globuläre Proteine [28]. Der hohe Anteil an Leuzin und Tryptophan und die hydrophobe Wirkregion des Hemmstoffes in den Fe–S-Proteinen von Komplex I und Komplex II der Atmungskette deuten auf besonders hydrophobe Bereiche im Hemmstoffprotein hin.

Mit dem Nachweis von Kohlenhydraten im Hemmstoff kann die Huminbildung bei der sauren Hydrolyse des Proteins und die unvollständige Aminosäurewiederfindung (bezogen auf Einwaage) erklärt werden.

Mit der Reindarstellung des Hemmstoffs RF und der Möglichkeit der Antikörpererzeugung ist das Feld weiterer eiweißchemischer Untersuchungen für Studien zur subzellulären Lokalisation und quantitativen RF-Bestimmung eröffnet worden, die Gegenstand anderer Publikationen sein werden.

Literatur

[1] RAPOPORT, S., u. W. GERISCHER-MOTHES: Hoppe-Seyler's Z. physiol. Chem. **302**, 167 (1955)
[2] RAPOPORT, S., u. CH. NIERADT: Hoppe-Seyler's Z. physiol. Chem. **302**, 179 (1955)
[3] RAPOPORT, S., u. C. WAGENKNECHT: Naturwissenschaften **43**, 582 (1956)
[4] WAGENKNECHT, C., u. S. RAPOPORT: Hoppe-Seyler's Z. physiol. Chem. **308**, 127 (1957)
[5] WAGENKNECHT, C., u. S. RAPOPORT: Hoppe-Seyler's Z. physiol. Chem. **309**, 201 (1957)
[6] SCHEWE, T., CH. HIEBSCH u. S. RAPOPORT: Acta biol. med. germ. **29**, 189 (1972)
[7] ALTENBRUNN, H.-J., u. S. RAPOPORT: Acta biol. med. germ. **2**, 599 (1959)
[8] WIESNER, R., u. S. RAPOPORT: Acta biol. med. germ. **31**, 289 (1973)
[9] RAPOPORT, S., und W. GERISCHER-MOTHES: Acta biol. med. germ. **3**, 450 (1959)
[10] SCHEWE, T., CH. HIEBSCH u. S. RAPOPORT: Unveröffentlichte Ergebnisse
[11] RAPOPORT, S., W. GERISCHER-MOTHES, u. CH. NIERADT-HIEBSCH: Hoppe-Seyler's Z. physiol. Chem. **304**, 207 (1956)
[12] RAPOPORT, S., u. W. GERISCHER-MOTHES: Hoppe-Seyler's Z. physiol. Chem. **315**, 38 (1959)
[13] WIESNER, R., CH. HIEBSCH u. S. ROSENTHAL: Acta biol. med. germ. **30**, 631 (1973)
[14] WIESNER, R., T. SCHEWE, CH. TANNERT u. S. RAPOPORT: In Vorbereitung
[15] VESTERBERG, O.: Meth. Enzymol. **28**, 389 (1972)
[16] WEBER, K., u. M. OSBORN: J. biol. Chem. **244**, 4406 (1969)
[17] GLOSSMANN, H., u. D. M. NEVILLE: J. biol. Chem. **246**, 6339 (1971)

[18] Backhausz, R.: Immunodiffusion und Immunoelektrophorese. Fischer, Jena 1967, S. 127
[19] Gros, C., u. B. Labonesse: Eur. J. Biochem. **7**, 463 (1969)
[20] Hirs, C. H. W.: J. biol. Chem. **219**, 611 (1956)
[21] Woods, K. R., u. K. T. Wang: Biochim. biophys. Acta **133**, 369 (1967)
[22] Edelhoch, H.: Biochemistry **6**, 1948 (1967)
[23] Roe, J. H.: J. biol. Chem. **212**, 335 (1955)
[24] Spiro, R. G.: Meth. Enzymol. **8**, 3 (1900)
[25] Robinson, H. K., u. D. H. Leaback: Biochem. J. **143**, 143 (1974)
[26] Mörikofer-Zwez, S., M. Cantz, H. Kaufmann, J. P. von Wartburg u. H. Aebi: Eur. J. Biochem. **11**, 49 (1969)
[27] Reeck, G. R., u. L. Fisher: Int. J. Peptide Protein Res. **5**, 109 (1973)
[28] Capaldi, R. A., u. G. Vanderkooi: Proc. natn. Acad. Sci, U.S.A. **69**, 930 (1972)
[29] Gottschalk, A. in: Glycoproteins. (= B.B.A. Library Vol. 5) A. Gottschalk (Hrsg.). Elsevier Publishing Company, Amsterdam, London, New York 1966, S. 96

Summary

R. Wiesner, Ch. Tannert, G. Hausdorf, T. Schewe, and S. Rapoport: Purification and characterization of the respiratory inhibitor RF from rabbit reticulocytes

In the stroma-free hemolysate of rabbit reticulocytes there exists a respiratory inhibitor (RF) of protein nature. It could be purified by the following procedure: $(NH_4)_2SO_4$-precipitation → DEAE-Sephadex A-50-chromatography → isoelectric focusing → gelfiltration on Sephadex G-200.
In addition to the activity the identity of the inhibitor could be proved immunologically at all steps of the purification. A molecular weight of about 80,000 was determined by gelelectrophoresis in the SDS-mercaptoethanol-system. The protein has an IP of 5.55—5.65 and consists of one polypeptide chain. This could be shown both by SDS-gelelectrophoresis and by estimation of only one N-terminal amino acid residue (glycine). Only after oxidation with performic acid the amino acid analysis of the protein gives reproducible values. The sum of weight-% is about 80 (without Tyr). The polarity of the protein with 41 mol-% agrees with the polarity of soluble globular proteins investigated by other authors. Acid hydrolysis without preceeding performic acid oxidation gives a sum of amino acid residues of only 35—45 weight-%, the glycoprotein nature of the inhibitor could be shown by staining the SDS-gels with Schiff's reagent and by carbohydrate determination with anthrone reagent.

Acta biol. med. germ., Band 36, Seite 405—410 (1976)

Institut für Physiologische und Biologische Chemie der Humboldt-Universität,
104 Berlin, DDR

Some properties of the lipoxygenase from rabbit reticulocytes

W. HALANGK, T. SCHEWE, CH. HIEBSCH, and S. RAPOPORT

Summary

A lipoxygenase was enriched from the stroma-free supernatant of rabbit reticulocytes. The enzyme causes drastic deterioration of mitochondrial membranes. The release of matrix enzymes is paralleled by formation of products of lipid peroxidation. The enzyme reacts with isolated phospholipids and free cis-unsaturated fatty acids. Some properties were determined: molecular weight, isoelectric point, temperature and *p*H-dependence and K_m value for linoleic acid. The enzyme is inhibited by reaction products and a variety of inhibitors, especially antioxidants and chelating agents.

Introduction

A few years ago a protein factor which causes lysis of mitochondria was found during studies of breakdown of reticulocyte mitochondria by SCHEWE et al. [1, 2]. First results showed that this factor is not identical with the known inhibitor proteins RF and RC from rabbit reticulocytes [3, 4].

On account of its action the factor was called mitochondria lysis factor. After exclusion of protease or phospholipase actions the peroxidation of the membrane phospholipids was tested.

The enzyme was identified as a soluble lipoxygenase which causes lysis of intact mitochondria by peroxidation of mitochondrial phospholipids. Soluble lipoxygenases have been described in a number of plant materials [5].

Lipoxygenases were not detected in animal tissues until recently. NUGTEREN [6] was the first to describe an arachidonate lipoxygenase in blood platelets.

In this paper our results concerning the identification and characterization of the reticulocyte mitochondria lysis factor as a true lipoxygenase are summarized.

Materials and methods

The lipoxygenase was enriched from stroma-free supernatant of rabbit reticulocytes obtained by bleeding anaemia. The osmotically haemolyzed reticulocytes were centrifuged after adjusting to *p*H 6.0 with 1 N HCl. To the resulting supernatant ammonium sulphate was added to a final concentration of 0.55 saturation ($AS_{0,55}$). After dialysis against 0.01 M Tris-HCl buffer, *p*H 7.2, an anion exchange chromatography on DEAE-Sephadex A 50 was carried out. The lipoxygenase was eluted at a NaCl-concentration of 0.1 M. The active fractions were collected and further purified by isoelectric focusing in Ampholine, *p*H 5—8.

The obtained enzyme fraction, while not pure, was free of haem-containing proteins.

Rat liver mitochondria were prepared by the method of DE DUVE et al. [7]. Isolated mitochondria were suspended in a medium containing 125 mM mannitol, 60 mM KCl, 60 mM Tris,

Abbreviations: MDH = malate dehydrogenase; MDA = malonyl dialdehyde; TBA = thiobarbituric acid; $AS_{0,55}$ = protein fraction at 0.55 saturation of ammonium sulphate; EDTA = ethylene diaminotetraacetic acid

10 mM KH_2PO_4, 5 mM $MgCl_2$ and 0.5 mM EDTA at *p*H 7.4. The final concentration was about 3 mg mitochondrial protein per ml.
Egg lecithin was prepared as methanolic solution according to Rhodes and Lea [8].
The tested polyunsaturated fatty acids linoleic acid ($C_{18:2}$), linolenic acid ($C_{18:3}$), 8,11,14-eicosatrienoic acid ($C_{20:3}$) and arachidonic acid ($C_{20:4}$) were a gift from Dr. Nugteren (Vlaardingen, Netherlands). We thank also Dr. Reichmann (Charité, Berlin) for linoleic acid and methyl linoleate and for gaschromatographic analyses.

Assay methods

The release of mitochondrial matrix enzymes was calculated as percentage of the total release after treatment with 0.2% Triton X-100. Malate dehydrogenase was assayed according to Siegel and Englard [9]. The formation of malonyl dialdehyde was determined with the thiobarbituric acid-method. The formation of fatty acid hydroperoxides was followed spectrophotometrically according to the method of Holman [10]. The reaction mixture contained fatty acid, 0.2% sodium cholate, 5% ethanol in 0.1 M phosphate buffer, *p*H 7.4.

Results

The lipoxygenase from rabbit reticulocytes causes drastic damage to membranes of isolated rat liver mitochondria [1, 2].
The lysis of mitochondria can be demonstrated by release of MDH. Fig. 2 shows the time dependent release of MDH and the formation of MDA from rat liver mitochondria. There can be observed a parallelism of membrane lysis and peroxidation of membrane phospholipids.
In Fig. 1 the temperature dependence is shown. Both processes, MDH release and MDA formation, show an optimum at 37 °C. A parallel behaviour with respect to the dependence on *p*H and the amount of enzyme was also observed.
The lipoxygenase is a protein with a molecular weight of 120,000 as determined by gel chromatography on Sephadex G-200 with catalase, inhibitor RF and haemo-

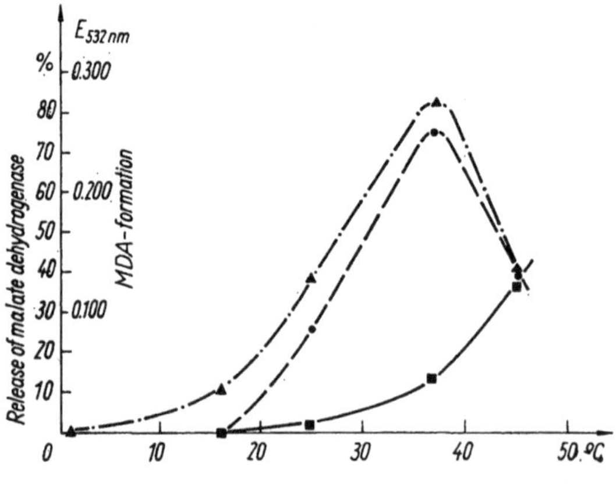

Fig. 1

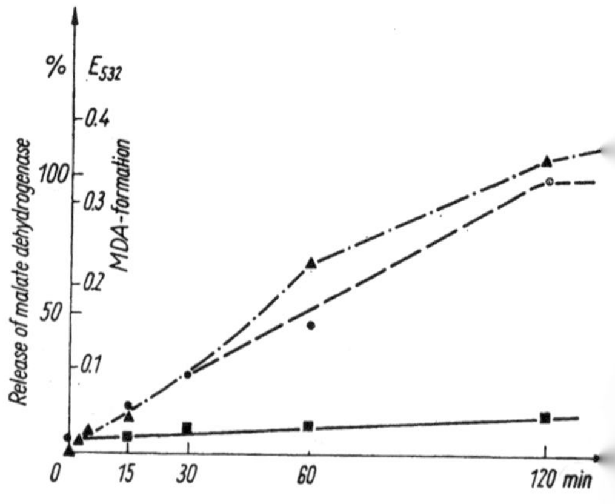

Fig. 2

Fig. 1. Temperature-dependent release of MDH by lipoxygenase (● ------ ●) and without lipoxygenase (■——■) and formation of MDA (▲ —·— ▲) from intact rat liver mitochondria (3 mg protein/ml). Incubation time 60 min with 3 mg $AS_{0,55}$-protein fraction

Fig. 2. Time-dependent release of MDH and MDA formation. Same conditions as Fig. 1. Temperature 37 °C globin as markers. The isoelectric point was determined to be 5.5 by isoelectric focusing in Ampholine, *p*H 5—8.

In addition to intact mitochondria the enzyme also peroxidizes purified mitochondrial phospholipids [11], egg lecithin and free cis-unsaturated fatty acids.

Some properties of the lipoxygenase were determined with linoleic acids as substrate by following the formation of linoleic acid hydroperoxide. Fig. 3 shows the temperature dependence of hydroperoxide formation. It is seen that the optimum of this reaction is at 20 °C. This is different from the results with mitochondria as substrate, but in the case of isolated phospholipids the MDA formation is parallel to the behaviour with linoleic acid. After a small increase in activity up to 20 °C a strong decrease follows so that at 37 °C only about 25% of the activity remains.

The *p*H-dependence of hydroperoxide formation shows an optimum at *p*H 7.4 to 7.8. At *p*H values smaller than 6.5 the enzyme is not active (Fig. 4).

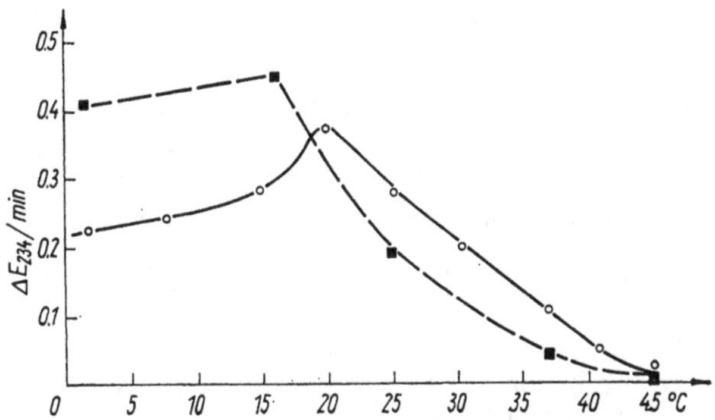

Fig. 3. Temperature dependence of linoleic acid hydroperoxide formation (o——o) and MDA formation from egg lecithin (1 mg/ml; ■ — — ■). Start with 7 µg enzyme (fraction of isoelectric focusing)

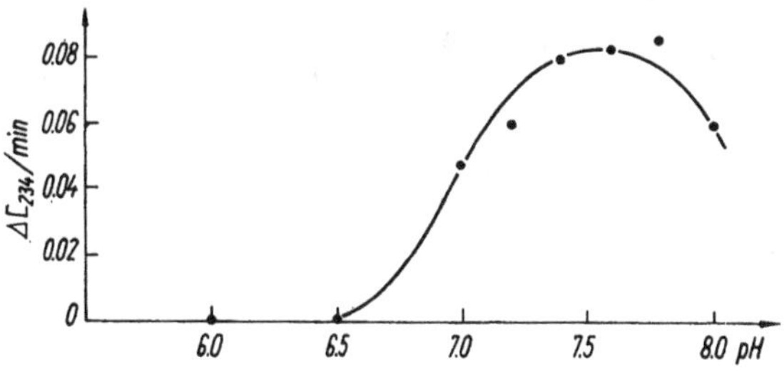

Fig. 4. *p*H-dependence of linoleic acid hydroperoxide formation. 7 µg enzyme (fraction of isoelectric focusing); reaction in 0.1 M phosphate buffers at 2 °C

For linoleic acid a K_m-value of about 0.1 mM is determined both at 37° and 2 °C. During reaction with free fatty acids the lipoxygenase is inactivated. It could be shown that this inactivation by peroxidized products is irreversible.

To investigate the substrate specificity of the enzyme a number of polyunsaturated fatty acids was tested. Fig. 5 shows the rate of peroxidation as per cent of V_{max} at certain concentrations of each fatty acid.

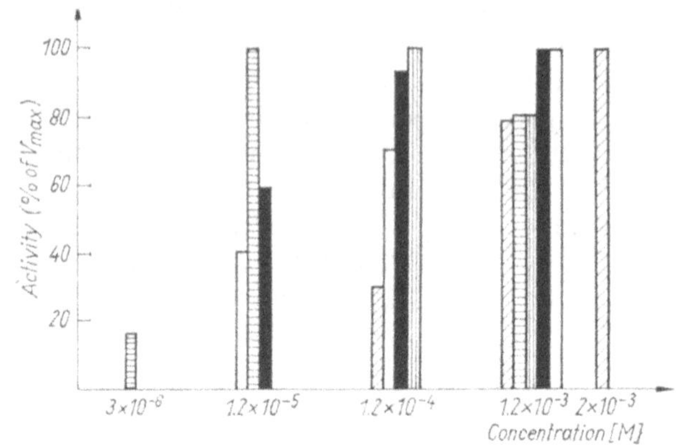

Fig. 5. Hydroperoxide formation from arachidonic acid (▤), 8,11,14-eicosatrienoic acid (▥), linolenic acid (▮), linoleic acid (□) and methyl linoleate (▧). Reaction with 7 μg enzyme at 2 °C

Certain trends both of the V_{max}- and K_m-values are apparent:

$$\text{Trend of } V_{max}.: C_{20:3} \geqq C_{18:2} > C_{18:3} \geqq C_{20:4} > C_{18:2}\, CH_3$$

$$\text{Trend of } K_m: C_{20:4} < C_{20:3} = C_{18:3} < C_{18:2} < C_{18:2}\, CH_3$$

A number of effectors such as antioxidants, chelating agents, and SH-reagents were tested.

In Table 1 the effects of some antioxidants are listed.

The results show an inhibition of the hydroperoxide formation by classical antioxidants, but with different efficiency.

In Table 2 the effects of some chelating agents are demonstrated.

The action of Tiron was slow; 50 per cent inhibition were obtained after 42 hrs incubation. p-Chloromercuribenzoate at a final concentration of 10 μM had a low effect of 16 per cent inhibition. Haemoglobin inhibits linoleic acid hydroperoxide formation strongly; 0.4 μM of haemoglobin inhibit the reaction by 54%. Serum-albumin (0.1%) in the reaction mixture inhibits the reaction totally by binding the free fatty acids.

Table 1

Effect of antioxidants on the lipoxygenase of reticulocytes

Antioxidant	Final concentration [μM]	Inhibition [%]
Propyl gallate	10	56
α-Naphthol	10	27
α-Tocopherol	100	53
Ubiquinone 50	100	69

The reaction mixture contained 0.6 mM linoleic acid, 0.2% sodium cholate, 5% ethanol in 0.1 M phosphate buffer, *p*H 7.4, at 2 °C plus the antioxydants. The reaction was started by the enzyme.

Table 2
The effect of chelating agents on the lipoxygenase of reticulocyte

Agent	Final concentration [mM]	Inhibition [%]
KCN	0.05	47
Tiron	0.01	50
EDTA	1	0
o-Phenanthroline	0.1	0

The reaction mixture contained 0.6 mM linoleic acid, 0.2% sodium cholate and 5% ethanol in 0.1 M phosphate buffer, *p*H 7.4, plus chelating agent in a 100 fold higher concentration than the final one. Preincubation time was 15 min at 37 °C. The reaction was started by the enzyme.

Discussion

The lipoxygenase from rabbit reticulocytes causes rapid and drastic damage of mitochondrial membranes in consequence of a peroxidation of their phospholipids. This is indicated by the parallelism between MDH release and formation of MDA as a function of time, temperature and other parameters (see Fig. 1 and 2). A haem-catalyzed autoxydation could be excluded by using haemoproteid-free enzyme preparations and by the heat lability of the enzyme. According to Tappel [12] it is possible to distinguish a lipoxygenase reaction from the haem-catalyzed peroxidation by using a fatty acid concentration smaller than 1 mM.

The lipoxygenase from reticulocytes shows some similarities to the lipoxygenase from soybeans with respect to the molecular weight [13], the slow inactivation by the chelator Tiron [14] and the insensitivity against p-chloromercuribenzoate. But there is also a variety of differences. The reticulocyte lipoxygenase is sensitive against cyanide and against the reaction products of lipid peroxidation. Soybeans lipoxygenase is activated by the 13-linoleic acid hydroperoxide [15].

We found a *p*H-optimum in the range between *p*H 7.4 and 7.8 in the presence of detergents such as cholate. Soybean lipoxygenase I possesses an optimum at *p*H 9.0, but other isoenzymes from this source possess an optimum at *p*H 6.5. The lipoxygenase from blood platelets is most active between *p*H 8.0 and 9.0 [6]. The temperature optima differ with mitochondria or fatty acids as substrates. The biological role of the reticulocyte lipoxygenase consists in the degradation of mitochondria. Therefore one would expect the physiological temperature optimum to be at 37 °C (see Fig. 3). Judging from the hydroperoxide formation from free fatty acids secondary reactions take place at temperatures higher than 20 °C. As shown in Fig. 5 all tested fatty acids are peroxidized by the reticulocyte lipoxygenase. We found differences in the rate of oxidation and in the affinity of the fatty acids to the enzyme. The highest rate was measured with 8,11,14-eicosatrienoic acid, but the highest affinity was observed in the case of arachidonic acid. Methyl linoleate is also a substrate for the enzyme, which is different from the arachidonate lipoxygenase described by Nugteren [6]. Antioxidants are inhibitors of the lipoxygenase reaction. The most powerful inhibitor was propyl gallate.

There were some differences as compared with the efficiency of classical antioxidants in autoxidative systems. First tocopherol was efficient only in a tenfold higher concentration with respect to propyl gallate and α-naphthol and secondly the molar ratio between antioxidants and fatty acid was relatively small, i.e. 1:60 in the case of propyl gallate. Among the iron chelating agents we found an inhibition of the reticulocyte lipoxygenase by cyanide and Tiron while EDTA and o-phenanthroline were ineffective. It is possible that the enzyme contains iron in the ferric state. Studies of the soybean lipoxygenase I showed a possible role of ferrous iron which is converted to the ferric state during the reaction with linoleic acid [16]. Other investigators detected ferric iron by the use of chelating agents [14].

In summary, it may be stated that the reticulocyte lipoxygenase possesses some similar properties, but also a variety of differences as compared with other known lipoxygenases. The biological importance of the enzyme is assumed to consist in triggering the breakdown of mitochondria during the maturation of reticulocytes [17]. The susceptibility of mitochondrial membranes to reticulocyte lipoxygenase is also dependent on the functional state of the mitochondria [18].

References

[1] SCHEWE, T., W. KRAUSE, W. HALANGK, and S. RAPOPORT: VII. Internationales Symposium über Struktur und Funktion der Erythrozyten, Berlin 1973. S. RAPOPORT and F. JUNG (eds). Akademie-Verlag, Berlin 1975, S. 607

[2] KRAUSE, W., T. SCHEWE, and D. BEHRISCH: Acta biol. med. germ. **34**, 1609 (1975)

[3] RAPOPORT, S., and W. GERISCHER-MOTHES: Hoppe-Seylers Z. physiol. Chem. **302**, 167 (1955)

[4] ALTENBRUNN, H. J., and S. RAPOPORT: Acta biol. med. germ. **2**, 599 (1959)

[5] BERGSTRÖM, S., and R. T. HOLMAN: Adv. Enzymol. **8**, 441 (1948)

[6] NUGTEREN, D. H.: Biochim. biophys. Acta **380**, 299 (1975)

[7] DE DUVE, C., B. C. PRESSMAN, R. GIONETTO, R. WATTIAUX, and F. APPELMANS: Biochem. J. **60**, 604 (1955)

[8] RHODES, D. N., and C. H. LEA: Biochem. J. **65**, 526 (1957)

[9] SIEGEL, L., and S. ENGLARD: Biochim. biophys. Acta **54**, 67 (1961)

[10] HOLMAN, R. T.: Archs Biochem. Biophys. **10**, 519 (1946)

[11] SCHEWE, T., W. HALANGK, CH. HIEBSCH, and S. RAPOPORT: FEBS Lett. **60**, 149 (1975)

[12] TAPPEL, A. L.: Fd Res. **18**, 104 (1953)

[13] STEBENS, F. C., D. M. BROWN, and E. L. SMITH: Archs Biochem. Biophys. **136**, 413 (1970)

[14] PISTORIUS, E. K., and B. AXELROD: J. biol. Chem. **249**, 3183 (1974)

[15] SMITH, W. L., and W. E. M. LANDS: J. biol. Chem. **247**, 1038 (1972)

[16] DE GROOT, I. I. M. C., G. A. VELDINCK, I. F. G. VLIEGENTHART, J. BOLDINGH, R. WEVER, and B. F. VAN GELDER: Biochim. biophys. Acta **377**, 71 (1975)

[17] SCHEWE, T., W. HALANGK, CH. HIEBSCH, and S. RAPOPORT: Acta biol. med. germ. **36**, 363 (1977)

[18] KRAUSE, W., and W. HALANGK: Acta biol. med. germ. **36**, 381 (1977)

Acta biol. med. german., Band 36, Seite 411—414 (1977)

Abteilung Neonatologie der Kinderklinik, Bereich Medizin (Charité) und
Institut für Physiologische und Biologische Chemie, Humboldt-Universität zu Berlin,
104 Berlin, DDR

Creatine during bleeding anemia of rabbits

I. SYLLM-RAPOPORT, E. DUMDEY, and S. RAPOPORT

Summary

In bleeding induced anemia of rabbits an increase of creatine in the red cells occurred, which in some cases amounted to more than 5-times the normal concentration. In all but one case plasma creatine rose as well. Reticulocytosis, increased O_2-consumption of the red cells and an increased level of erythropoietin indicated active erythropoiesis and a shift towards a younger cell population.
Whether the rise of red cell creatine is solely due to a higher proportion of young cells remains an open question.

Introduction

Even though creatine is present in red blood cells in relatively high concentrations (about 0.5 mmoles/l cells), its functional role and mode of entry into the cell are as yet unknown. GRIFFITHS and FITZPATRICK [1] pointed out that young red cells have a higher creatine content than old ones.
This was confirmed by VALERI and FORTIER [2] and by OPALINSKI and BEUTLER [3]. Anemias and other hypoxic states with active erythropoiesis result in a rise of red cell creatine concentration [1—4]. It is not clear whether this rise is caused solely by the higher proportion of young cells.

Material and methods

In 4 adult rabbits an anemia was produced by daily removal of 45—50 ml blood from the ear vein. The lowest hematocrit values reached were 12—18%.
Venous blood was obtained in heparinized glass tubes. The samples were kept in ice water and centrifuged at 4 °C for the separation of the cells from the plasma. The buffy coat was removed; the cells were not washed. Their creatine concentration was corrected by means of the cell sample hematokrit and the creatine concentration of the plasma determined separately.
Creatine in red cells and plasma was determined according to the method of EGGLETON et al. [5] modified by GRIFFITHS [6]. All determinations were done in duplicates.
The O_2-consumption of the red cells was determined by the Warburg technique at 37 °C.
The erythropoietin level in the plasma was estimated by means of the ^{59}Fe-incorporation into mouse erythrocytes after injecting the rabbit plasma into supertransfused mice[1].
In 2 animals the experiment was stopped after 14 days, 2 rabbits died on the 6th and 8th day after beginning of bleeding. Their hematocrit values were 20% and 12%, respectively.

Results

In Fig. 1—4 the behaviour of the creatine concentration in red cells and blood plasma are shown for 4 rabbits during the course of the bleeding anemia.

[1] We thank Mrs. R. POHLE for the performance of the tests.

A rise of creatine concentration in the red cells (in some cases amounting to more than 5 times the normal concentration) occurred within the first 6—10 days, declining thereafter in spite of continued daily removal of blood. In all but one animal the concentration of plasma creatine increased as well. Reticulocytosis,

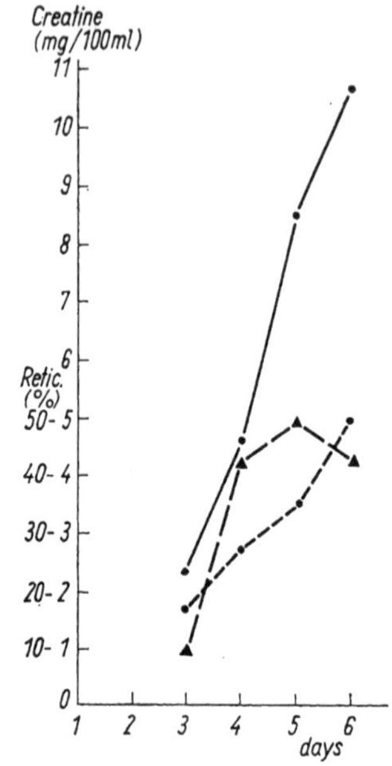

Fig. 1

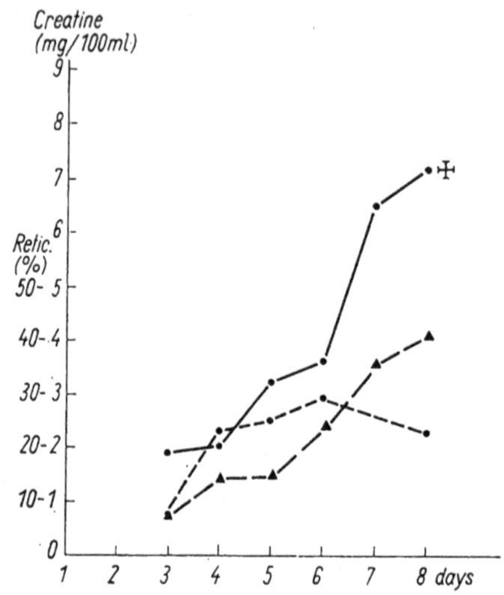

Fig. 2

Fig. 1 and 2. Creatine concentration in red blood cells and plasma, reticulocyte counts during bleeding anemia.

●——●, erythrocytes; ▲ — — ▲, plasma; ● ------ ●, % reticulocytes.

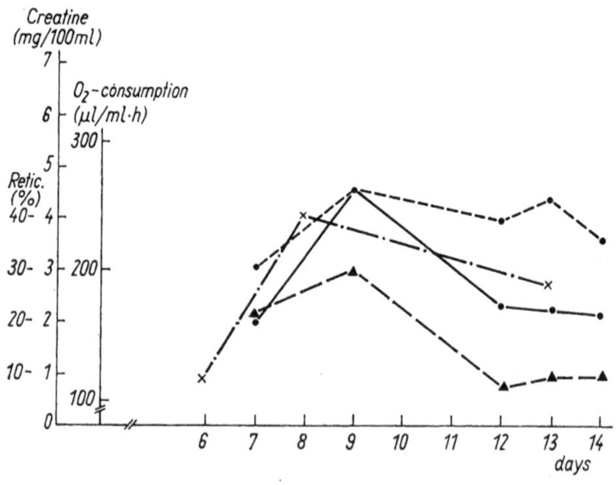

Fig. 3. Creatine concentration in red blood cells and plasma, reticulocyte counts and O_2-consumption of red cells during bleeding anemia.

Symbols cf. Fig. 1; × —·— ×, O_2-consumption

increased O_2-consumption of the red cells (Fig. 3) and an elevated erythropoietin level in the plasma (Fig. 4) indicates a shift towards a younger cell population. However, the rise in cell creatine seemed to exceed the number of reticulocytes. In Fig. 5 the calculated concentrations of creatine in reticulocytes, i.e. related to 100 %reticulocytes, are plotted against the days of the experimental anemia for all 4 rabbits. It is shown that the reticulocytes which appear in the blood stream during the course of bleeding anemia do not seem to be furnished with a constant creatine content.

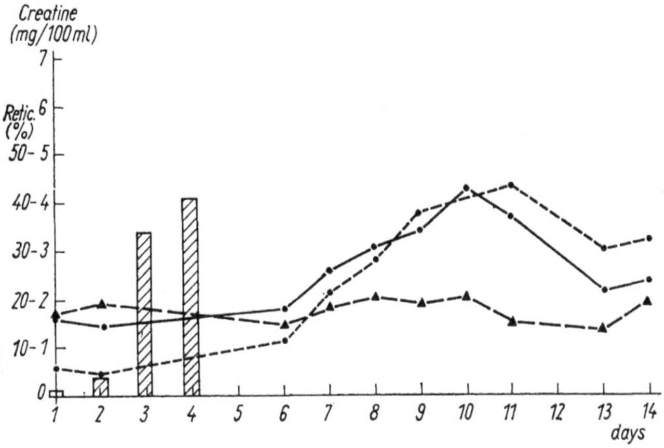

Fig. 4

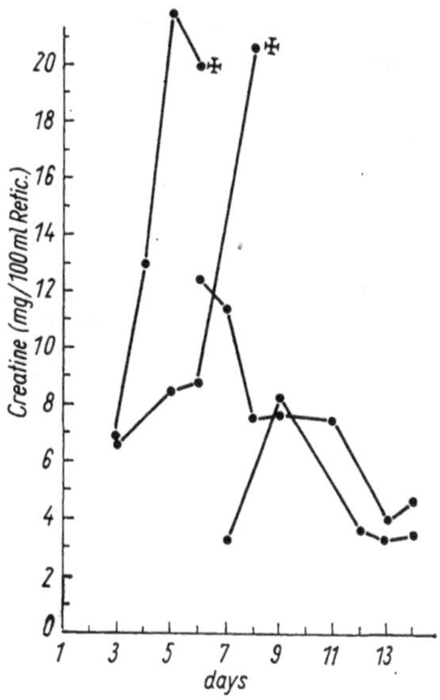

Fig. 5

Fig. 4. Creatine concentration in red blood cells and plasma, reticulocyte counts and level of plasma erythropoietin (^{59}Fe-incorporation) during bleeding anemia Symbols cf. Fig. 1; ////, Fe-incorporation.

Fig. 5. Calculated creatine concentrations of reticulocytes during the course of bleeding anemia

Discussion

Our experiments confirm the previous observations on the increase of red cell creatine in clinical cases of anemic hypoxia [1—4] and also in experimental hemolytic anemia of rabbits caused by phenylhydrazine as reported by GRIFFITHS and LOTHIAN [4]. Our findings, however, contradict those of GRIFFITHS and LOTHIAN regarding the results on bleeding anemia in rabbits, where they found no rise in red cell creatine. The reason for this contradiction is most probably the fact that they removed only 10—20 ml of blood for 5 days and never reached the point of sufficient hypoxic stress to induce active erythropoiesis as borne out by the low reticulocyte count, even though the hemoglobin dropped to 6 g/100 ml.

Our results do not allow a definite answer to the question whether the rise in red cell creatine is due solely to active erythropoiesis (i.e. a greater percentage of young cells with high creatine content) or whether in addition creatine may be taken up from the plasma either by reticulocytes or erythrocytes in the blood stream.

Griffiths [7] observed no overall rise of creatine content of human red cells incubated in plasma with creatine concentrations which even exceeded the normal cellular level. We have confirmed this result. Whether the same holds for rabbit red cells, which maintain a lower concentration gradient between cells and plasma (rabbits: 3:1; humans: 10:1) has yet to be shown.

Moreover, the question remains open whether a young red cell population, especially reticulocytes, might have the capacity for enriching their creatine by transfer from the plasma.

References

[1] Griffiths, W. J., and M. Fitzpatrick: Br. J. Haemat. **13**, 175 (1967)
[2] Valeri, C. R., and N. L. Fortier: New Engl. J. Med. **281**, 1452 (1969)
[3] Opalinski, A., and E. Beutler: New Engl. J. Med. **285**, 483 (1971)
[4] Griffiths, W. J., and Elizabeth J. Lothian: Br. J. Haemat. **17**, 477 (1969)
[5] Eggleton, P., S. R. Elsden, and N. Gough: Biochem. J. **37**, 526 (1943)
[6] Griffiths, W. J.: J. clin. Path. **21**, 412 (1968)
[7] Griffiths, W. J.: Br. J. Haemat. **15**, 389 (1968)

Acta biol. med. germ., Band 36, Seite 415—416 (1977)

Abteilung Neonatologie der Kinderklinik, Bereich Medizin (Charité) and
Institut für Physiologische und Biologische Chemie, Humboldt-Universität zu Berlin,
104 Berlin, DDR

Creatine transport into human red blood cells

I. SYLLM-RAPOPORT, E. DUMDEY, and S. RAPOPORT

GRIFFITHS [1] showed that even very high creatine concentrations of the surrounding plasma did not result in an overall increase of cellular creatine content of human red cells. We could confirm these observations. GRIFFITHS concluded that the human red cell is highly impermeable to creatine, the cell membrane "offering a barrier to the free exchange of creatine with the plasma".
However, by incubating red cells with tracer amounts of 1-[^{14}C]-creatine we could demonstrate an active transport into the cell. Fig. 1 and 2 show a definite increase in specific activity in the red cells and a fall of activity in the plasma, while no transport takes place at 0 °C.

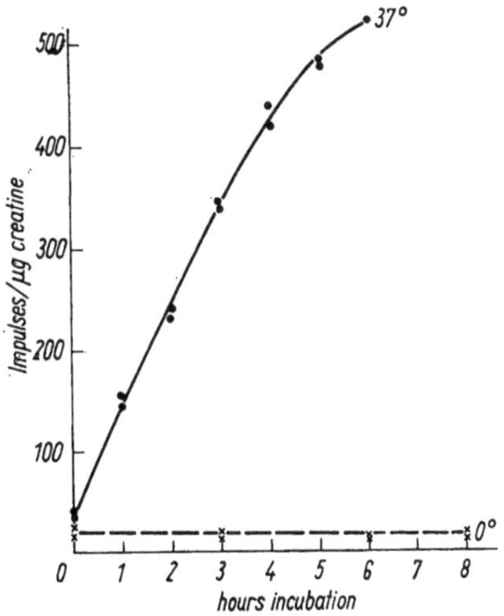

Fig. 1

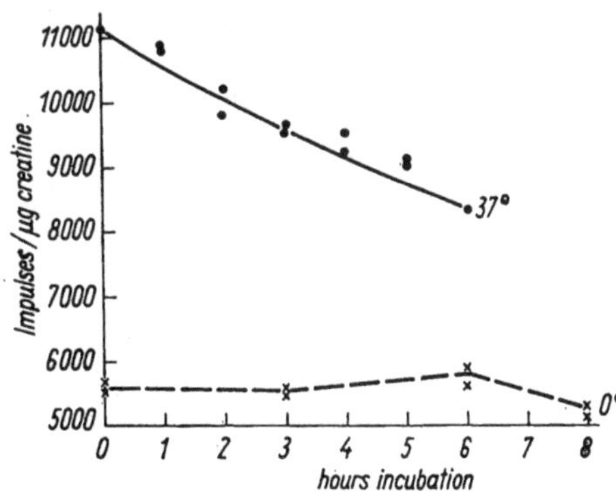

Fig. 2

Fig. 1. Specific 1-[^{14}C]-creatine activity in red blood cells during incubation at 37 °C and at 0 °C

Fig. 2. Specific 1-[^{14}C]-creatine activity in the plasma during incubation at 37 °C and 0 °C

The renewal rates of erythrocyte creatine with normal and increased creatine are shown in Table 1. At normal plasma creatine levels the half renewal time of cell creatine is $2^1/_2$ days. The renewal rate increases in a saturable manner with increasing plasma creatine concentrations (Fig. 3).

Table 1

Renewal rates of erythrocyte creatine with normal and increased creatine plasma concentration

Date	calculated from	Renewal rate (% creatine/6 hrs)		
		— Creatine[1]	+ Creatine	Plasma conc. (mg/100 ml)
24. 4. 76	Plasma	4.2		
4. 5. 76	Plasma	5.9	10.9	3.8
	Erythrocyte	6.1	—	
20. 5. 76	Plasma	4.0	11	} 4.3
	Erythrocyte	4.0	12	
1. 6. 76	Plasma	5.0	—	
	Erythrocyte	5.3	—	

[1] values 0.61—0.85 mg/100 ml Plasma

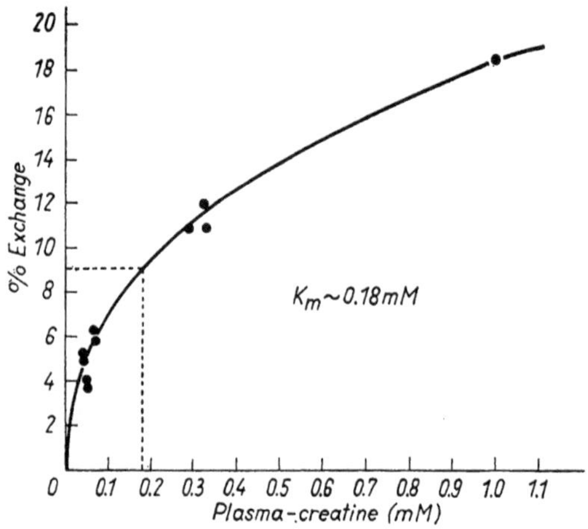

Fig. 3. Creatine exchange of human red cells with increasing plasma creatine concentration

References

[1] Griffiths, W. J.: Br. J. Haemat. 15, 389 (1968)

Acta biol. med. germ., Band 36, Seite 417—419 (1977)

Katedra patologické fyziologie fakulty všeobecného lékařství University Karlovy v Praze, ČSSR

The transplantation of bone marrow cells with a change of hemoglobin pattern induced by bleeding

K. Šulc, J. Neuwirt, T. Trávníček, and E. Radikovská

Introduction

The hypoxia evoked by blood loss induces a change in the hemoglobin type which seems to be related to the production of erythropoietin.

The change in the hemoglobin type is most significant in sheep in which great amounts of Hb C are produced after bleeding or erythropoietin administration [1].

Hb C is normally present in extremely small amounts in all sheep and its appearance probably represents an increased rate of Hb C synthesis and not the activation of a completely repressed gene. The switch-over is usually complete within 10 days of the onset of severe anemia.

In our laboratory changes in the hemoglobin type after bleeding were also found in rats [2—5].

It is still a matter of dispute whether the change of the hemoglobin type after bleeding occurs at the transcriptional or at the translational level.

In our experiments we tried to prove whether the change of the hemoglobin type appearing in rats after bleeding was transplantable and whether this change could be also detected in radiation chimeras.

Materials and methods

The experiments were carried out on Wistar inbred male rats (type A). In these rats four main hemoglobin fractions are to be found. The rats were bled and their bone marrow cells were used for transplantation. The amount of blood withdrawn was about 1/3 of circulating blood volume. The rats were sacrified 24 hrs after bleeding.

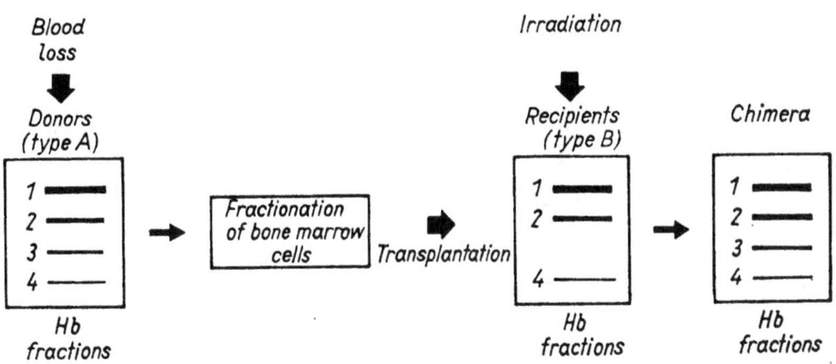

Fig. 1. Scheme of the experiments

$1 \cdot 10^6$ bone marrow cells were transplanted to a genetically different strain of rats (type B) which has only 3 hemoglobin fractions. The recipient rats (type B) were irradiated with a dose of 850 Rads and the transplantation was performed 3 hrs after irradiation. The bone marrow cells were fractionated prior to the transplantation. The fractionation procedure was modified according to the method used by Peterson and Ewans [6] and Miller and Phillips [7]. The fractionation of bone marrow cells was performed on the column with a linear gradient (1%–5%) of polymerized sucrose (Ficoll) with the addition of a constant amount of 0.1% of bovine serum albumin. The analysis of hemoglobin types in peripheral blood was carried out on a thin layer starch gel electrophoresis according to Trávníček and Šulc [8]. On the starch plate the protein compounds and the hemoglobin compounds were detected. As a criterion of the production of the radiation chimera in rats the appearance of the third main hemoglobin fraction, which is normally missing in rats of the strain B, was used (Fig. 1).

Results

The analysis of the hemolysate of peripheral blood was performed 7, 14, 21 and 28 days after irradiation and transplantation. The irradiated rats (type B) were divided into 5 groups. The control animals were in the first group; in the second group the animals were transplanted with unfractionated bone marrow cells. The second, third and fourth fractions obtained after cell separation were administered to animals in groups 3, 4 and 5, respectively. The highest content of bone marrow stem cells was found in the 3rd fraction. The results of the experiments are summarized in Table 1. The percentage of fraction which appears as a result of the production of the radiation chimera corresponds to the percentage of this fraction in donor rats after bleeding.

Table 1

Relative amount of haemoglobin fractions in per cent in control and experimental animals

Hb fractions	Control animals (type A)	9 days after blood loss (type A)	Control recipient animals (type B)	Chimera (recipient — type B) 21 days after transplantation
1	52.5 ± 3.1	41.4 ± 4.1*	52.5 ± 2.2	38.5 ± 5.8*
2	19.0 ± 2.7	23.2 ± 1.3	22.8 ± 2.5	22.1 ± 2.4
3	16.5 ± 3.1	20.1 ± 2.3*		21.8 ± 2.9*
4	10.7 ± 3.6	15.4 ± 2.6*	24.7 ± 2.3*	17.1 ± 2.7*

* = $p < 0.05$

Discussion

According to these experiments it may be concluded that the change of the hemoglobin type occurring in rats after bleeding is transplantable to recipient rats. This indicates that the change in the hemoglobin type occurs at the transcriptional level in the pluripotent hemopoietic stem cells.

The pattern of hemoglobin fractions of donor rats after bleeding progressively increases in the peripheral blood of recipient animals. The appearance of the specific hemoglobin types cannot be explained on the basis of the supply of 1 million cells per rat only. If this were the case, the hemoglobin pattern typical for rats after bleeding would be detected immediately after transplantation. In our experiments the changes were detectable from the 14th day which shows that the radiation chimera was produced.

From some experiments with hemoglobin formation during chick embryo development it was suggested that transcription of hemoglobin genes may precede the onset of hemoglobin synthesis quite considerably [9]. If these experiments can be extrapolated to mammalian erythropoiesis then the transcription of hemoglobin genes may precede by some time the initiation of hemoglobin synthesis. It seems that the selective transcription of hemoglobin genes may occur very early during differentiation already at the level of the pluripotent hemopoietic stem cells. Since the changes in hemoglobin phenotype occur also after the administration of erythropoietin and are partially abolished by anti-erythropoietin, it has been suggested that this hormone may affect the pattern of genome activation for globin chain production [10]. However, it was proposed that the erythropoietin-responsive stem cells are the site of action of erythropoietin and not the pluripotent stem cells which are the only cells producing the radiation chimera. Our results outline an interesting possibility that erythropoietin influences the genome derepression already at the level of the pluripotent hemopoietic stem cells.

References

[1] van Vliet, G., and T. H. J. Huisman: Biochem. J. **93**, 401 (1964)

[2] Trávníček, T., E. Trávníčková, and K. Šulc: Physiologia bohemoslov. **15**, 175 (1966)

[3] Trávníček, T., E. Trávníčková, and K. Šulc: Physiologia bohemoslov. **16**, 160 (1967)

[4] Trávníček, T., Z. Vodrážka, J. Borová, J. Čejka, J. Salák, and K. Šulc: Physiologia bohemoslov. **16**, 543 (1967)

[5] Trávníček, T., J. Borová, and K. Šulc: Physiologia bohemoslov **20**, 27 (1971)

[6] Peterson, E. A., and W. H. Ewans: Nature, Lond. **214**, 824 (1967)

[7] Miller, R. G., and R. A. Phillips: J. cell. Physiol. **73**, 191 (1969)

[8] Trávníček, T., and K. Šulc: Physiologia bohemoslov **16**, 167 (1967)

[9] Wilt, F. H.: Adv. Morphogen. **6**, 89 (1967)

[10] Lewis, J. P., and T. H. J. Huisman in: The Regulation of Erythropoiesis and Haemoglobin Synthesis. Proceedings of the International Symposium Erythropoieticum, Praha 1970. T. Trávníček and J. Neuwirt (eds). University Karlova, Praha 1971, p. 391

Acta biol. med. germ., Band 36, Seite 421—424 (1977)

Department of Biology, Faculty of General Medicine, Charles University,
Prague, Czechoslovakia

Haemoglobin as a marker in bone marrow transplants of rats

R. Brdička and V. Křen

Summary

A pair of congenic strains has been produced, which differs only in haemoglobin phenotype and allows long-term observation after bone marrow transplantation without immuno-histocompatibility reactions.

Introduction

On the basis of genetic Hb-polymorphism of the rat (Rattus norvegicus) we have tried to breed animals offering several advantages for the analysis of the erythron function:

1. Elimination of GVHR after bone marrow transplantation;
2. unequivocal distinguishing of the activity of "own" and transplanted cells;
3. possibility of long-term observation;
4. suppression of the so-called allogenic inhibition phenomenon.

Material and methods

The *congenic strain production* that would differ in haemoglobin phenotype only was supposed to yield the above mentioned advantages. Breeding of the congenic strain started by crossing

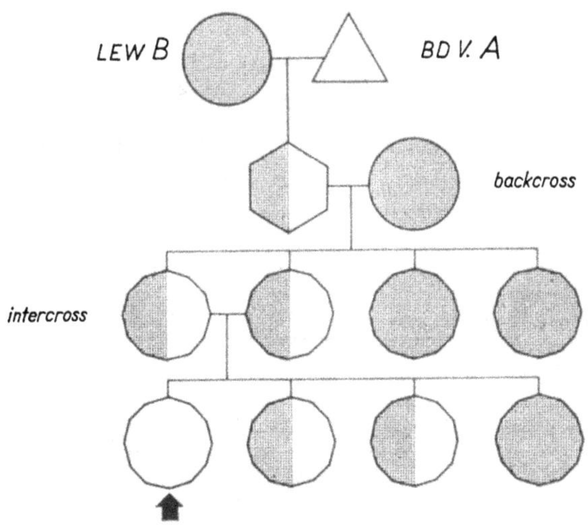

Fig. 1. The way of rendering animals more and more similar to the original strain LEW, which is simulated by the change of the shape, but bearing the Hb allele of the strain BD V. The arrow indicates "animals" homozygous for the Hb allele A used as donors

Abbreviations: Hb = haemoglobin; GVHR = graft versus host reaction; MHS = main histocompatibility system of the highly inbred strain LEW/Cub (Hb phenotype B) with the coloured diagnostic strain BD V/Cub (Hb phenotype A). In generations 1—4 the simple backcross system of mating was used but in the following generations it was changed in the cross-intercross system to get homozygous animals (Fig. 1). Moreover selections against heterogeneity in the MHS [1] and chromogen locus were made. The probability of homozygous mating in a locus genetically independent of the Hb allele transferred has been calculated to be more than 95% in animals used for the experiments presented here [2].

Bone marrow transplantation

Bone marrow cells were obtained by aspiration from trepanated femurs into chilled solution of 1 part of donor's serum and 4 parts of saline and applied via tail vein in a dose of $50 \cdot 10^6$ living cells. Recipients not older than 100 days were irradiated with a total whole-body dose of 900 R (15 R/min) from ^{60}Co source no more than 6 hrs. before the bone marrow application

Detection of the erythron's function

As an indicator of the erythron's function the Hb electrophoretic pattern was used. Haemolysates prepared from 0.2 ml of blood were analyzed at 7 day intervals by electrophoresis on Cellogel strips (Cellogel, Chemetron, Italy) under conditions suggested by the producer. Strips stained with amidoblack 10 B were transparentized and checked densitometrically with an integrating densitometer (Vernon, France). Genetic polymorphism of rat Hb is related to two out of four main components (Fig. 2), e.g. to number 3, 5 and 4, 6 as described in detail elsewhere [3]. Under commonly used conditions the mobility of components 4 and 5 is very close which is not suitable for the quantitative evaluation and for this reason component 3 was chosen as a gauge and its relative quantity was compared with artificially prepared blood mixtures (Fig. 3).

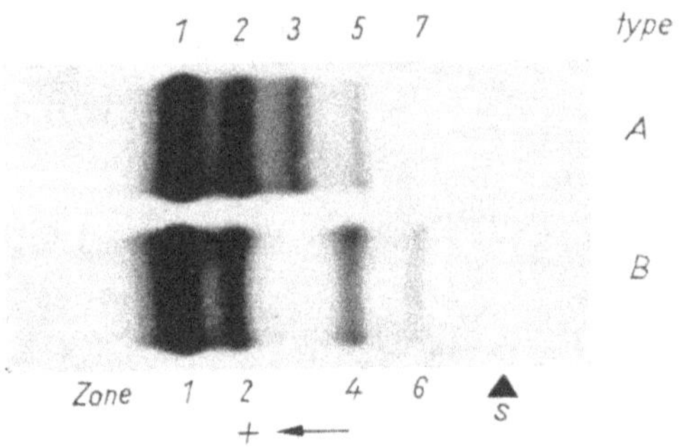

Fig. 2. Electropherogram of rat Hb on Cellogel RS
A) Elfo pattern of Hb phenotype A,
B) Elfo pattern of Hb phenotype B.

Relative amounts of Hb fractions

	1	2	3	5		(7)	
	53.4±2.6	22.2±1.1	15.7±1.3	6.7±1.0		2.1±1.2	A
	58.5±3.3	21.4±1.3		14.6±2.3		5.6±1.1	B
	1	2		4		6	

Fig. 3. Relative quantities of electrophoretic components calculated from 10 densitometrically evaluated elfo patterns.
A) Means and S.E. of components 1, 2, 3, 5 and 7 of the phenotype A, B) Means and S.E. of components 1, 2, 4 and 6 of the phenotype B

Results

It was shown in our preliminary experiments, the results of which are to be published in Folia biologica (Praha), that the combination of LEW/Cub donor and LEW · HbA recipient was not favourable for quantitative measurements and therefore a reciprocal combination was preferred in the present study. From the point of view of the change of recipient's Hb for that of the donor, both experiments gave similar results (Fig. 4). The quantity of recipient's Hb continuously decreases

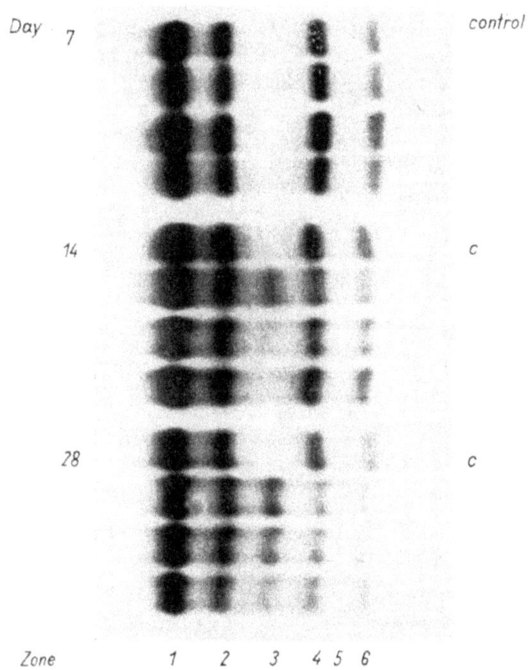

Fig. 4. Elfo Hb patterns of 3 individual rats obtained at the 7, 14 and 28th day after bone marrow transplantation LEW HbA → LEW Hb B. c: Neither irradiated, nor transplanted control

and is practically not observable after the 5th week from transplantation. Within day 14—21 the presence of donor's Hb becomes evident and its relative quantity grows until the recipient's Hb has vanished. Transplanted animals have survived to date (more than 200 days) and in some of them the gradual restitution of the original recipient's Hb pattern has been observed.

Discussion

Our results confirmed the suitability of our experimental material for further studies and are in good agreement with the known data on rat red blood cell life span, on the onset and the course of Hb synthesis after transplantation of the bone marrow into lethally irradiated recipients [4, 5]. The reappearance of recipient's Hb after a period when donor's Hb prevailed indicates that the dose of 900 R did not eradicate the erythropoietic capacity absolutely and shows that the genetic similarity between the original strain LEW/Cub and its congenic strain LEW · HbA is not yet sufficient since "own" cells are preferred. These results have inspired experiments which have already been started, concerning the acuity of distinction between "mine" and "other's" on the level of genes.

Acknowledgement

We would like to mention gratefully the skillful assistance of Mrs. V. Bernardová and Mr. V. Vojčik.

References

[1] Křen, V., D. Křenová, and O. Štark: Folia biol., Praha **16**, 305 (1970)
[2] Green, S. L., and D. P. Doolittle in: Methodology in mammalian genetics. W. J. Burdette (Ed.). Holden-Day, San Francisco 1963, p. 341
[3] Brdička, R.: Xe Congrès européén sur les groupes sanguines et le polymorphisme biochimique des animaux, Paris 1966. Institute National de la Recherche Agronom, Paris 1967, p. 407
[4] Blackett, N. M., P. J. Roylance, and K. Adams: Br. J. Haemat. **10**, 453 (1964)
[5] Constable, T. B., and N. M. Blackett: Expl Haemat. **2**, 131 (1974)

Acta biol. med. germ., Band 36, Seite 425—432 (1977)

[1] Zentralinstitut für Molekularbiologie der Akademie der Wissenschaften der DDR, Bereich Bioregulation, 1115 Berlin-Buch, DDR
[2] Bezirksinstitut für Blutspende- und Transfusionswesen, 701 Leipzig, DDR
[3] Bezirksinstitut für Blutspende- und Transfusionswesen, 112 Berlin, DDR

Untersuchungen zur Vitalität von Human-Knochenmark bei der Gewinnung und Tieftemperaturkonservierung

A. Daene[1], M. Holle[2], K. Stolle[2], A. Weissflog[2], M. Ehrenberg[2], V. Thierbach[2], H. Grützmann[1], Ch. Coutelle[1], H. Roigas[3] und S. Rosenthal[1]

Zusammenfassung

Die kryoprotektive Wirkung von Dimethylsulfoxid, Glyzerin und Polyäthylenglykol bei der Tieftemperaturkonservierung von Human-Knochenmark wurde mit Hilfe des Eosin-Farbstoffausschlußtests, eines Fluoreszenzmikroskopietests mit Akridinorange sowie eines RNS- und eines DNS-Synthesetests beurteilt. Das angewandte Verfahren der Entnahme, Aufbereitung und Konservierung von Human-Knochenmark liefert in den angegebenen Testen bei Dimethylsulfoxid 50%, bei Glyzerin 30% und bei Polyäthylenglykol 10% vitale kernhaltige Zellen, bezogen auf das aufgearbeitete Knochenmark ohne Kryoprotektivum vor Tieftemperaturkonservierung. Der Verlust an vitalen kernhaltigen Zellen wird bei Dimethylsulfoxid und Glyzerin überwiegend durch den Zusatz des Kryoprotektivums hervorgerufen, während bei Polyäthylenglykol auch der Einfrier- und Auftauprozeß zu einer starken Verringerung der Anzahl vitaler kernhaltiger Zellen führt. Dimethylsulfoxid ist mit 50% Verlust an vitalen kernhaltigen Zellen als Kryoprotektivum unter den angewandten Bedingungen der Entnahme, Aufbereitung und Konservierung von Human-Knochenmark am besten geeignet.

Einleitung

Mit der zunehmenden Bedeutung von Transplantationen vitaler Gewebe und Organe in der Medizin ergibt sich die Notwendigkeit der Entwicklung optimaler Entnahme-, Aufbereitungs- und Konservierungsverfahren. Für die Beurteilung solcher Verfahren ist der Aufbau gut handhabbarer, standardisierter Testsysteme, die Aussagen über den Grad einer Zell- bzw. Organschädigung zulassen, erforderlich.

Für die Ausarbeitung eines optimierten Abnahme-, Aufbereitungs- und Konservierungsverfahrens von Human-Knochenmark wird versucht, durch die parallele Testung wichtiger zellulärer Funktionen möglichst umfassende Kenntnisse über Vitalität und Proliferationsfähigkeit der isolierten Knochenmarkzellen bzw. ihre Schädigung zu gewinnen.

Neben den schon lange in der Praxis angewandten Farbstofftesten erscheinen Teste zur Erfassung von zellulären biochemischen und biologischen Grundfunktionen gut geeignet, um Aussagen über den Grad einer Zellschädigung zu erhalten. Hierzu gehören die Nukleinsäuresynthese, der Energiestoffwechsel und — als besonders wichtig für Knochenmark — Teste zum Nachweis der Stammzellproliferation.

Im Rahmen der Kooperation mehrerer Forschungsgruppen werden gegenwärtig derartige Teste entwickelt und erprobt. Die Ergebnisse der verschiedenen Testsysteme werden korreliert, um Aussagen für die Anwendung einfacher Teste in der Praxis zu erhalten. Die nachfolgend dargelegten Resultate sind der erste Versuch einer Zusammenführung, Abgleichung und Korrelation der bereits in Erprobung an Human-Knochenmark befindlichen Teste.

Methoden und Material

Bisher wurden folgende Vitalitätstests standardisiert und zur Beurteilung des entwickelten Verfahrens der Aufbereitung und Konservierung von Human-Knochenmark eingesetzt:

Eosin-Farbstoffausschlußtest

Die Knochenmarkzell-Suspension wurde in einer Leukozyten- bzw. Erythrozytenpipette mit 0,1%iger Eosinlösung (Chemapol, Prag/ČSSR) in isotonem Phosphatpuffer verdünnt. In einer Zählkammer wurden die vitalen (ungefärbten) und avitalen (rotgefärbten) Zellen pro µl getrennt gezählt.

Fluoreszenzmikroskopie-Test mit Akridinorange

Die Knochenmarkzell-Suspension wurde in einer Leukozyten- bzw. Erythrozytenpipette mit 0,01%iger Akridinorangelösung (Akridinorange für die Mikroskopie, Merck AG) in isotonem Phosphatpuffer verdünnt. Ein Tropfen dieser Verdünnung wurde auf einen Objektträger gegeben und mit einem Deckglas abgedeckt. Es wurde mit dem Durchlicht-Mikroskop Amplival, der Fluoreszenzleuchte HBO/200, der Filterkombination Erregerfilter BG 12/ Sperrfilter OG 1, dem Okular PK 20 und dem Objektiv Apochromat 40/0,95 (Geräte von VEB Carl Zeiss, Jena) gearbeitet.
Pro Versuchsansatz wurden 2 × 100 kernhaltige Zellen ausgewertet.
Die Blaulichtexposition wurde so kurz wie möglich gehalten, da sie zu einer Änderung der Zellfärbung führt.
Für die einzelnen Stufen der Zellschädigung wurde das in Tab. 1 wiedergegebene empirische Schema aufgestellt.

RNS-Synthese-Test

Er wurde entsprechend einem standardisierten Testverfahren durchgeführt [1]. Der Test beruht auf dem Einbau von [^{3}H]-Uridin in die RNS. $8 \cdot 10^5$ Knochenmarkzellen wurden 80 min bei 37 °C mit $8{,}3 \cdot 10^{-6}$ M [^{3}H]-Uridin (UVVVR Prag, ČSSR, spez. Aktivität = 24,2 Ci/mMol) in 500 µl modifiziertem Eagle-Medium inkubiert. Nach Auswaschen des nicht eingebauten [^{3}H]-Uridins wurden die Zellen mit Hyamin aufgeschlossen und die Radioaktivität mit Toluolszintillator im Flüssigkeitsszintillationszähler gemessen.

DNS-Synthese-Test

1 ml Knochenmarksuspension ($5-15 \cdot 10^6$ kernhaltige Zellen) wurde 1:10 mit modifiziertem Eagle-Medium (10,2 g Eagle-Basalmedium, Staatliches Institut für Immunpräparate und Nährmedien, Berlin-Weißensee, 14 mM NaÄDTA · 2 H_2O, 15 mM $MgCl_2 \cdot 6\,H_2O$, ad 1000 ml aqua dest., *p*H 7,4, Osmolarität 300 mosmolar) verdünnt, 8 min bei 1500× *g* abzentrifugiert und mit modifiziertem Eagle-Medium auf das Ausgangsvolumen aufgefüllt. Die Inkubation erfolgte für 40 min bei 37 °C mit 5 µCi [^{3}H]-Thymidin (Akademie der Wissenschaften der DDR, Zentralinstitut für Kernforschung, Dresden, spez. Aktivität 5,6 Ci/mMol) und 10^{-5} M 5-Fluordesoxyuridin (Ferak, Berlin). Der Ansatz wurde nach der Inkubation mit 5%iger Trichloressigsäure gefällt, gewaschen und hydrolysiert [2]. Zur Messung der in die DNS eingebauten Radioaktivität wurden 0,2 ml Trichloressigsäurehydrolysat mit 2 ml Methanol und 10 ml Toluolszintillator im Flüssigkeitsszintillationszähler ausgewertet.
Die quantitative DNS-Bestimmung erfolgte mittels Diphenylaminreaktion [3].
Als Maß des Einbaues von [^{3}H]-Thymidin wird die spezifische Aktivität der DNS (dpm/mg DNS) angegeben.

Tabelle 1

Farb- und morphologische Veränderungen der Zellbestandteile im Fluoreszenz-Mikroskopie-Test mit Akridinorange in Abhängigkeit vom Grad der Zellschädigung

Veränderung der Zellbestandteile	IV (vitale Zellen)	III		II			I		0	
		a	b	a	b	c	a	b	a (avitale Zellen)	b (avitale Zellen)
Kern Farbe	grün	grün	gelblich	grün	gelb	dunkel	grün	gelb bis rötlich	grün	intensiv rot
morphologische Struktur	unverändert	geschwollen	unverändert	verwaschen	unverändert	keine Struktur	homogen	unverändert	keine Struktur	unverändert
Membran	unverändert	unverändert	unverändert	unverändert	unverändert	grün	defekt	unverändert	defekt	unverändert
Zytoplasma Grundfärbung	grün	grün	grün	grün	grün	grün	grün	grün	nicht sichtbar	nicht sichtbar
morphologische Struktur	rote Granula	rote Granula	rote Granula	vergröberte rote Granula	vergröberte rote Granula	vergröberte rote Granula	verklumpte rote Granula Vakuolen	verklumpte rote Granula Vakuolen	nicht sichtbar	nicht sichtbar
Zellmembran	intakt	intakt	intakt	teilweise defekt	teilweise defekt	teilweise defekt	defekt	defekt	defekt	defekt

Material

Das Knochenmark wurde von gesunden Spendern durch dorsale und ventrale Beckenkammpunktion mit Klima-Rosegger-Kanülen unter Allgemeinanästhesie (Intubationsnarkose mit Lachgas/Sauerstoff/Halan) aus 8—10 Punktionsstellen gewonnen. Das Absaugen erfolgte mit einem Pumpenvakuum (0,8—0,9 at) über Schläuche von Blutübertragungsgeräten direkt in 120 ml-Blutkonservenflaschen, in denen je 50 ml modifiziertes Zolipk 76 (20 mM Glukose, 130 mM Saccharose, 3 mM NaÄDTA, 30 mM Na-Zitrat, 67 ml Infukoll M 40 ad 100 ml aqua dest.) vorgelegt waren. Im Mittel wurden 400—500 ml Blutmarkgemisch pro Spender entnommen.

Die Aufbereitungszeit von 5 Std. beinhaltete eine Sedimentation bei $1 \times g$ zur Trennung von Erythrozyten und kernhaltigen Zellen und eine Zentrifugation des Überstandes mit den kernhaltigen Zellen ($700 \times g$, 30 min) zur Volumeneinengung. Die Präparation erfolgte bei +4 °C. Als Kryoprotektiva wurden Dimethylsulfoxid, 5%ig über Aktivkohle vakuumdestilliert, (E_{275}^{1cm}. < 0,1), Polyäthylenglykol, MG = 400, 10%ig (Versuchsprodukt des VEB Chemische Werke Buna) und Glyzerin, 10%ig (doppelt destilliert, Ferak, Berlin) eingesetzt. Die Knochenmarkproben wurden in 60 ml-Aluminium-Kontainer (UdSSR) überführt. Das Einfrieren erfolgte in 3 Phasen: Die im Eisbad aufbewahrten 4—6 Kontainer (je Spender) wurden in ein Alkoholbad von —15 °C (Ultrakryostat MK 70, VEB MLW-Prüfgerätewerk Medingen) gebracht. Der Temperaturabfall erfolgte bis —13 °C mit ca. 2 °C/min. Anschließend wurden die Kontainer in dem auf —170 °C temperierten, luftgefüllten Rezipienten des Ultrakryostaten N 180 (VEB MLW-Prüfgerätewerk Medingen) unter Verwendung von flüssigem Stickstoff mit ca. 7 °C/min bis —140 °C abgekühlt. Danach erfolgte das Tauchen der Kontainer in flüssigen Stickstoff und Lagerung im Biogefäß Bo 200 (VEB Kältetechnik, Niedersachswerfen).

Der Temperaturverlauf während des Einfrierungsprozesses wurde in einem getrennten Kontainer (Blutplasma mit 5% Dimethylsulfoxid) unter Verwendung von Platin-Widerstands-Thermometern (Pt 100, VEB Spezialmeßgerätewerk Leipzig) gemessen und mittels des Kompensationsbandschreibers MK VW 3 (Meßgerätewerk Erich Weinert, Magdeburg) registriert.

Zum Auftauen wurden die Kontainer ca. 45 s bei +40 °C im Wasserbad unter Temperaturkontrolle bis 0 °C geschüttelt.

Die beschriebenen Vitalitätsteste wurden zur Einschätzung der schädigenden Wirkung der Kryoprotektiva auf das Knochenmark vor Tieftemperaturkonservierung und zur Beurteilung des Einfrier- und Auftauschadens eingesetzt.

Ergebnisse

Wirkung von Kryoprotektiva

Die Vitalität der Knochenmarkzellen wurde nach Zugabe von Dimethylsulfoxid, Polyäthylenglykol und Glyzerin in den Farbstoff- und Nukleinsäuresynthese-

Tabelle 2
Wirkung von Kryoprotektiva auf Knochenmarkzellen

Teste	5% Dimethyl-sulfoxid	10% Polyäthylen-glykol	10% Glyzerin
Kernhaltige Zellen	92	95	81
Eosin-Farbstoffausschlußtest	102	107	45
Fluoreszenzmikroskopie mit Akridinorange, Stufe IV	89	72	36
RNS-Synthese	54	41	26
DNS-Synthese	64	48	65

Angaben in % bezogen auf Ergebnisse an aufgearbeitetem Knochenmark ohne Kryoprotektiva = 100% für $n = 4$. Es sind die Mittelwerte angegeben.

Tests vor Tieftemperaturkonservierung bestimmt (Tab. 2). Die durchschnittliche Einwirkungszeit der Kryoprotektiva betrug 30 min.
Tab. 2 zeigt, daß durch die Kryoprotektiva keine Abnahme der Gesamtzahl kernhaltiger Zellen verursacht wird, so daß auf die Ausgangszellzahl (mit Volumenkorrektur für die zugesetzte Menge an Kryoprotektiva) bezogen werden kann. Im Eosin-Farbstoffausschlußtest und bei der Fluoreszenzmikroskopie mit Akridinorange ist nur nach Einwirkung von 10% Glyzerin ein deutlicher Abfall des Anteils vital reagierender Zellen festzustellen. Die übrigen Kryoprotektiva vermindern den Anteil vital reagierender Zellen hingegen kaum.
Sowohl RNS- als auch DNS-Synthese fallen in Abhängigkeit von den einwirkenden Kryoprotektiva auf 60 bis 30% der Ausgangsaktivität ab.

Wirkung des Einfrier- und Auftauprozesses

In Tab. 3 ist der Abfall an vitalen kernhaltigen Zellen in den Farbstoff- und Nukleinsäuresynthese-Tests durch die Tieftemperaturkonservierung dargestellt. Durch die Tieftemperaturkonservierung erfolgt keine Verringerung der Gesamtzahl kernhaltiger Zellen im Vergleich zu der entsprechend der zugesetzten Kryoprotektivamenge volumenkorrigierten Ausgangszellzahl.

Tabelle 3

Wirkung des Einfrier- und Auftauprozesses auf Knochenmarkzellen

Teste	ohne Kryoprotektivum	5% Dimethylsulfoxid	10% Polyäthylenglykol	10% Glyzerin
Kernhaltige Zellen	94	100	87	92
Eosin-Farbstoffausschlußtest	27	63	50	70
Fluoreszenzmikroskopie mit Akridinorange, Stufe IV	16	41	55	104
RNS-Synthese	17	93	11	113
DNS-Synthese	10	52	20	40

Angaben in % bezogen auf Ergebnisse vor Tieftemperaturkonservierung = 100% für $n = 4$, DNS-Synthese $n \geqq 2$. Es sind die Mittelwerte angegeben.

In allen vier Tests sinkt der Anteil vitaler Zellen bei der Knochenmarkprobe, die ohne Kryoprotektivum eingefroren und aufgetaut wurde, unter 30% ab.
Das mit Glyzerin eingefrorene Knochenmark zeigt nur eine Abnahme der DNS-Synthese-Aktivität. In den anderen Tests ist keine Veränderung gegenüber der Probe vor Tieftemperaturkonservierung festzustellen.
Mit Dimethylsulfoxid eingefrorenes Mark weist in den Farbstoff- und DNS-Synthese-Tests einen Vitalitätsabfall von rund 50% auf. Dagegen verändert sich die RNS-Synthese-Aktivität gegenüber der uneingefrorenen Probe nicht.
Im Eosin-Farbstoffausschlußtest und in der Fluoreszenzmikroskopie mit Akridinorange ist die Abnahme vital reagierender Zellen bei der Knochenmarkprobe mit Polyäthylenglykol mit rund 50% geringer als der Abfall der Nukleinsäuresynthese-Aktivität mit rund 80%.

Wirkung der Kryoprotektiva und des Einfrier- und Auftauprozesses

Tab. 4 zeigt den durch Kryoprotektiva und Tieftemperaturkonservierung hervorgerufenen Gesamtabfall an vitalen kernhaltigen Zellen.

Tabelle 4

Wirkung der Kryoprotektiva und des Einfrier- und Auftauprozesses auf Knochenmarkzellen

Teste	ohne Kryo-protektivum	5% Dimethyl-sulfoxid	10% Polyäthylen-glykol	10% Glyzerin
Kernhaltige Zellen	94	92	82	74
Eosin-Farbstoffausschlußtest	27	64	48	33
Fluoreszenzmikroskopie mit Akridinorange, Stufe IV	16	36	40	23
RNS-Synthese	17	47	5	28
DNS-Synthese	10	52	9	27

Die Angaben erfolgen in % bezogen auf die Ergebnisse an aufgearbeitetem Knochenmark ohne Kryoprotektiva vor Tieftemperaturkonservierung = 100% für $n = 4$, DNS-Synthese $n \geqq 2$. Es sind die Mittelwerte angegeben.

Während bei den Proben ohne Kryoprotektivum, mit Dimethylsulfoxid und mit Polyäthylenglykol keine kernhaltigen Zellen verlorengehen, deutet sich bei Glyzerin ein geringer Verlust an, der in den Ergebnissen aber nicht berücksichtigt wurde. Der Mittelwert der Ergebnisse der vier zur Beurteilung der Knochenmarkzellen herangezogenen Tests zeigt bei der Probe ohne Kryoprotektivum einen Rest vitaler Zellen von 18%, bei Dimethylsulfoxid von 46% und bei Glyzerin von 28%. Dabei liegen die Ergebnisse im Eosin-Farbstoffausschlußtest durchgehend über dem sich aus den vier Testen ergebenden Mittelwert.

Bei der Knochenmarkprobe mit Polyäthylenglykol ist der Abfall an vital reagierenden Zellen mit rund 50% sehr viel geringer als die Abnahme der Nukleinsäuresynthese mit 90%.

Dimethylsulfoxid besitzt unter der angewandten Methode der Gewinnung, Aufbereitung, des Einfrierens und Auftauens von Knochenmark mit rund 50% vital reagierenden Zellen bzw. Nukleinsäuresyntheseaktivität die geringste schädigende Wirkung und die beste kryoprotektive Eigenschaft von den getesteten Substanzen.

Diskussion

Um das zur Aufbereitung und Konservierung von Human-Knochenmark entwickelte Verfahren beurteilen und verbessern zu können, wurden die bereits standardisierten Farbstoff- und Nukleinsäuresynthese-Tests eingesetzt. Die dargestellten Ergebnisse sind ein erster Versuch der Zusammenführung, Abgleichung und Korrelation der genannten Tests.

Grundsätzlich muß zwischen den Schäden durch das Kryoprotektivum und Schäden durch die Tieftemperaturkonservierung unterschieden werden. Die unterschiedliche Beanspruchung der Zellen infolge der verschiedenen methodischen Anforderungen der einzelnen Tests könnte eine Ursache für abweichende

Ergebnisse bei den einzelnen Tests (Tab. 2 u. 3) sein. So spiegeln die Farbstofftests die durch die Kryoprotektiva hervorgerufenen Schäden nicht wider. Jedoch scheint jede zusätzliche Belastung, wie die zur Messung der Nukleinsäuresynthese notwendigen Prozeduren (Zentrifugation, Inkubation), diese latenten Schäden zu manifestieren und zu einem Abfall in der Nukleinsäuresynthese zu führen. Hieraus könnten sich die positiven Ergebnisse in den Farbstofftests bei Dimethylsulfoxid und Polyäthylenglykol (Tab. 4) erklären. Dagegen scheint der Schädigungsgrad bei Glyzerin so hoch zu sein, daß er bereits in den Farbstofftesten manifest wird. Erst nach Einfrieren und Auftauen (Tab. 3) erscheinen die durch Dimethylsulfoxid bereits partiell geschädigten, jedoch noch vital reagierenden Zellen (Tab. 2) in den Farbstofftests als avital und täuschen damit einen Einfrier- und Auftauschaden vor, während sie im RNS-Synthese-Test bereits als Kryoprotektivaschaden erfaßt wurden und in Tab. 3 nur die Schädigung durch Frieren und Tauen dargestellt wird. Für die Richtigkeit dieser Überlegungen spricht die recht gute Übereinstimmung der Ergebnisse der Farbstoff- und Nukleinsäuresynthesetests in den Versuchen ohne Kryoprotektivum, mit Dimethylsulfoxid und mit Glyzerin für die Überalles-Reaktion auf Kryoprotektiva und Tieftemperaturkonservierung (Tab. 4).

Ein ähnliches Verhalten wie für Dimethylsulfoxid wäre für Polyäthylenglykol zu erwarten. Wahrscheinlich beeinflußt aber die Kombination Polyäthylenglykol/Dextran (modifiziertes Zolipk 76) den Farbstofftransport und damit die Anfärbbarkeit der Zellen. Für die Nukleinsäuresynthesemessung ist mit keiner Behinderung des [^{3}H]-Thymidin- bzw. [^{3}H]-Uridintransports zu rechnen, da die Zellen vorher mit Eagle-Medium gewaschen und inkubiert wurden. Hieraus könnten die positiven Ergebnisse der Farbstoffteste bei Polyäthylenglykol (Tab. 2—4) erklärt werden.

Aus den aus der Diskussion ersichtlichen methodisch bedeutungsvollen Differenzen zwischen den einzelnen Tests ergibt sich die Notwendigkeit der weiteren Angleichung.

Unter Berücksichtigung der methodisch bedingten Diskrepanzen liefert das vorliegende Verfahren der Aufbereitung und Konservierung von Human-Knochenmark mit Dimethylsulfoxid rund 50%, mit Glyzerin rund 30% und mit Polyäthylenglykol rund 10% vitale kernhaltige Zellen, bezogen auf das aufgearbeitete Knochenmark ohne Kryoprotektivum vor Tieftemperaturkonservierung. Die erreichten Resultate stimmen mit den in der Literatur beschriebenen gut überein [4, 5]. Der Abfall an vitalen kernhaltigen Zellen wird bei Dimethylsulfoxid und Glyzerin fast ausschließlich durch die Zugabe des Kryoprotektivums hervorgerufen. Der Schaden durch den Einfrier- und Auftauprozeß ist minimal. Durch Versuche mit Rattenknochenmark konnte nachgewiesen werden, daß neben der Temperatur die Zeitdauer der Aufbreitung entscheidend für den Schädigungseffekt der Kryoprotektiva ist (Daene, unveröffentlichte Versuchsergebnisse). Um die vorliegenden Ergebnisse zu verbessern, muß ein beschleunigtes Verfahren zur Trennung von Erythrozyten und kernhaltigen Zellen entwickelt werden. Es sollte darüber hinaus beide Zellarten schärfer als die bisherige Methode trennen und so die Ausbeute an kernhaltigen Zellen erhöhen.

Bei Polyäthylenglykol setzt sich der Gesamtverlust an vitalen kernhaltigen Zellen aus dem Schaden durch die Zugabe des Kryoprotektivums und dem durch den

Einfrier- und Auftauprozeß zusammen. Die Knochenmarkprobe mit Polyäthylenglykol verhält sich während des Frierens und Tauens wie die Probe ohne Kryoprotektivum. Offensichtlich besitzt das verwendete Polyäthylenglykol M 400 in der Endkonzentration von 10% bei dem vorliegenden Einfrier- und Auftauschema für kernhaltige Zellen keine bzw. nur eine dem im modifizierten Zolipk 76 enthaltenen Dextran entsprechende kryoprotektive Wirkung. Es erscheint daher als Kryoprotektivum für die Tieftemperaturkonservierung von Knochenmark nicht geeignet.

Unter den gegenwärtigen Bedingungen der Aufbereitung und Konservierung von Human-Knochenmark ist Dimethylsulfoxid mit 50% Abfall an vitalen kernhaltigen Zellen als Kryoprotektivum am besten geeignet.

Die für den Transplantationserfolg unter anderem entscheidende Frage der Proliferationsfähigkeit des Knochenmarks kann mit den vorliegenden Tests nicht geklärt werden. Daher müssen in Zukunft mit den sich zur Zeit in Erarbeitung befindlichen Proliferationstesten Überprüfungen der optimierten Aufbereitungs- und Konservierungsbedingungen und Korrelationsuntersuchungen gegenüber den anderen Testverfahren erfolgen.

Literatur

[1] Daene, A., Ch. Coutelle, V. Thierbach, M. Holle, C. Raderecht u. H. Roigas: Folia haemat. (im Druck)

[2] Schneider, W. C.: J. biol. Chem. **161**, 293 (1945)

[3] Dische, Z.: Mikrochemie **8**, 4 (1930)

[4] Mosimann, W., M. Furlan, E. A. Beck u. U. Bucher: Schweiz. med. Wschr. **102**, 1600 (1972)

[5] Pegg, P. E.: J. appl. Physiol. **19**, 123 (1964)

Summary

A. Daene, M. Holle, K. Stolle, A. Weissflog, M. Ehrenberg, V. Thierbach, H. Grützmann, Ch. Coutelle, H. Roigas, and S. Rosenthal: Investigations on the vitality of human bone marrow in the preparation and cryopreservation

The cryoprotective effect of dimethylsulfoxide, glycerol and polyethylenglycol during freezing and thawing of human bone marrow was investigated by eosin staining test, an acridinorange fluochrome staining test and by RNA- and DNA-synthesis tests.

In these tests the overall yield of vital nucleated cells, referred to the number in the absence of cryoprotectants and freezing and thawing, amounted to 50% with dimethylsulfoxide, 30% with glycerol, and 10% with polyethylenglycol. With dimethylsulfoxide and glycerol the loss of vital nucleated cells is almost entirely due to the addition of cryoprotectants. Polyethylenglycol freezing and thawing also leads to a great loss of vital nucleated cells.

The results with dimethylsulfoxide show that the currently employed techniques of punction, preparation, freezing and thawing of bone marrow are suitable for clinical application.

Acta biol. med. germ., Band 36, Seite 433—441 (1977)

Zentralinstitut für Molekularbiologie der Akademie der Wissenschaften der DDR, Bereich Bioregulation, 1115 Berlin-Buch, DDR

Some unusual properties of globin-mRNA isolated from rabbit reticulocytes

V. Hahn, H. D. Hunger, F. Hiepe, A. Ladhoff, H. Grützmann, and Ch. Coutelle

Summary

Globin mRNA isolated from rabbit reticulocytes showed an unusually high molecular weight for the β-chain mRNA (280000 daltons), a large difference between the molecular weight of the α-chain mRNA (230000 daltons) and the β-chain mRNA and a dense and compact structure in the electron microscope (average length of the molecules 0.14 µm), which can be transformed to a linear structure (average length of the molecules 0.44 µm corresponding to a molecular weight of about 420000 daltons). The poly(A)-tail of the mRNA estimated by polyacrylamide gel electrophoresis with poly(A)-markers as reference substances has a size between 12 and 52 nucleotides. The mRNA works as a good template in the cDNA-synthesis with reverse transcriptase and full length copies of the mRNA were obtained by this reaction. The cDNA hybridizes to an amount of 70% with the globin mRNA at a 1:4 cDNA/mRNA ratio.

Introduction

The proteins made in immature mammalian blood cells are almost exclusively hemoglobins [1], so that the messenger RNA population of these cells will be largely globin mRNA. According to this and to the fact that the reticulocytes have a low ribonuclease activity, which minimize the problems of nuclease contamination usually encountered in RNA purification, it was not surprising that the globin mRNA has been one of the first mRNA isolated in a pure form and characterized to direct the synthesis of the corresponding protein in cell-free systems [2—9]. In connection with our work concerning the isolation of precursor mRNA, which have been predicted by several authors [10—12], it was one of our tasks to prepare globin mRNA. The isolated mRNA showed some unusual properties, which we want to describe and discuss here.

Material and methods

Globin mRNA preparation

High anemic rabbits, which received 40 mg phenylhydrazine per kg body weight per day over a period of 4 days, were sacrified by heart puncture at the 7th day. The polysomes, prepared according to Williamson et al. [13], were suspended in 0.01 M Tris, *p*H 7.4; 0.01 M NaCl; 0.01 M EDTA; 0.2% sodium dodecylsulfate; 0.1% deoxycholate and diluted with 0.1 M Tris *p*H 9.0, to an $A_{260} < 0.100$. The solution was made 0.5% in sodium dodecylsulfate and total polysomal RNA was extracted by 1 volume of phenol/chloroform/isoamylethanol (50/50/2; v/v/v). The buffer phase was removed by centrifugation and the nonaqueous phase was extracted twice with 1 volume phenol/chloroform (50/50; v/v) and several times with chloroform (0.5 volume) until all turbidity had been removed. The RNA was precipitated from the clear buffer phase with 2.5 volumes ethanol after addition of sodium acetate to 0.1 M at −20 °C and was reprecipitated twice. The mRNA was isolated from the total polysomal RNA by binding the poly(A)-carrying molecules to poly(U)-sepharose in 0.01 M Tris, *p*H 7.5; 0.1 M NaCl; 0.01 M EDTA; 0.2% sodium dodecylsulfate and elution with bidistilled water.

Electrophoresis

The electrophoresis of the mRNA, cDNA and the poly(A) under denaturating and nondenaturating conditions was performed in vertical polyacrylamide slab gels, according to the methods described by Loening [14], Staynov et al. [15], Gould and Hamlyn [16] and Peacock and Dingman [17]. A selfmade apparatus was used, which allows the separation of eight different samples at the same time on one flat-bed gel (15 × 15 cm, 5 mm thick). This equipment showed great accuracy, if the migration distances of the same substance layered onto the different slots are compared. Pure 28 S RNA, 18 S RNA, 5 S RNA and 4 S RNA were used as molecular weight markers. After the separation the slabs were scanned directly in the UV-light using the Hitachi spectrofluorometer MPF-2A or the Gilford spectrophotometer 2400-2.

Electron microscopy

For the electron microscopic investigations a very simple protein-free preparation method was used [12], which is based on the adsorption of nucleic acids at the air-water interphase [18].

Estimation of poly(A)

300 μg globin mRNA were digested with RNase A and RNase T_1 for 1 h at 37 °C (1 ml 2× SSC).
After deproteinization the reaction mixture was passed through a poly(U)-sepharose column. The poly(A) was eluated at 50 °C with distilled water.
The electrophoresis was performed in 8% slab gels for 4 hrs 90 mA.
Marker 4S, 5S, 18S RNA and poly(A)-marker was from Miles.

Enzymatic synthesis of cDNA

The synthesis of DNA complementary to the globin mRNA was made by Frolova et al. [19, 20].

Hybridization

Appropriate amounts from solutions of globin mRNA, the corresponding cDNA and tRNA were mixed and lyophilized. The dry mixture was solved in 4 μl hybridization buffer (0.5 M NaCl, 0.025 M Tris, *p*H 7.5, 0.01 M EDTA, 50% formamide), transfered in a capillary, which was then closed on both ends by melting and incubated at 42 °C for 70 hrs. After the incubation the capillary was rinsed with 146 μl nuclease buffer (0.07 M sodium acetate, *p*H 4.5, 2.8 mM zinc sulfate, 0.14 M NaCl, 10 μg/ml denatured E. coli DNA). Nuclease S_1 was added and the solution incubated for 2 hrs at 37 °C. Then 50 μl of this solution were given to GFA-filter, dried, transfered to scintillation vials, scintillation fluid added and counted for activity. To the remaining 100 μl of the mixture 25 μl 30% TCE and 25 μl 0.1% bovine serum albumin solution were added; the whole mixture was incubated for 30 min in an ice-bath and then centrifuged for 2 min. From the supernatant 100 μl were taken and given to GFA-filters, washed, dried and then counted for non-acid precipitable activity.

Results and discussion

The globin mRNA isolated as described in methods gives normally after binding to and elution from the poly(U)-sepharose in the polyacrylamide gel electrophoresis under denaturing conditions a pattern shown in Fig. 1. If there still exists some material in the region corresponding to 18 S RNA or 4 S RNA (marked by the arrows), then the mRNA was further purified by sucrose density gradient centrifugation or by a second binding to poly(U)-sepharose.

The poly U-sepharose was a kind gift from Dr. L. Fisher, Pharmacia Fine Chemicals AB, Sweden

The globin mRNA showed under non-denaturating conditions in the polyacrylamide gel electrophoresis at two different acrylamide concentrations (2.6% and 3.5%) only a single peak. No splitting in α- and β-chain mRNA could be seen. For these single peaks a molecular weight of 230000 daltons (2.6% gel) or 240000 daltons (3.5% gel) can be calculated from the calibration curves (Fig. 2). These values are in good agreement with the values reported by other authors (see Table 1) and support the fact that the molecular weight of the globin mRNA is about 80000—90000 daltons higher than what one would expect from the size of the globin chains. That means there are between 240 and 260 nucleotides in the mRNA which are noninformative. A part of these (about 50 nucleotides) belongs to the poly(A)-tail at the 3'-end [22, 23, 30—32], and the other part (about 100 nucleotides) to the unique noninformative sequence [21], located between the informative part of the mRNA and the poly(A)-tail at the 3'-end. The remaining 100 nucleotides should then be localized at the 5'-end. The function of the noninformative nucleotides are still unknown.

In a second experiment the globin mRNA was separated by polyacrylamide gel electrophoresis under denaturating conditions. Three different systems were used, as shown in Fig. 3 [15—17]. A separation into two bands took place, if the formamide systems was used. Only a single band could be seen, if urea was used as denaturating agent, but an unusually high molecular weight of 300000 daltons was observed. The splitting of the globin mRNA into two bands during the form-

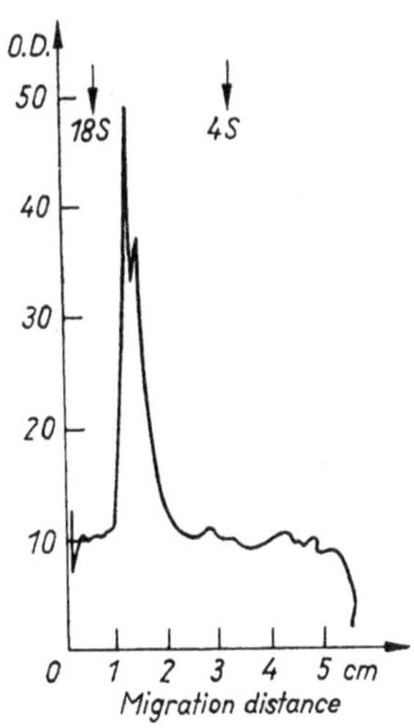

Fig. 1

Gelsystem	Flat bed of PPA-Gel according to LOENING	
	2.6%	3.5%
Conditions of electrophoresis	150min 80mA	180min 65mA
Estimation method	Staining with stains all	
Estimated MW	230.000	240.000

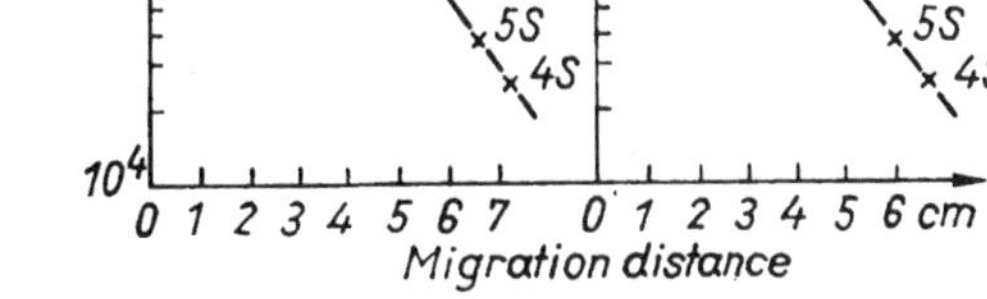

Fig. 2

Fig. 1. Electrophoretic pattern of globin-mRNA in 5,2% PAA-gel with formamide, scanned at 265 nm

Fig. 2. Electrophoresis of globin-mRNA under nondenaturating conditions. Parameters of the electrophoresis and calibration curves. The open circle shows the MW of the globin-mRNA

Table 1
Comparison of the molecular weights of globin-mRNA of different species

Estimation method PAA-gel electrophoresis conditions	Species	Number of bands	Estimated molecular weight
Non denaturating	Rabbit	1	230000–240000
	22.	1	230000
	23.	1	230000
	24.	1	210000–235000
	Duck 25.	1	180000–220000
	26.	1	210000
	Chicken 27.	1	260000 ± 20000
	Mouse 13.	1	225700 ± 13700
Denaturating			
a) with urea	Rabbit	1	300000
b) with formamide			
I non buffered	Rabbit	2	α-chain 220000
			β-chain 260000
II buffered	Rabbit	2	α-chain 230000–240000
			β-chain 280000–290000
	16, 28.	2	α-chain 202000
	29.		β-chain 227000
		2	α-chain 210000
			β-chain 240000

Gelsystem	5.2% Flat-bed-Gel with formamide accordding to STAYNOV	5.2% Flat-bed-Gel with buffered formamide acc. to GOULD a. HAMLYN	2.5% Flat-bed-Gel with Agarose and 6M Urea acc. to PEACOCK a. DINGMAN
Electrophoretic conditions	285 min 75 mA 25 °C	35 min 50 mA 120 min 80 mA 25 °C	180 min 50 mA 25 °C
Estimation method	Staining with stains all	Scanning with UV-light	
Estimated MW	260000 240000	280000 230000	300000

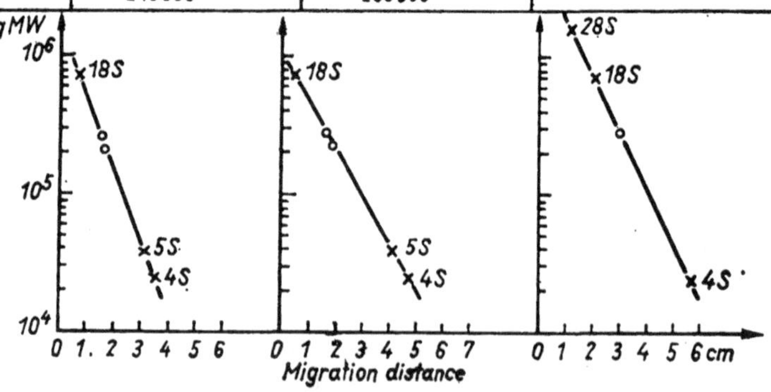

Fig. 3. Electrophoresis of globin-mRNA under denaturating conditions. Parameters of the electrophoresis and calibration curves. The open circles show the MW of the globin-mRNA

amide electrophoresis is well known [16] and it has been shown that the two bands correspond to α-chain mRNA and β-chain mRNA [28, 29, 33, 34]. Surprising is the observation that in our globin mRNA the second band has a molecular weight of 280000—290000 daltons.

Because it was shown (PONKA, personal information) that our globin mRNA directs the synthesis of α- and β-globin in a cell-free system, we can assume that the second band corresponds to the β-globin chain. The explanation for the high value of the molecular weight of the β-chain could be that the β-globin mRNA has a larger noninformative part at the 5′-end as the α-globin mRNA.

The mRNA was analyzed for its poly(A)-tail by polyacrylamide electrophoresis after digestion with pancreas RNase and RNase T_1 and binding of the nondigested poly(A) to poly(U)-sepharose. As can be seen from Fig. 4, no discrete poly(A)-tail exists. The values reach from oligo$(A)_{12}$ up to poly$(A)_{52}$. It is important to calculate the poly(A)-length by using pure poly(A)-markers [35, 36], otherwise one gets an overestimation of the poly(A)-tail (cf. Fig. 4). If 5 S RNA and 4 S RNA were used for calibration, then one would estimate between 35 and 110 nucleotides for the poly(A)-part. From this point of view all estimations of the poly(A)-tail of globin mRNA must be seen, which had been made up to now. Therefore the values between 100 and 150 nucleotides are very unlikely and the values estimated by using poly(A)-markers [31, 32] seems to be more accurate.

In the electron microscope the globin mRNA showed a very dense and compact structure (Fig. 5). If the contour length of these types of molecules was measured, the distribution shown in Fig. 6 is found. An average length of 0.14 μm was estimated, corresponding to a molecular weigth of about 150000 daltons under the

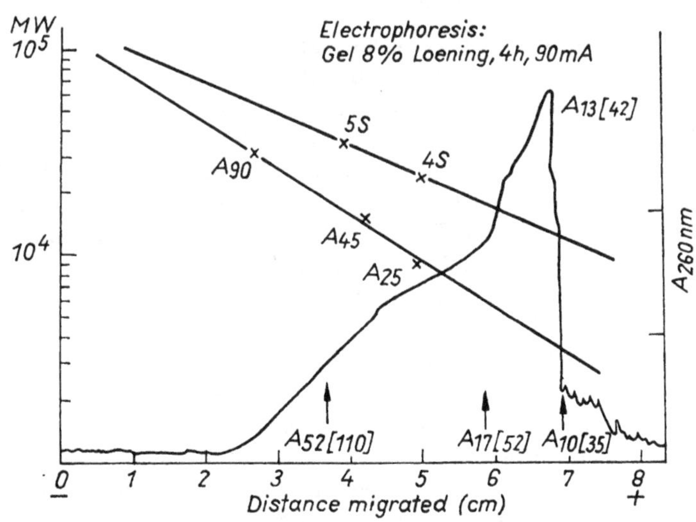

Fig. 4. Electrophoresis of the poly(A)-part of the globin-mRNA.
Pattern of the gel scanned at 265 nm.
Calibration curves with heteropolymer and homopolymer polynucleotides as standards for MW-estimation.
The arrows show distinct poly(A)-values estimated by using the poly(A)-markers as standards.
The numbers in parenthesis are calculated on the basis of heteropolymer RNAs as standards.

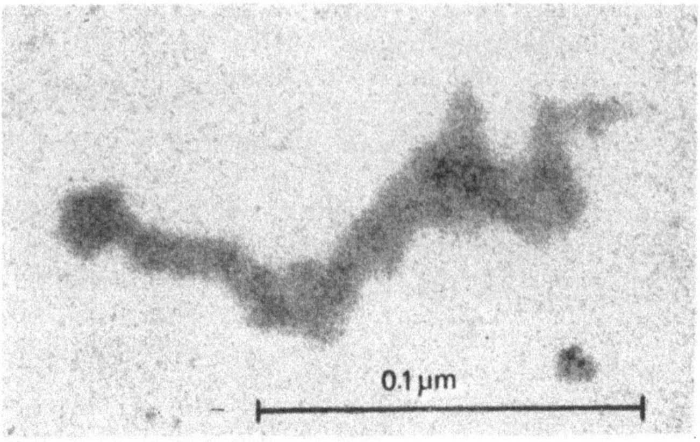

Fig. 5. Electron microscopic picture of the globin-mRNA.

assumption of a nucleotide distance of 3.4 A. But as shown by LADHOFF et al. [37] the non-linear compact form can be transformed under certain conditions into a linear form, which is about three times longer. On the basis of a mean value of 0.44 µm a molecular weight of about 420000 daltons can be calculated. That is nearly two times that molecular weight which has been calculated on the basis of the results from the polyacrylamide gel electrophoresis. If the mRNA was treated in the same way as done for the electron microscopic investigations (heating for 5 min at 90 °C in 80% formamide containing 4 M urea), no change in their electrophoretic mobility could be observed.

Recently for the ovalbumin mRNA similar discrepancies were reported [35]. There is no correspondence of the molecular weights estimated by different methods (sedimentation analysis, electrophoresis and electron microscopy). Similar observations have been made for HeLa cell ribosomal RNA [38—40], for globin and myosin mRNA [42, 13] and immunoglobulin K chain mRNA [42—45]. It was discussed that the reason for these discrepancies might be the larger G + C content of the ribosomal RNAs used as markers, considering the fact that secondary structure in 28 S RNA is still evident even after spreading in 80% formamide/4 M urea [46]. Therefore it can be possible that according to the higher G + C content of the ribosomal markers and some residual secondary structure these RNAs have a somewhat higher mobility in the electrophoresis than the mRNAs, which would lead to an overestimation of their molecular weight. Similarly, cytoplasmic rRNAs can be expected to sediment faster in sucrose gradients than do mRNAs as a result of their more condensed structure. This effect would lead to an underestimation of the molecular weights of mRNAs by means of the sedimentation analysis.

A second fact may further contribute to these discrepancies. As discussed by McLEOD [47], poly(A)-carrying mRNA molecules have under the conditions used for estimation of the sedimentation constants a larger radius of gyration as the more base-paired ribosomal RNA markers of the same molecular weight and one gets therefore by this method an underestimation of the S-values. In the polyacrylamide gel electrophoresis one can expect the opposite behaviour. According to the larger radius of gyration the mRNA molecules would have a slower apparent electrophoretic mobility and that would lead to an overestimation of their molecular weight by this method. Therefore it seems to be essential to determine mRNA molecular weights by several independent methods using appropriate standards.

But nevertheless there still exists the fact that the estimation of the molecular weight of the globin mRNA by electron microscopy gives a value of about 420000 daltons corresponding to 1250 nucleotides, whereas the highest molecular weight estimated by polyacrylamide gel electrophoresis is only about 300000 daltons corresponding roughly to 900 nucleotides. For this difference we have up to now no suitable explanation.

The globin mRNA was analyzed for its template activity for reverse transcription by revertase from AMV. Our rabbit globin mRNA showed ten times more activity than a preparation of pigeon globin mRNA under the same experimental conditions. The resulting cDNA has been analyzed by polyacrylamide gel electrophoresis for its size. As shown in Fig. 7 about 6 to 7 discrete peaks can be clearly

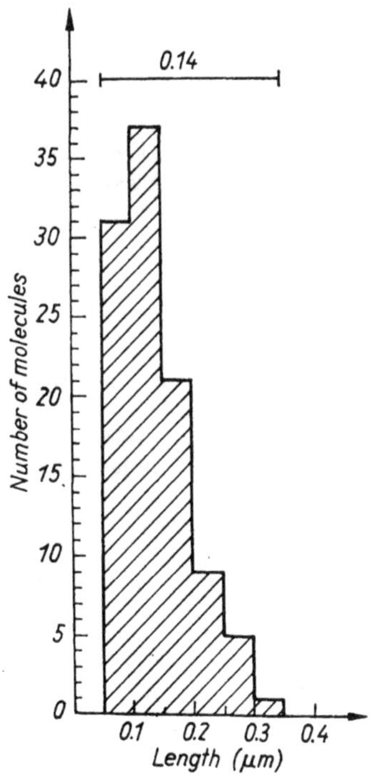

Fig. 6

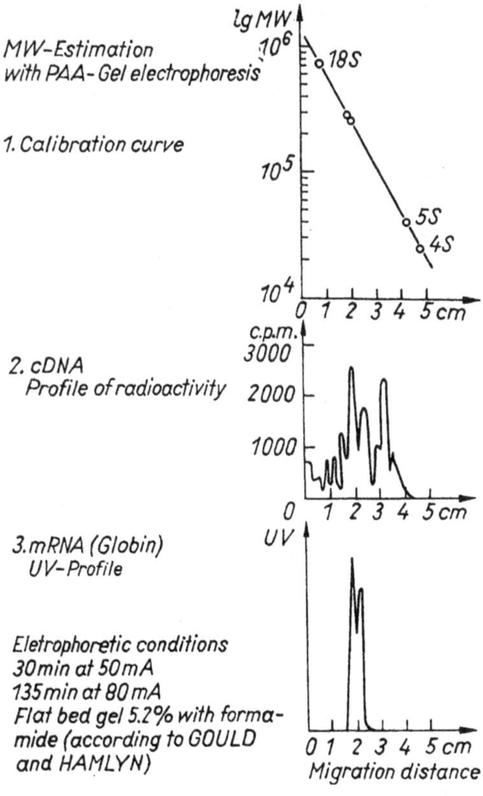

Fig. 7

Fig. 6. Size distribution of the conture length of nonlinear compact globin-mRNA molecules

Fig. 7. Electrophoresis and MW-estimation of globin-mRNA and the corresponding DNA under denaturating conditions distinguished. Two of them are in good correspondence with the peaks for the α- and β-chain of globin mRNA separated at the same flat bed gel. Therefore one can assume that full transcripts of the mRNA have been made under the selected conditions. Besides there exist 3 peaks with higher molecular weights (roughly 400000, 500000 and 600000 daltons) corresponding to 1200, 1500 and 1800 nucleotides and two or three with lower molecular weights (between 90000 and 120000 daltons) corresponding to 270 and 360 nucleotides. This observation has been analyzed in more detail by FROLOVA et al. [20].

The prepared cDNA was hybridized with different amounts of globin mRNA and the titration curve shown in Fig. 8 was obtained. The mRNA hybridizes with the corresponding cDNA to an amount of 70%. The plateau is reached at a 1:4 cDNA/mRNA ratio. The value of 70% hybridization must be seen in connection with the fact that cDNA alone shows about 30% resistance to S_1 nuclease treatment. Our further work with the mRNA will be focused on the open questions, concerning the high secondary structure, the differences in the molecular weight estimations and the limited hybridization.

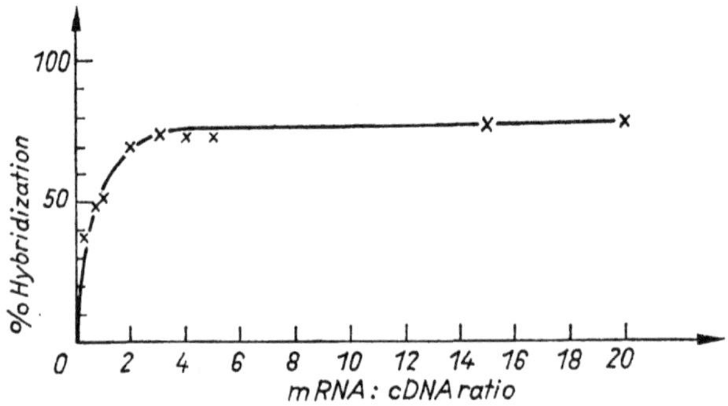

Fig. 8. Titration of globin-cDNA with mRNA excess

References

[1] DINTZIS, H. M., H. BORSOOK, and J. VINOGRAD in: Microsomal Particles and Protein Synthesis. R. B. ROBERTS (Ed.). Pergamon, London 1958, p. 95

[2] MARBAIX, G., and A. BURNY: Biochem. biophys. Res. Commun. **16**, 522 (1964)

[3] CHANTRENNE, H., A. BURNY, and G. MARBAIX: Prog. Nucl. Acid Res. molec. Biol. **7**, 173 (1967)

[4] SCHAPIRA, G., J. C. DREYFUS, and N. MALEKNIA: Biochem. biophys. Res. Commun. **32**, 558 (1968)

[5] LOCKARD, R. E., and J. B. LINGREL: Biochem. biophys. Res. Commun. **37**, 204 (1969)

[6] LAYCOCK, D. G., and J. A. HUNT: Nature, Lond. **221**, 1118 (1969)

[7] LABRIE, F.: Nature, Lond. **221**, 1217 (1969)

[8] MATHEWS, M. B., M. OSBORN, and J. B. LINGREL: Nature, Lond. **233**, 206 (1971)

[9] NIENHUIS, A. W., D. G. LAYCOCK, and W. F. ANDERSON: Nature New Biol. **231**, 205 (1971)

[10] SCHERRER, K., H. LATHAM, and J. E. DARNELL: Proc. natn. Acad. Sci. U.S.A. **49**, 240 (1963)

[11] GEORGIEV, G. P., O. P. SAMARINA, M. J. LERMAN, and M. N. SMIRNOV: Nature, Lond. **200**, 1291 (1963)

[12] LADHOFF, A. M., B. J. THIELE, and CH. COUTELLE: Eur. J. Biochem. **58**, 431 (1975)

[13] WILLIAMSON, R., M. MORRISON, G. LANYON, R. EASON, and J. PAUL: Biochemistry **10**, 3014 (1971)

[14] LOENING, U. E.: Biochem. J. **102**, 251 (1967)
[15] STAYNOV, D. Z., J. C. PINDER, and W. B. GRATZER: Nature New Biol. **235**, 108 (1972)
[16] GOULD, H. J., and P. H. HAMLYN: FEBS Lett. **30**, 301 (1973)
[17] PEACOCK, A. C., and C. W. DINGMAN: Biochemistry **7**, 668 (1973)
[18] FROMMER, M. A., and I. R. MILLER: J. phys. Chem. **72**, 2862 (1968)
[19] FROLOVA, L. YU., A. V. TENNOV, K. G. GASARYAN, V. Z. TARANTUL, YU. N. BARANOV, CH. COUTELLE, H. GRÜTZMANN, V. HAHN, F. HIEPE, N. A. GRAEVSKAJA, and L. L. KISSELEV: Molek. Biol. USSR **10**, 944 (1976)
[20] FROLOVA, L. YU., A. V. TENNOV, N. A. SCOBELEVA, L. L. KISSELEV, V. HAHN, CH. COUTELLE, and H. GRÜTZMANN: Acta biol. med. germ. **36**, 323 (1977)
[21] PROUDFOOT, N. J., and G. G. BROWNLEE: Nature, Lond. **252**, 359 (1974)
[22] SOREQ, H., U. NUDEL, R. SALOMON, M. REVEL, and U. Z. LITTAUER: J. molec. Biol. **88**, 233 (1974)
[23] HUEZ, G., G. MARBAIX, E. HUBERT, M. LECLERCQ, U. NUDEL, H. SOREQ, R. SALOMON, R. LEBLEU, M. REVEL, and U. Z. LITTAUER: Proc. natn. Acad. Aci. U.S.A. **71**, 3143 (1974)
[24] GASKILL, P., and D. KABAT: Proc. natn. Acad. Sci. U.S.A. **68**, 72, (1971)
[25] MOREL, C., B. KAYIBANDA, and K. SCHERRER: FEBS Lett. **18**, 84 (1971)
[26] PEMBERTON, R., D. HOUSMAN, H. F. LODISH, and C. BAGLIONI: Nature New Biol. **235**, 99 (1972)
[27] MACLEAN, N., and W. MADGWICK: Cell Different. **2**, 271 (1973)
[28] SHEARMAN, J. J., P. H. HAMLYN, and H. J. GOULD: FEBS Lett. **47**, 171 (1974)
[29] BERNS, A., P. JANSEN, and H. BLOEMENDAL: FEBS Lett. **47**, 343 (1974)
[30] LIM, L., and E. S. CANELLAKIS: Nature, Lond. **227**, 710 (1970)
[31] HUNT, J. A.: Biochem. J. **131**, 327 (1973)
[32] KAUFMANN, S. J., and K. GROSS: Biochim. biophys. Acta **353**, 133 (1974)
[33] KAZAZIAN, H. H. JR., P. G. SNYDER, and TU-CHEN CHENG: Biochem. biophys. Res. Commun. **59**, 1053 (1974)
[34] MORRISON, M. R., S. A. BRINKLEY, J. GORSKI, and J. B. LINGREL: J. biol. Chem. **249**, 5290 (1974)
[35] WOO, S. L. C., J. M. ROSEN, C. D. LIARAKOS, Y. C. CHOI, H. BUSCH, A. R. MEANS, B. W. O'MALLEY, and D. L. ROBBERSON: J. biol. Chem. **250**, 7027 (1975)
[36] BURNESS, A. T. H., I. U. PARDUE, and N. O. GOLDSTEIN: Biochem. biophys. Res. Commun. **67**, 1408 (1975)
[37] LADHOFF, A. M., B. J. THIELE, CH. COUTELLE, and S. ROSENTHAL: Acta biol. med. germ. **36**, 305 (1977)
[38] VESCO, C., and S. PENMAN: Proc. natn. Acad. Sci. U.S.A. **62**, 218 (1969)
[39] ATTARDI, G., Y. ALONI, B. ATTARDI, D. OJALA, L. PICAMATTOCIA, D. L. ROBBERSON, and B. STORRIE: Cold Spring Harb. Symp. quant. Biol. **35**, 599 (1970)
[40] ATTARDI, B., B. CAAVIOTO, and G. ATTARDI: J. molec. Biol. **44**, 47 (1969)
[41] MORRIS, G. E., E. A. BUZASH, A. W. ROURKE, K. TEPPERMAN, W. C. THOMPSON, and S. M. HEYWOOD: Cold Spring Harb. Symp. quant. Biol. **37**, 535 (1972)
[42] BROWNLEE, G. G., E. M. CARTWRIGHT, N. J. COWAN, J. M. JARVIS, and C. MILSTEIN: Nature New Biol. **244**, 236 (1973)
[43] STAVNEZER, J., R. C. C. HUANG, E. STAVNEZER, and J. M. BISHOP: J. molec. Biol. **88**, 43 (1974)
[44] SWAN, D., H. AVIV, and P. LEDER: Proc. natn. Acad. Sci. U.S.A. **69**, 1967 (1972)
[45] DELOVITCH, T. L., and C. BAGLIONI: Proc. natn. Acad. Sci. U.S.A. **70**, 173 (1973)
[46] WELLAUER, P. K., I. B. DAWID, D. E. KELLEY, and R. P. PERRY: J. molec. Biol. **89**, 397 (1974)
[47] MCLEOD, M. C.: Anal. Biochem. **68**, 299 (1975)

Discussion

In the discussion to the report of GAZARYAN it was emphasized that skipped divisions probably occur in the anemic bone marrow of pigeons; however at least one division may occur in the peripheral blood. The significance of the [^{3}H]-thymine incorporation is unknown. With reference to the report of LADHOFF et al. two questions were raised: 1) whether the pulse-labelled RNA was identical with that studied by electronomicroscopy and 2) with respect to the oligo(U)-sequence. LADHOFF answered that the sequence contains about 80% residues and is about 25 nucleotides long. It is absent in globin mRNA. With respect to the report of MARBAIX it was suggested (PONKA) to inject oocytes with a heme-binding protein together with poly(A) free mRNA in order to demonstrate complete inhibition of α-mRNA translation and inhibition of its breakdown. SCHERRER referred to recent work on capping which is found apparently both in the nucleus and in the cytoplasm and thus may occur on initial transcripts and during the processing. MARBAIX told that the poly(A) segment decreases from 60 to 10 nucleotides within 40 hrs in the blood stream. LIMBORSKA et al. supplemented their report as follows: the concentration of the d-nucleotide triphosphates was 0.2 mM; the bloods used originated from patients with nonthalassemic hemolytic anemia. The "heavy" fraction of the alkaline sucrose gradient was 80% resistant to S_1 nuclease, specific for single strands. The yield of products of 1200—1500 nucleotides was much greater if both DNA-polymerase and revertase were present. The discussion to the report of MÜLLER elicited the following points: The presence of glucose diminishes liberation of ammonia. In the absence of glucose, ribose originating from the increased breakdown of nucleotides yield the acceptor for the NH_2 group. Serine is the only amino acid which yields CO_2 in mature erythrocytes. The transport mechanism responsible for the high ratios of amino acids between mitochondria and cytosol is unkown. So is the exact rate of breakdown of the stroma. With respect to the presentation of PONKA it was discussed that the reason for a lesser inhibition by INH of 9 S RNA formation as compared with that of heme synthesis may be the presence of a pool of "free" heme. The possibility of an effect of heme on translation was suggested (ROSENTHAL), while PONKA maintained that it is unlikely that untranslatable globin mRNA exists in the early stages of erythropoesis. The report of SCHEWE elicited an extensive discussion. Among others the following points were made: 1) Hemoglobin inhibits lipoxygenase and has a possible role in the regulation of mitochondrial breakdown. 2) The occurrence of respiratory inhibitors in other tissues has been established; that of lipoxygenase is being looked for. 3) Lipoxygenase alone is sufficient for lysis but not for the inhibition of respiration. 4) Lipoxygenase does not seem to act on the plasma membrane. 5) It is reported that peroxidized phospholipids are more susceptible to the action of phospholipases.

Workshop

Two main topics were discussed:

1. Pre-mRNA: Dr. Scherrer pointed out that the basic techniques of pre-mRNA isolation established by GEORGIEV, namely stepwise temperature extraction of the RNA from cells or nuclei by water/phenol, has undergone several modifications (*p*H according to SABATINI; combination with protease treatment according to SCHERRER; combination with additional hydrophobic compounds). Another principle of extraction was inaugurated by BONNER and HOLMES, who used SDS in combination with urea. Dr. SCHERRER reported on his own experiences. The main disadvantage of Georgiev's technique consists in the formation of stable aggregates in the cold during the first steps of extraction. In order to avoid them, high temperatures have to be applied. On the other hand, high temperatures lead to a rapid decomposition of the high molecular weight RNA of the nucleus. The main advantage of Georgiev's method consists in the purity of the pre-mRNA fractions. The formation of the aggregates can be diminished and the extraction at +80 °C can be avoided by

- extraction procedures with a highly diluted cell supension (vol buffer: vol cells = 100:1);
- DNase treatment of the cell or nucleus lysates;
- solution of the RNA-protein complex after the +65 °C extraction in SDS;
- enzymatic degradation of protein by protease K (other proteases — pronase — cannot be used because of their RNase impurities).

Dr. SCHERRER recommended nuclei as the source of pre-mRNA. With total cells rRNA impurities are likely.
The procedure of BONNER and HOLMES has the disadvantage of a high protein impurity of the pre-mRNA fraction. THIELE proposed to combine this procedure with a protease treatment. Addition of hydrophobic compounds (chloroform, cresol, naphthaline, urea) to the biphasic water/phenol system reduces the solubility of the proteins, so that there is no reason to add such compounds (SCHERRER).
Dr. GAZARYAN reported on high molecular weight RNA prepared by extraction of the total RNA and a separation by gradient centrifugation.
The following discussion dealt with the avoidance of the aggregation of high and low molecular weight RNA by centrifugation in organic media. Because of the fact that DMSO gradients are tricky and might be employed for small amounts only, gel electrophoresis in formamide was proposed (SCHERRER). A low *p*H value during the extraction avoids DNA impurities, however, under the conditions of low *p*H values the poly (A)-tail of the pre-mRNA is degraded (SCHERRER). At *p*H 8 a quantitative preparation of poly $(A)^+$-pre-mRNA can be carried out (MARBAIX).
Isolation of specific pre-mRNA by affinity chromatography (cDNA-cellulose columns) might be a very promising technique (SCHERRER, GAZARYAN, THIELE). SCHAFFRITZ and ROSS have obtained high molecular weight globin pre-mRNA by using a column with oligo-I-cellulose: poly $(C)^+$-cDNA. It is now generally accepted that the affinity chromatography with mercuriated cDNA and –SH columns gives good results. For investigations with the electron microscope this method

cannot be used, as a high degree mercuriation leads to a degradation of cDNA and mRNA.
The doubts concerning the solidity of the pre-messenger concept in animal cells are probably due to methodical reasons. They might be overcome by the isolation of specific genes, back hybridizing those genes with cDNA and the eventual finding of additionally transcribed sections (SCHERRER).

2. mRNA, especially globin mRNA: Dr. MARBAIX pointed out that the choice of the isolation procedure for obtaining pure globin mRNA depends on the RNase content of the tissue in question. In case of reticulocytes, which possess small amounts of RNase only, one might use polysomes. After treatment with EDTA they dissociate into mRNP and ribosomal subunits. The mRNP complex is purified by repeated sucrose gradient centrifugations. The mRNA is made soluble by SDS and proteinase K treatment. This procedure is followed by an affinity chromatography with poly(U)-sepharose. About 70—80% of the RNA is bound and consists mainly of globin mRNA. About 20—30% of the RNA do not bind to the column and consits mainly of non-globin mRNA. In order to avoid the degradation of the RNA in case of the presence of higher amounts of RNase, the tissue is frozen and homogenized in liquid nitrogen. After isolation of the total RNA the mRNA fraction is obtained by running the total RNA through a poly(U)-sepharose column.

This method has been used successfully for obtaining poly(A)$^+$ RNA from carp islets (T. RAPOPORT).
Dr. MARBAIX then reported on a stepwise elution of RNA fractions with different poly(A)-tails from the poly(U)-sepharose column by increasing the temperature between +4 and +50 °C. The excess of poly(U) is critical and might lead to an unspecific binding of RNA.

Dr. SCHERRER reported on his experiences with tissues rich in RNase. The procedure includes a gentle lysis of the cells in order not do destroy the lysosomes, which are then separated quantitatively. He prefers additional purification with formamide gel electrophoresis. For testing the degree of purity one has to look for the specific products of the translation in vitro. An additional possibility is offered by the synthesis of cDNA of a mRNA fraction.
It is now generally accepted that wheat germ represents a cell free system of protein synthesis with a low endogenous background, though re-initiation rate is low. For long messages the reticulocyte system seems to be useful. However, its background is high. The cell free liver system according to STAEHELIN is complicated and does not permit routine work. The oocyte system permits the checking of mRNA stability and the processing of the translation product; however, its background is also very high.
Dr. FROLOVA discussed the conditions for obtaining full copies of cDNA from RNA and from its hairpins. What seems to be important is the avoidance of hidden breaks within the template, a long poly(A)-tail and the concentration of nucleoside triphosphates. The number of hairpins can be increased by using a combination of reverse transcriptase and DNA polymerase I. The length of the poly(A)-tail of the globin message was reinvestigated by several groups using either the poly(A)-homopolymer as a marker (MARBAIX, HAHN), or intercalation as a tool for titration

of the poly(A) length. The poly(A)-tail is about 40—60 nucleotides long. While Dr. MARBAIX did not find readenylation of the 3′ end or RNA pieces in red cells, Dr. SCHERRER said he could.
Dr. LADHOFF showed preliminary data of globin mRNA: poly(U)-hybrids as seen in the electron microscope. Only 10% of the mRNA molecule hybridize with poly(U), and within these sections there is no condensed structure. The hybrid sections are longer than the poly(A)-tail.
Only very little information is available at present on the occurrence of additional mRNA types from red cells. From the literature (TIEDEMANN) a histone mRNA is known running within the 9S fraction. This histone mRNA does not possess a poly(A)-tail.
Dr. NEUWIRT and Dr. SCHERRER discussed different forms of related leukemias in animals, e.g. Friend leukemia in mice and erythromyeloblastosis in birds. Both types of leukemias are induced by viruses. In human beings such acute leukemias have not been observed.
Dr. MARBAIX reported on the work of the group of Dr. BURNY (Brussels). It was shown that the bovine leukemia is caused by an oncornavirus. They have made use of the methods of molecular biology to isolate the virus, to synthesize the cDNA of the viral RNA, to isolate viral specific proteins including revertase etc. and to check for the presence of viral proteins by immunological methods.

Section II

Regulation of Energy Metabolism

Acta biol. med. germ., Band 36, Seite 451—460 (1977)

[1] Department of Biochemistry, Kyushu University School of Medicine, Fukuoka, Japan, and
[2] Kliniskt-kemiska centrallaboratoriet, Akademiska sjukhuset, Uppsala, Sverige

Thermodynamics of red cell glycolysis[3]

S. MINAKAMI[1] and C.-H. de VERDIER[2]

Summary

1. Changes of Gibbs energy, enthalpy and entropy in red cells during steady-state glycolysis were discussed.
2. The heat production of red cells at various metabolic conditions was measured on a flow microcalorimeter with simultaneous analyses of lactate and other metabolites. The results were discussed in relation to enthalpy changes of the different metabolic steps in the glycolytic pathway.

Professor YOSHIKAWA and one of the authors have calculated the Gibbs energy change (ΔG) of human red cells during steady-state glycolysis, from the standard Gibbs energy changes ($\Delta G^{\circ\prime}$) and the mass action ratios (Γ) of the metabolic steps, taking advantage of the characteristics of red cells that the cells have no oxidative metabolism and no compartmentation of metabolites [2]. The thermodynamic treatment has given useful information on the regulation of glycolytic metabolism. We planned to extend the study further by compiling the enthalpy values (ΔH) of glycolytic reactions and by calculating the entropy changes (ΔS), which will be discussed in this communication. A calorimetric study was also made to show the heat production of red cells at various metabolic conditions.

Thermodynamic parameters of glycolytic reactions

Table 1 shows the Gibbs energy, enthalpy and entropy changes of glycolytic reactions in red cells at steady-state. The Gibbs energy changes are calculated from the standard Gibbs energy changes and mass action ratios of the reactions ($\Delta G = \Delta G^{\circ\prime} + RT \ln \Gamma$). The entropy changes are those calculated from ΔH and ΔG ($T\Delta S = \Delta H - \Delta G$).

Unfortunalety, few reliable enthalpy changes of glycolytic reactions are found in the literature. Few calorimetric studies on glycolytic reaction have recently been carried out and earlier data are liable to have large errors. The errors in earlier data may be exemplified by the enthalpy change of ATP hydrolysis which was given by MEYERHOF and LOHMANN [3] as —50 kJ and was

[3] A detailed account of this work is being published elsewhere [1].

Abbreviations used in the table and the figures:
As: arsenate; Fru-6-**P**: fructose 6-phosphate; Fru-1,6-$\mathbf{P}_2$: fructose 1,6-bisphosphate; Glc: glucose; Glc-6-**P**: glucose 6-phosphate; Gra: glyceraldehyde; Gra-**P**: glyceraldehyde 3-phosphate; 2-**P**-Gri: 2-phosphoglycerate; 3-**P**-Gri: 3-phosphoglycerate; 2,3-$\mathbf{P}_2$-Gri: 2,3-biphosphoglycerate; Grk: dihydroxyacetone; Grk-**P**: dihydroxyacetone phosphate; Ino: inosine; Lac: lactate; MIA: monoiodoacetate; Prv: pyruvate; **P**-Prv: phosphoenolpyruvate

Table 1
Thermodynamic parameters of glycolytic reactions at physiological conditions in red cells

Reactions	ΔH/kJ	ΔG/kJ	$T\Delta S$/kJ
1. 1/2 Glc + 1/2 ATP → 1/2 Glc-6-**P** + 1/2 ADP	−13.8 [5]*	−20.9	7.4
2. 1/2 Glc-6-**P** → 1/2 Fru-6-P	4.8 [18]	0.2	4.6
3. 1/2 Fru-6-**P** + 1/2 ATP → 1/2 Fru-1,6-$\mathbf{P}_2$ + 1/2 ADP	−13.8 [5]	−13.0	−0.8
4. 1/2 Fru-1,6-$\mathbf{P}_2$ → Gra-**P**	20.9 [19]	−0.4	21.3
5. Gra-**P** + NAD + Pi + ADP → 3-**P**-Gri + NADH + H+ + ATP	−2.9 [20]	−1.0	−1.9
6. 3-**P**-Gri → 2-**P**-Gri	0 [21, 22]	−0.6	0.6
7. 2-**P**-Gri → **P**-Prv + H_2O	14.6 [23]	−1.3	15.9
8. **P**-Prv + ADP → Prv + ATP	0 [24]	−15.5	15.5
9. Prv + NADH + H+ → Lac + NAD+	−44.4 [25]	0	−44.4
10. 1/2 Glc + ADP + Pi → Lac + ATP (Sum: 1 to 9)	−34.5	−52.6	18.0
11. ATP → ADP + Pi	−20.0 [4]	−59.8	39.7
12. 1/2 Glc → Lac (Sum: 10 + 11)	−55.7	−112.5	57.8
13. NADH → NAD+ + H+ + 2e	39.7 [25]		
14. 2,3-$\mathbf{P}_2$-Gri → 3-**P**-Gri + Pi	−5.4 [6]		

* Numbers in brackets are the related references on enthalpy changes. Some of them are used with modifications.

later corrected to —20 kJ by BENZINGER and HEMS [4]. The data calculated from the temperature dependencies of equilibrium constants by Van't Hoff's equation are usually less precise than calorimetric data and no combustion data of organic phosphate compounds are available. We tried to compile enthalpy values of glycolytic reactions with some modification of reported data, with some preliminary calorimetric measurements (Fig. 1 is given as an example) and with estimation from the data of analogous reactions (e.g. the phosphofructokinase reactions from hexokinase reaction [5] and the hydrolysis of 2,3-bisphosphoglycerate from that of glycerol 3-phosphate [6]). The overall enthalpy change during glycolysis can be calculated both from the heat of formation of glucose and lactate [7] and from the sum of the enthalpy changes of glycolytic reactions shown in Table 1.

A graphical representation of the energy changes during glycolysis (Fig. 2) demonstrates following thermodynamic properties of red cells. Gibbs energy decreased mainly at three steps, i.e. at the hexokinase, phosphofructokinase and pyruvate kinase steps, while the changes at other steps are essentially negligible. This situation indicates that the steps of latter groups are not the committed steps in red cell glycolysis, which we discussed previously [2]. On the contrary, entropy is increasing gradually until the pyruvate kinase step and a large decrease of entropy is observed at the last step of glycolysis, the lactate dehydrogenase, while the changes of enthalpy have no special tendency. It should be mentioned that the last lactate dehydrogenase step is highly exothermic with a marked decrease of the entropy. This shows that the glycolysis is driven mainly by the entropy term ($T\,\Delta S$) of the Gibbs energy change ($\Delta G = \Delta H - T\,\Delta S$)

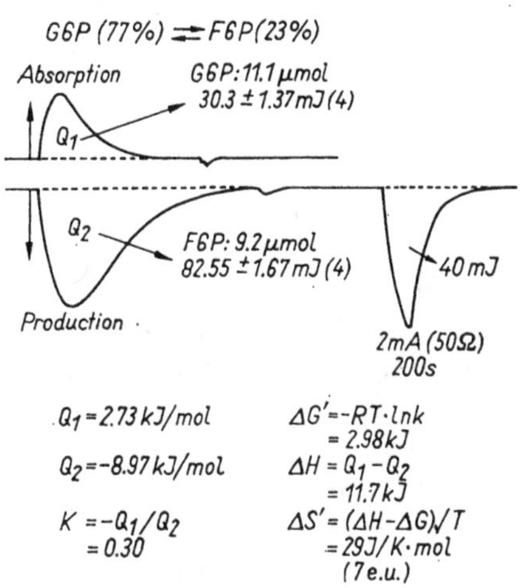

Fig. 1

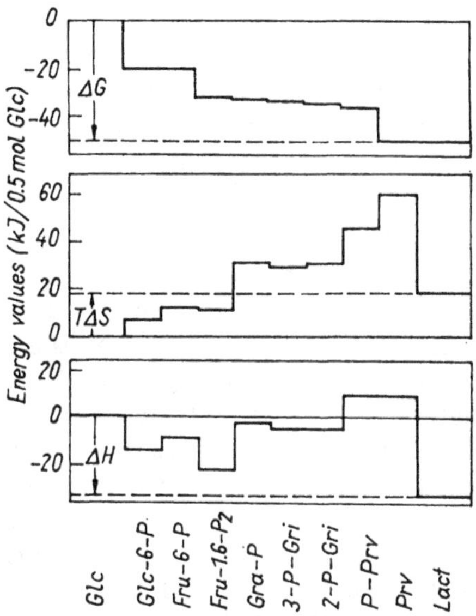

Fig. 2

Fig. 1. Determinations of enthalpy, Gibbs energy and entropy of phosphoglucose isomerase reaction.
The reaction was started by the addition of known amount of the substrate, glucose 6-phosphate or fructose 6-phosphate. The heat absorbed or produced by the reaction from one substrate to the equilibrium mixture was measured by an LKB batch microcalorimeter. The calculation procedure was given under the trace

Fig. 2. Changes of Gibbs free energy, entropy and enthalpy during steady-state glycolysis in red cells at 37 °C. The level of 1/2 Glc + 1/2 ATP is set as zero and the vertical changes correspond to the values for half a mole of glucose

until the pyruvate kinase step and the reaction is finally driven by the enthalpy term at the last lactate dehydrogenase step, accompanying a large decrease in entropy. If we may be allowed to speculate on these marked changes of entropy and enthalpy at the lactate dehydrogenase step, the step is regulating the heat and entropy production of cells, because pyruvate is permeable to the cellular membrane.

Heat production of red cells

Calorimetry is a technique suitable for the study of energy metabolism in intact cells and tissues, which can be exemplified by the classical works of MEYERHOF, HILL and their colleagues on muscle. The technique has, however, several disadvantages which have made its application limited.

The development of sensitive commercial calorimeters opens up new possibilities for the application of calorimetry on studies of cellular metabolism. LEVIN [8] and MONTI and WADSÖ [9] measured the heat production of human blood cells from normal and pathologic subjects and indicated diagnostic applications. Although this approach seems to be promising for metabolic studies of intact cells, fundamental knowledge of the origin of heat production is a prerequisite for the application, since heat production is non-specific and can be observed with most chemical and physical reactions.

Accordingly, we studied the heat production of red cells with simultaneous measurement of metabolite changes. The heat production was discussed in relation to the enthalpy changes of glycolytic reactions.

Calorimetric procedure

Thermograms are shown to explain the procedure (Fig. 3). Red cells suspended in buffered saline were incubated in a thermostated vessel, the *p*H of the suspension being kept constant by the addition of 0.3 M NaOH. The cell suspension was pumped with a flow rate of about 17 ml/h into an LKB flow microcalorimeter which had been fed with the buffered saline giving a base-line on the chart. After about 5 min lag time for the suspension to pass through a dead space, the shift from the base was observed, which indicates the rate of heat production (heat effect) of the suspension. When the heat production rate became steady, a compound was added into the incubation vessel. Sometimes, several compounds were added in succession. When an experiment was finished, the calorimeter cell were flushed again with the buffered saline to give the base-line and then calibration was carried out with an electric current. The calibration in Fig. 3 corresponds to 50 μW (180 mJ/h or 43 mcal/h). Aliquots of samples were taken out from the vessel at fixed intervals and were used for enzymic metabolite assays.

Effect of pH on the heat production

Since the proton concentration of the suspension is one of the most influential factors for red cell glycolysis [10], the heat production rate of red cells is supposed to depend on the *p*H of the suspension. An increase in the heat production rate was observed when the *p*H of the suspension increased, in agreement with that observed by LEVIN [8]. The thermogram as well as changes of major metabolites are shown in Fig. 4. The heat production is essentially correlated with the lactate formation. Furthermore, an overshoot of the heat formation rate was observed when the *p*H of the suspension changed from 7.2 to 7.6, which is supposedly connected with the reduction of pyruvate and the formation of 2,3-bisphosphoglycerate.

Heat production associated with glycolysis

One milliliter of human red blood cells produces heat at a rate of about 400 mJ/h under physiological conditions. About half of this heat production is apparently not associated with glycolysis, which can be shown by the heat at low *p*H or in the presence of sodium fluoride shown in Fig. 4, because essentially no lactate formation was observed in these conditions.

The amount of heat produced by glycolytic reactions can be estimated by the titration of glucose-starved red cells with glucose, as shown in Fig. 5. When a limited amount of glucose was added to well-washed glucose-starved cells, a transient increase in heat production was observed because of the high affinity of the hexokinase to glucose. The area below the curve and the above the glucose-

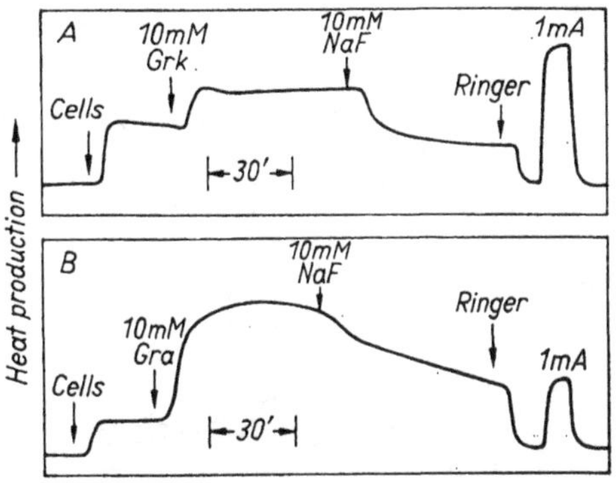

Fig. 3

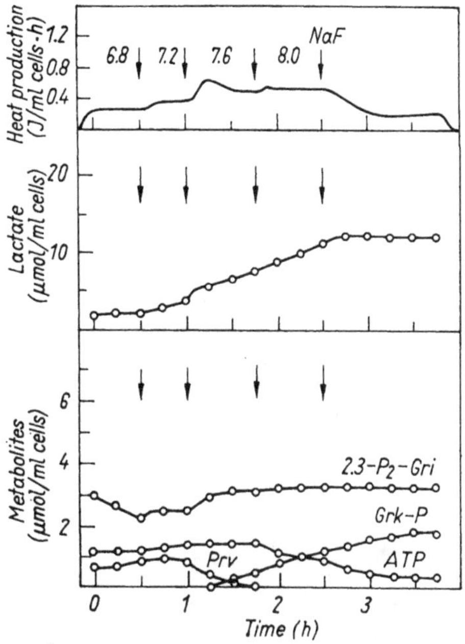

Fig. 4

Fig. 3. Thermograms showing the effect of (A) dihydroxyacetone and (B) D-glyceraldehyde on red cells.

Red cells were suspended in an isotonic saline containing 5 mM KCl, 5 mM inorganic phosphate, 1 mM $MgCl_2$ and 5 mM glucose and were incubated at *p*H 7.4 at 37 °C. The hematocrit value was 30%. (A) The lactate formation rate before and after the addition of Grk were 3.3 and 5.5 μmol/h per ml cells, respectively. (B) The lactate formation rate before and after the addition of Gra were 3.6 and 3.0 μmol/h per ml cells, respectively. Glycerol was formed after the addition, at the rate of 0.74 μmol/h per ml cells

Fig. 4. Effect of the *p*H of the red cell suspension on the heat production rate and the metabolite concentrations.

The conditions of the incubation were as in Fig. 3, except the glucose concentration (10 mM) and the *p*H. The *p*H of the suspension was changed at the points indicated by arrows

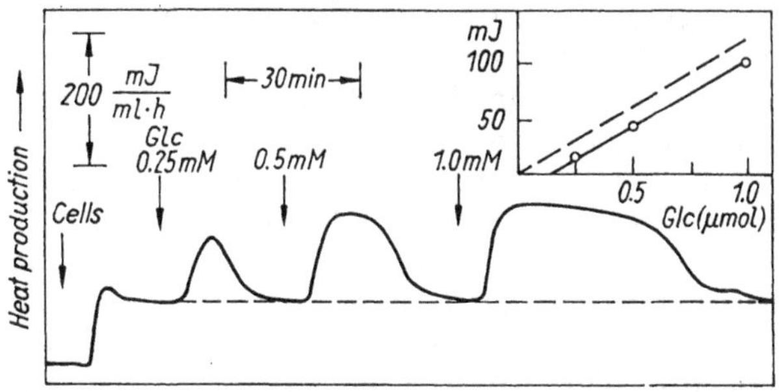

Fig. 5. Estimation of heat produced due to glycolysis.

Washed red cells (hematocrit 40%) were incubated in a buffered saline (15 mM Pi, *p*H 7.4). Glucose was added to the final concentrations indicated. The inserted figure shows the relation between the amount of glucose added and the amount of heat evolved per ml suspension, in which the broken line was drawn by adding an equal amount of heat produced assuming that the line should pass the origin

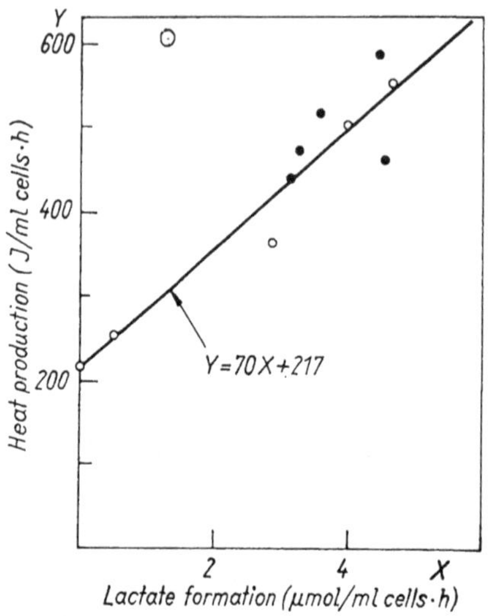

Fig. 6

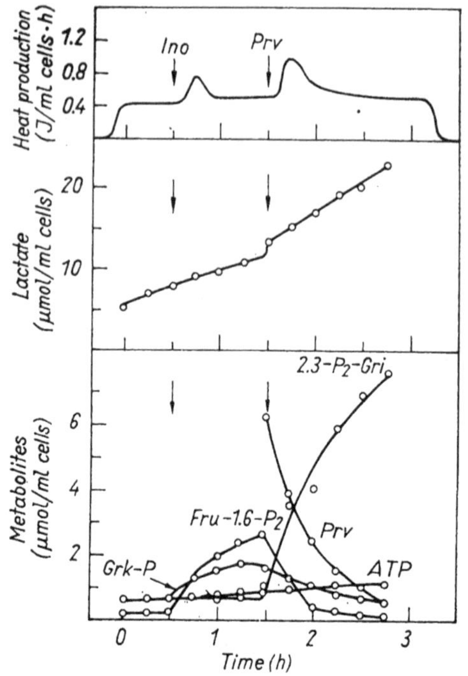

Fig. 7

Fig. 6. Relation between lactate formation and heat production in red cells. (○) Values from Fig. 4, (●) from steady-state values with glucose in Fig. 3, 4, 8 and 9

Fig. 7. Effect of inosine and subsequently of pyruvate.
Red cells were incubated in a buffered saline (15 mM Pi, *p*H 7.6) containing 10 mM glucose. The hematocrit value was 50%. Inosine was added to the vessel to a final concentration of 10 mM. Pyruvate was subsequently added to a final concentration of about 15 mM. To show the initial changes, pyruvate was added first and the sample was taken out without delay

free steady-state level is the heat which is formed in association with the utilization of a given amount of glucose. With a correction of the heat loss at dead space, the heat production per mol of glucose was calculated as about 130 kJ.

The glycolytic heat may be estimated more properly by comparison of the heat production and the lactate formation. The lactate production seems to be a better index than the glucose consumption to assess the glycolytic heat, because of a large contribution of the pyruvate reduction in the heat production. When 10 pairs of values of lactate formation (X) and the heat production (Y) were plotted, a regression straight line, $Y = 70X + 217$, was obtained by the method of least squares (Fig. 6). The intercept, about 200 kJ/ml cells per h, is the heat which is independent of glycolysis and the slope of the line gives an estimate of the amount of heat produced per mole of lactate formed, which is 70 kJ/mol lactate or 140 kJ/mol glucose.

The expected heat production due to glycolysis is the sum of the enthalpy change of the reaction and the neutralization heat of the acid formed. The former is about −55 kJ per mole lactate as shown in Table 1. The latter is difficult to assess because of the mixed buffer system of the cells. However, it may be appropriate to assume that most of the protons formed may be neutralized by imidazole groups of hemoglobin ($\Delta H = \sim -25$ -30 kJ [11]), which are the most abundant buffer groups in red cells. Other groups which may be concerned with the neutralization are carboxylic acid ($\Delta H = -4.6$ kJ), inorganic phosphate ($\Delta H = 17.6$ kJ) and phosphate groups of 2,3-bisphosphoglycerate, ATP and other organic phosphate compounds. The neutralization of phosphate groups is usually endothermic: e.g. the values for ATP und ADP are given as $\Delta H = 5.4$ kJ [12].The heat production rate obtained in the present experiment can be explained by a neutralization heat of about $\Delta H = -15$ kJ, which is well in accordance with the above consideration.

Heat production associated with particular steps of glycolysis

In the above calculation, we assumed that the cells were in a metabolic steady-state. Since, however, it is almost impossible to have a complete steady-state of metabolites in vitro, heat production due to changes of metabolite concentrations should be taken into consideration. This necessitates the measurement of intermediates simultaneously with the calorimetry. The accumulation or exhaustion of a metabolites will cause the change in the heat production, which may be predicted by the enthalpy of the metabolite.

When inosine was added to the cell suspension, an increase in heat formation with an overshoot was observed, although the rate of the lactate formation remained essentially constant (Fig. 7). The increased heat production rate can be considered to be due to the formation of triose phosphates through the pentose phosphate cycle and the reverse reaction catalyzed by the aldolase to form fructose 1,6-bisphosphate. Because of highly endothermic nature of the aldolase reaction, the formation of triose phosphates without passing through the step may cause the increase in heat production and the reversal of the step causes an overshoot of the heat production. The subsequent addition of pyruvate initiated a rapid utilization of pyruvate coupled with the formation of 2,3-bisphosphoglycerate. The transient increase in the rate of heat production observed by the addition of pyruvate is the reflection of the highly exothermic nature of the pyruvate reduction. The coupled reaction, glyceraldehyde 3-phosphate + pyruvate + Pi → 2,3-bisphosphoglycerate + lactate, is highly exothermic with ΔH of -10.9 kJ.

Dihydroxyacetone and glyceraldehyde have also caused increases in the heat formation (Fig. 3) but to a different extent depending on the substrate used: the former caused only 50% increase whereas the latter almost 400%. This may be explained by the difference in their metabolism. Dihydroxyacetone is phosphorylated to dihydroxyacetone phosphate [13], whereas glyceraldehyde is metabolized through several pathways, of which the reduction to glycerol by the NADP-aldose reductase seems to be of special importance, as it supplies $NADP^+$ to the phosphogluconate oxidative pathway [14]. The effect of fluoride on the heat production was also different: the increased heat formation due to dihydroxyacetone was inhibited by fluoride whereas the one induced by glyceraldehyde was only slightly influenced by the inhibitor.

Three kinds of glycolytic inhibitors, fluoride, monoiodoacetate and arsenate, were also used to study the relationship between metabolic changes and the heat production. Fluoride inhibits the supply of pyruvate so that the lactate productions stops. This caused the decrease in the heat production. Monoiodoacetate inhibits the lactate formation, though pyruvate is supplied by the degradation of 2,3-bisphosphoglycerate. The decrease of ATP was also observed. The degradation of 2,3-bisphosphoglycerate and ATP may be reflected on a rather gradual decrease of the heat production rate observed (Fig. 8) because both ATP hydrolysis and the conversion of 2,3-bisphosphoglycerate into pyruvate are exothermic reactions, the enthalpy of the former reaction being —20 kJ and that of the latter —10.9 kJ. The origin of the initial overshoot after the addition of monoiodoacetate is not clear. It might be due to the interaction of the inhibitor and proteins. Arsenate at first accelerates the lactate formation by uncoupling the glyceraldehyde-phosphate dehydrogenase and thereby ATP and 2,3-bisphosphoglycerate decrease [15]. A big overshoot of the heat production was observed before reaching a steady state (Fig. 9).

A preliminary experiment was carried out on the heat production in hemolyzate (Fig. 10), in order to study the heat production associated with partial reactions

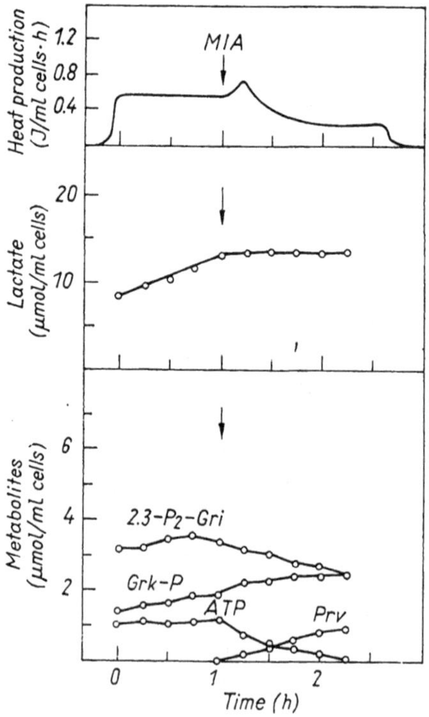

Fig. 8

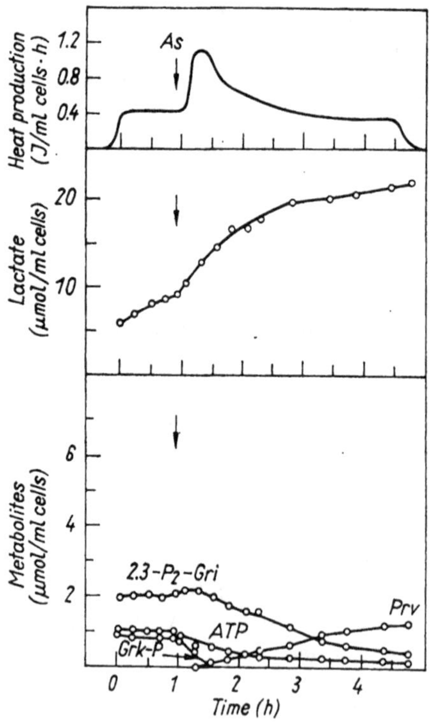

Fig. 9

Fig. 8. Effect of monoiodoacetate.
The conditions were essentially the same as in Fig. 2, except the *p*H was kept at 7.6 and the hematocrit value was 31%. Monoiodoacetate (2 mM) was added at the point indicated by an arrow

Fig. 9. Effect of arsenate.
The conditions were essentially the same as in Fig. 2, except the *p*H was kept at 7.4 and the hematocrit value was 35%. Sodium arsenate was added to a final concentration of 10 mM

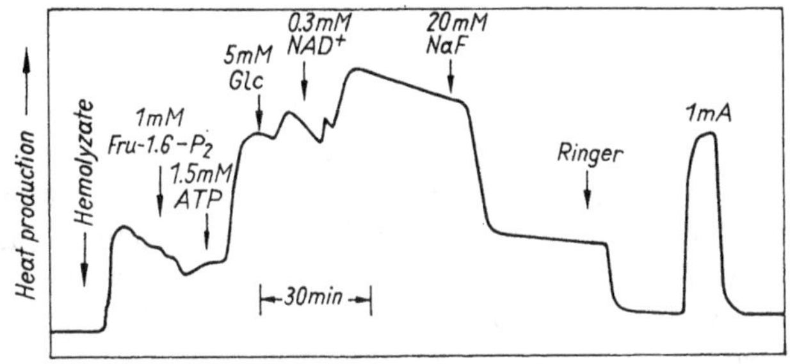

Fig. 10. Thermogram showing the effect of substrates and coenzymes on hemolyzate Red cell hemolyzate (hemoglobin concentration 116 g/l, 50 mM KCl, 2 mM NaCl, 2 mM Pi, 1 mM $MgCl_2$, *p*H 7.6) was incubated at 37 °C. No lactate formation was observed before the addition of NAD^+. After the NAD^+ addition, the rate of lactate formation became 3.3 μmol/h per ml cells

of glycolysis. Hemolyzate also showed an appreciable liberation of heat without the addition of substrates and coenzymes, i.e. without the lactate formation. The addition of fructose 1,6-bisphosphate caused some transient absorption of heat which can be ascribed to the conversion of the hexose bisphosphate to triose phosphates. The addition of ATP showed an increase in the heat production rate, which is connected with ATP hydrolysis. Subsequent addition of glucose caused a transient increase of heat production which may be ascribed to the phosphorylation of the sugar. Finally, the lactate formation was initiated by the addition of NAD^+ and the heat production rate increased. The heat production was inhibited by fluoride.

Heat production associated with oxidative metabolism:

Although mature human red cells lack the capacity of respiration, oxygen uptake is induced by the addition of methylene blue [16]. Table 2 shows the heat production and oxygen uptake of red cells in the presence of methylene blue. By increasing methylene blue concentration, the heat production rate and the oxygen

Table 2

Heat production and O_2 uptake of red cells in presence of methylene blue.

Ten days old red cells were suspended in buffered saline (50 mM Pi, *p*H 7.0 at 37 °C). The hematocrit value was 30%. Respiration was measured with a Warburg apparatus at 37 °C. The values of the heat production rate used for the calculation of heat/O_2 are those in excess of the control values without methylene blue

Methylene blue concentration (μM)	Heat production rate (J/ml cells · h)	O_2 uptake (μmol/ml cells · h)	Heat/O_2 (kJ/mol)
0	0.35	<0.1	—
1.9	1.0	1.3	500
3.8	1.58	1.66	741
9.4	1.92	2.16	727
18.8	2.04	2.23	758
37.5	2.28	3.35	547
93.8	2.48	4.16	502

uptake increased and they seemed to be correlated with each other, giving 500 to 700 kJ per mole oxygen consumed. The enthalpy change calculated from the reaction, 1/6 glucose + $O_2 \rightarrow CO_2 + 2\,H_2O$, is —488 kJ and that of the NADH oxidation by mitochondrial respiratory chain [17] is —500 kJ per mole oxygen. The heat production in the presence of methylene blue is somewhat higher than these values, which may be due partly to other exothermic reactions including the neutralization of CO_2 and partly to the difference in conditions employed in calorimetry and manometry. Anyhow, the heat production associated with oxidative metabolism is extremely higher than the one associated with glycolysis. A part of the heat production in the absence of glycolytic reaction may be ascribed to oxidative reactions in the cells.

Concluding remarks

Thermodynamic treatment of red cells is expected to yield a good model for cellular metabolism on the one hand and a theoretical background for practical application of calorimetry, on the other. Further calorimetric investigation will be necessary to have an insight in the metabolic regulations of intact cells.

Acknowledgement

Calorimetric work was carried out at the Department of Clinical Chemistry, University Hospital, Uppsala, while one of the authors (S. M.) was a recipient of fellowship from the Swedish Medical Research Council. This work was supported by the Swedish Medical Research Council (B76-13X-152).

References

[1] MINAKAMI, S., and C.-H. de VERDIER: Eur. J. Biochem. **65**, 451 (1976)
[2] MINAKAMI, S., and H. YOSHIKAWA: Biochem. biophys. Res. Commun. **18**, 345 (1965); MINAKAMI, S., and H. YOSHIKAWA: J. Biochem., Tokyo **59**, 139 (1966)
[3] MEYERHOF, O., and K. LOHMANN: Biochem. Z. **253**, 431 (1932)
[4] BENZINGER, T. H., and R. HEMS: Proc. natn. Acad. Sci. U.S.A. **42**, 896 (1956)
[5] MCGLOTHLIN, C. D., and J. JORDAN: Anal. Chem. **47**, 786 (1975)
[6] OHLMEYER, P.: Hoppe-Seyler's Z. physiol. Chem. **282**, 37 (1945)
[7] WILHOIT, R. C. in: Biochemical Microcalorimetry. H. D. BROWN (Ed.). Academic Press, New York 1969, p. 305
[8] LEVIN, K.: Scand. J. clin. Lab. Invest. Suppl. no. 135 (1973)
[9] MONTI, M., and I. WADSÖ: Scand. J. clin. Lab. Invest. **32**, 47 (1973)
[10] MINAKAMI, S., and H. YOSHIKAWA: J. Biochem., Tokyo **59**, 145 (1966)
[11] WYMAN, J., Jr.: J. biol. Chem. **127**, 1 (1939)
[12] IZATT, R. M., and J. J. CHRISTENSEN: J. phys. Chem. **66**, 1030 (1962)
[13] BEUTLER, H., and E. GUINTO: Blood **41**, 559 (1974)
[14] BEUTLER, H., and E. GUINTO: J. clin. Invest. **53**, 1528 (1974)
[15] SAITO, T., and S. MINAKAMI: J. Biochem., Tokyo **67**, 211 (1967)
[16] HARROP. G. A., and E. S. G. BARRON: J. exp. Med. **48**, 207 (1928)
[17] POE, M., H. GUTFREUND, and R. W. ESTBROOK: Archs Biochem. Biophys. **122**, 204 (1967)
[18] KAHANA, S. E., O. H. LOWRY, D. W. SCHULZ, J. V. PASSONNEAU, and E. J. CRAWFORD: J. biol. Chem. **235**, 2178 (1960)
[19] MEYERHOF, O., and K. LOHMANN: Biochem. Z. **275**, 430 (1935)
[20] MEYERHOF, O.: Ann. N. Y. Acad. Sci. **45**, 377 (1944)
[21] MEYERHOF, O., and W. SCHULZ: Biochem. Z. **297**, 60 (1938)
[22] PIZER, L. I. in: The Enzyme. **6**, 179. 2nd Ed. P. D. BOYER, H. LARDY, and K. H. MYRBÄCK (Eds). Academic Press, New York 1962
[23] WOLD, F., and C. E. BALLOU: J. biol. Chem. **227**, 301 (1957)
[24] MEYERHOF, O., and W. SCHULZ: Biochem. Z. **281**, 292 (1935)
[25] KATZ, S.: Biochim. biophys. Acta **17**, 226 (1955)

Acta biol. med. germ., Band 36, Seite 461—468 (1977)

Zentralinstitut für Molekularbiologie der Akademie der Wissenschaften der DDR,
1115 Berlin-Buch, DDR
Institut für physiologische und biologische Chemie der Humboldt-Universität,
104 Berlin, DDR

An extended model of the glycolysis in erythrocytes

T. A. RAPOPORT, M. OTTO, and R. HEINRICH

Summary

An extension of a previous model [2] is proposed of the glycolysis of erythrocytes which includes realistic rate laws for the hexokinase-phosphofructokinase system and for the 2,3-P_2G phosphatase. Whereas most conclusions previously drawn are reinforced, the mechanism of ATP regulation is different in the present model. The ATP concentration is mainly regulated by the inhibitory action of ATP and the activating effect of AMP on the phosphofructokinase. The role of the 2,3-P_2G bypass as a buffer of changes in the ATP demand is of lesser significance than previously thought. Besides the feedback action of the adenine nucleotides on the hexokinase-phosphofructokinase system in the quasi-steady state the role of 2,3-P_2G as an energy source is important since it can yield ATP for a certain period of time. The present version of the model describes qualitatively the experimental data on the modulation of Na^+-K^+-ATPase.

Introduction

Two models of the glycolysis of erythrocytes have been proposed previously. The first model did not include the adenine nucleotides as variables and represented therefore essentially a linear chain of enzymic reactions [1]. The extended model included the adenine nucleotides [2]. It could also describe time dependent processes. Although the model was in agreement with a multitude of experimental data, two recent kinds of experiments were not consistent with the predictions. First, the 2,3-P_2G phosphatase was shown to be saturated with its substrate 2,3-P_2G [3, 4] in contrast to the assumption of the model of an almost first order reaction. Previous calculations have shown that no meaningful steady state of the system exists if saturation of the 2,3-P_2G phosphatase is taken into account while retaining all other model assumptions (unpublished results). The second discrepancy concerns the modulation of the Na^+-K^+-transport ATPase. Whereas the model predicted an increase of the glycolytic flux when the ATPase was inhibited, the reverse was actually observed [5—7]. This indicates that the net response to the adenine nucleotides of the hexokinase-phosphofructokinase system was not in accord with reality.

In this paper we propose a modified version of the previous model [2] which takes into account the recent experimental data. The concequences of the new model assumptions are shown for the conditions of steady states and quasi-steady states.

Abbreviations: G6P: glucose-6-phosphate; F6P: fructose-6-phosphate; 2,3-P_2G: 2,3-bisphosphoglycerate; 1,3-P_2G: 1,3-bisphosphoglycerate; P-Pyr: phosphoenolpyruvate; A: sum of the adenine nucleotides; HK: hexokinase; PFK: phosphofructokinase; PK: pyruvate kinase; P_2GM: bisphosphoglycerate mutase; P_2Gase: 2,3-bisphosphoglycerate phosphatase; Enol: enolase; PGM: phosphoglycerate mutase; PGK: phosphoglycerate kinase; ATPase: nonglycolytic ATP-consuming processes

Model assumptions

Most of the assumptions of the previous model have been retained [2]. For the reaction catalyzed by the 2,3-P_2G phosphatase the following rate law was employed

$$v_{2,3\text{-}P_2Gase} = \frac{k_{P_2Gase} \cdot [2{,}3\ P_2G]}{K_m + [2{,}3\ P_2G]}$$

K_m was experimentally determined to be 300 μM [4].

The response of the hexokinase-phosphofructokinase system to the ATP concentration was constructed as previously described. However, more realistic rate laws for the single enzymes were used and the existence of a conservation quantity for the adenine nucleotides and the adenylate kinase were taken into account. The construction is shown in Fig. 1 and in more detail in [8]. For the calculation of the ATP response of the hexokinase the rate law and the parameter values given by GERBER et al. [9] were employed. The hexokinase displays an approximate Michaelis-Menten function with respect to ATP up to 2.0 mM (water basis). The enzyme is inhibited by G6P. For the phosphofructokinase the following rate law was used [8, 10]

$$v_{PFK} = V_{max} \frac{\frac{0.4\,[G6P]}{K_F}}{1 + \frac{0.4\,[G6P]}{K_F}} \cdot \frac{\frac{[MgATP]}{K_{MgATP}}}{1 + \frac{[MgATP]}{K_{MgATP}}} \cdot \frac{1}{1 + L}$$

$$L = L_0 \frac{\left(1 + \frac{[ATP]_{free}}{K_{ATP}}\right)^4 \cdot \left(1 + \frac{[Mg]_{free}}{K_{Mg}}\right)^4}{\left(1 + \frac{0.4\,[G6P]}{K_F}\right)^4 \cdot \left(1 + \frac{[AMP]}{K_{AMP}}\right)^4}$$

The parameter values are given in the legend to Fig. 1. The phosphofructokinase is activated at low ATP concentrations and inhibited allosterically at higher ones. Owing to the conservation equation a change of the ATP concentration acts not only by a variation of its inhibitory effect but also by virtue of the complementary changes of AMP. F6P has an activating influence on the phosphofructokinase. For the construction of both the hexokinase and phosphofructokinase curves only the complex formation of Mg^{2+} with ATP was taken into account. The net flux through the hexokinase-phosphofructokinase system which is identical with the steady state glycolytic flux is given by the intersection of the individual curves (heavy line in Fig. 1). The net curve displays a maximum. Under in vivo conditions the flux is inhibited by an increase of the ATP concentration. The calculated values for the flux and metabolite concentrations at the in vivo point are given in the legend to Fig. 1 and compared with experimental data. Having obtained this agreement the net curve was fitted to the following descriptive rate law

$$v_{HK\text{-}PFK} = \frac{V_{max} \frac{[ATP]}{K_A}}{1 + \frac{[ATP]}{K_A} + \left(\frac{[ATP]}{K_i}\right)^n}$$

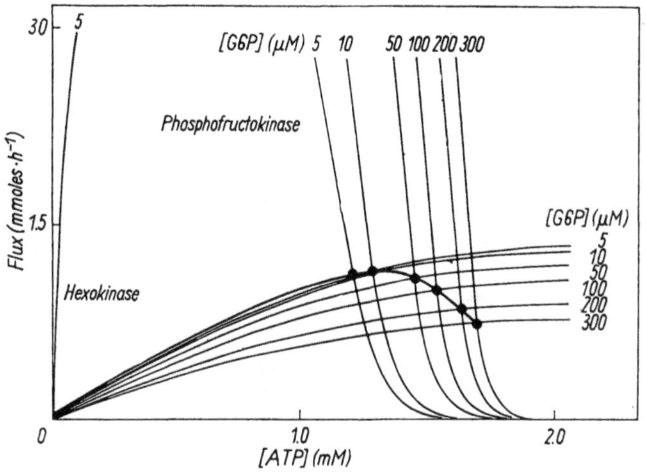

Fig. 1. Construction of the ATP response of the hexokinase-phosphofructokinase system. The concentrations in the figure are given per litre of cell water. For the calculation of the hexokinase curves the kinetic model and the parameter values given by GERBER et al. [9] were used. For the phosphofructokinase the rate law given in the text and the following parameter values were employed (given per litre water): $V_{max} = 250\ mM \cdot h^{-1}$, $K_{MgATP} = 68\ \mu M$, $K_F = 100\ \mu M$, $K_{ATP} = 10\ \mu M$, $K_{Mg} = 440\ \mu M$, $K_{AMP} = 33\ \mu M$, $L_0 = 0.05$. ΣA and Mg_{total} were held fixed at 20 mM. Only the complex formation of Mg^{2+} and ATP was taken into account ($K_{ass} = 12.3\ mM^{-1}$). At the in vivo point the following concentrations of the metabolites are calculated (experimental values in parenthesis, see also [18]): $[ATP]_{total} = 1.12$ mM (1.2), [MgATP] = 0.99 mM (1.05), $[ATP]_{free} = 0.13$ mM (0.13), [AMP] = 0.1 mM (0.05), [G6P] = 73 μM (44), flux = $1.05\ mM \cdot h^{-1}$ (1.1). These values are given per litre of cells. The heavy line in the figure corresponds to the net flux through the hexokinase-phosphofructokinase system and was fitted with the equation given in the text with the following parameter values: $V_{max} = 2.9\ mM\ h^{-1}$, $K_A = 1.168$ mM, $K_i = 1.241$ mM, $n = 12$ (values are given per litre of cells).

where the parameters have the values given in the legend to Fig. 1. Here ATP corresponds to ATP_{total} in the equations of the underlying detailed description of the enzymes.

Similar curves as shown in Fig. 1 were obtained with different total sums of adenine nucleotides [8].

Results

Properties of the steady states

In Fig. 2 are represented the steady state solutions of ATP and the fluxes versus the first order rate constant of the nonglycolytic ATP-consuming processes. The arrow in the figure indicates the situation in vivo. It corresponds to the parameter set which gives the best agreement with the experimental data (see legend to Fig. 2). The ATP curve is similar to that calculated with the previous model. However, the reason for the relative constancy of the ATP in the region of physiological interest is different. The previously proposed stoichiometric mechanism which was based on a simplified scheme of SEL'KOV [11] was connected with the 2,3-P_2G bypass and is of importance only for great activations of the ATP-con-

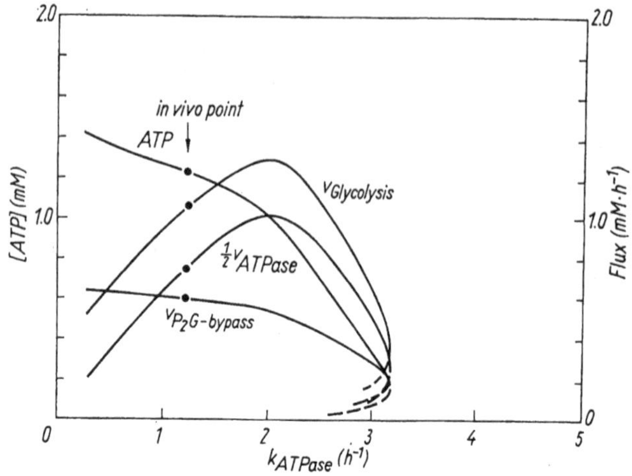

Fig. 2. ATP concentration and fluxes in the steady state as functions of the rate constant of the ATP-consuming processes (k_{ATPase}).

The v_i denote the fluxes specified by the superscript. The curves were calculated with the following parameter values: $k_{\text{P}_2\text{GM}} = 1.1 \cdot 10^5 \cdot \text{h}^{-1}$, $k_{\text{P}_2\text{Gase}} = 2.13 \cdot \text{h}^{-1}$, $k_{\text{PK}} = 0.67 \cdot \text{h}^{-1}$, $q_{\text{Enol}} \cdot q_{\text{PGM}} \cdot q_{\text{PGK}} = 300$, $K_{\text{P}_2\text{G}} = 40$ µM (inhibition constant for 2,3-P_2G of the 2,3-P_2G mutase), $\Sigma A = 1.4$ mM. The other parameter values are given in the text and in the legend to Fig. 1. At the in vivo point (arrow in the figure) the following concentrations of the metabolites are calculated (experimental values in parenthesis): [ATP] = 1.22 mM (1.2), [ADP] = 144 µM (185), [AMP] = 34 µM (50), [2,3 P_2G] = 4.49 mM (4.7), [P-Pyr] = 21 µM (26), [1,3 P_2G] = 0.58 µM (0.5), flux = 1.05 mM · h^{-1} (1.1). The dotted parts of the curves indicate unstable steady states. The rate of the nonglycolytic ATP-consuming processes is given by the equation $k_{\text{ATPase}} \cdot$ [ATP].

suming processes. Then, the increased ADP concentration leads to a lowered concentration of 1,3-P_2G and of the rate through the bypass. The compensatory increase of the flux through the phosphoglycerate kinase leads to an additional ATP production and therefore stabilization of the ATP concentration. In the present model, however, the flux through the bypass is quite constant due to the saturation of the 2,3-P_2G phosphatase with its substrate. Therefore the main mechanism for the stabilization of the ATP level is of nonstoichiometric nature. If k_{ATPase} is increased, the concentration of AMP rises and that of ATP falls. This leads to an activation of the phosphofructokinase and, via the release of the inhibition of the hexokinase by G6P, to an increased glucose consumption (Fig. 2). The higher flux produces more ATP in the lower part of the glycolytic chain and thus compensates for the increased rate of the ATP-consuming processes.

At variance with the previous model [2], the present extension explains the increase of the glycolytic flux with activation of the Na^+-K^+-ATPase [5—7]. The increase of AMP is the main factor for the activation of the glycolytic flux. In addition, the model is also in agreement with data on blood preservation where glycolytic flux and ATP change in parallel. Under these conditions the system is situated to the right of the maximum of the glycolytic flux in Fig. 2 [12].

Fig. 3 gives the recalculated plots of the ATP concentration versus the rate of the ATP consuming processes (not their rate constant) for erythrocytes of various animal species. The points in the figure give the in vivo situation in the cells. The interesting location of the in vivo points near to the maxima of the ATP consumption rates is in accord with the previous results [2]. The erythrocytes of the various species perform about 85% of the maximally possible work which seems to represent an evolutionary adaptation.

The control strengths are given in Table 1 of the nonequilibrium enzymes in the steady state (left column). The most pronounced influence on the flux have the nonglycolytic ATP-consuming processes (ATPases). The hexokinase-phosphofructokinase system and the 2,3-P_2G phosphatase have also significant flux control. On the other hand, pyruvate kinase and 2,3-P_2G mutase have no influence on the glycolytic steady state flux. The diminished flux control by the hexokinase-phosphofructokinase system compared to the previous model can be explained by the feedback action of the adenine nucleotides; if the system is activated the increase of ATP and the decrease of AMP counteract flux activation. The role of the 2,3-P_2G phosphatase is due to its strong influence on the ATP concentration. The activation state of this enzyme determines the flux through the bypass which does not contribute to ATP production.

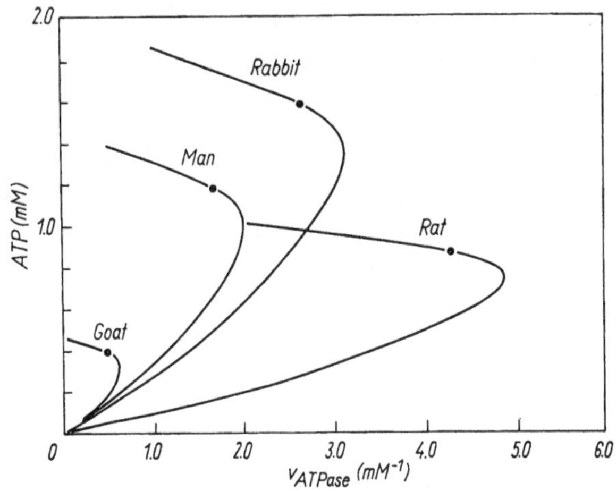

Fig. 3. ATP concentration as a function of the rate of the nonglycolytic ATP-consuming processes (v_{ATPase}) for erythrocytes of various species.
The parameters for the enzymes of the different species were taken from [18] and [2]. The points in the figure give the in vivo situation for the cells

Table 1

Control strengths of the nonequilibrium enzymes in the steady state and in the quasi-steady state at unrelaxed 2,3-P_2G concentrations

Enzyme	Steady state	Quasi-steady state
Hexokinase-phosphofructokinase system	0.24	0.78
ATP-consuming processes	0.55	0.25
Pyruvate kinase	−0.02	−0.16
2,3-P_2G mutase	0.02	0.17
2,3-P_2G phosphatase	0.21	−0.04

The influence of the conservation sum for the adenine nucleotides on the steady state of the glycolytic system is considered in Fig. 4. The net response of the hexokinase-phosphofructokinase system to the ATP concentration was constructed as in Fig. 1 for various values of the total sum of adenine nucleotides [8]. The total Mg^{2+} concentration was held constant in view of the very slow transport of Mg^{2+} through the cell membrane [13]. Then the various descriptive rate laws for the hexokinase-phosphofructokinase system were used for the overall model of the erythrocyte glycolysis (Fig. 4). The increase of the total sum of adenine nucleotides by a factor of 1.5 results in an approximately equal increase of ATP so that the ratio [ATP]/([ATP] + [ADP] + [AMP]) remains almost constant. The glycolytic flux is çhanged by only 15% and the flux through the 2,3-P_2G bypass remains practically constant. These results are explained by the simultaneous increase of ATP and AMP which compensate each other in their effects on the phosphofructokinase. The theoretical predictions are in agreement with experimental data. Human subjects with a maturation stop have higher total sums of adenine nucleotides in the transfused erythrocytes but the ratios between the adenine nucleotides and the glycolytic flux are unchanged [14, 15]. On the other hand, the model would predict a decrease of the 2,3-P_2G concentration owing to the higher ADP- and therefore lower 1,3-P_2G-levels. This was not observed experimentally. This is probably due to the neglect in the model of important factors such as variations in the interactions of ATP, 2,3-P_2G, Mg^{2+} and hemoglobin and of the influence of Mg^{2+} on the apparent equilibrium constants of the adenylate kinase and phosphoglycerate kinase reactions. Despite this discrepancy with the experimental data the model shows that the particular dependence of the velocity of the phosphofructokinase on the concentrations of ATP and AMP serves to keep constant the erythrocyte metabolism even in the case of great

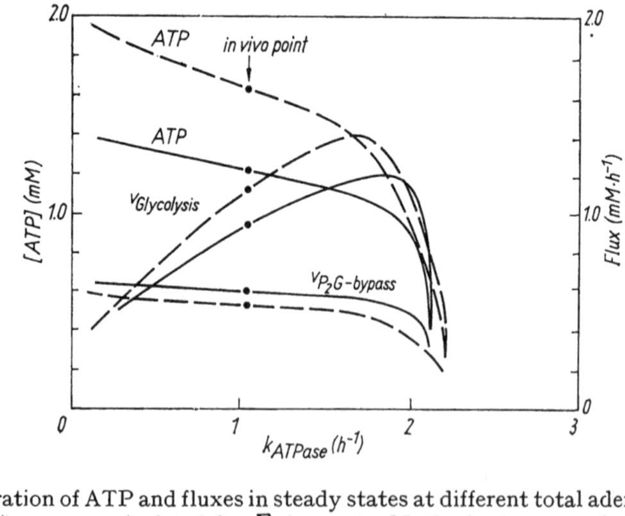

Fig. 4. Concentration of ATP and fluxes in steady states at different total adenine nucleotides. The continous lines were calculated for $\Sigma A = 1.4$ mM, the broken ones for $\Sigma A = 2.1$ mM (per litre cells). The net curves of the hexokinase-phosphofructokinase system given in [8] were employed. The curve at 1.4 mM does not entirely coincide with that given in Fig. 2 since the net curves of the system were fitted with different parameter values to simulate the dependence on ΣA. The arrow in the figure indicates the in vivo point of the system

variations in the individual concentrations. This may explain the greater constancy of the ATP level in the cells of one subject as compared to that among cells of different donors [16].

Properties of the quasi-steady states

The pronounced time hierarchy of the glycolytic reactions has been discussed previously in detail [2]. 2,3-P_2G is a slow variable with a relaxation time of more than 10 hours. A quasi-steady state of the system in erythrocytes is observed about half an hour after perturbation and is defined by the steady state assumption for all metabolites except for 2,3-P_2G. Fig. 5 shows the effect of k_{ATPase} on the ATP level in the quasi-steady state. For comparison the line of the true steady state is included in the figure. ATP is more constant in the quasi-steady state. This means that the glycolytic system can tolerate ATP overconsumption for a short time period better than for longer periods. In contrast to the steady state conditions the rate of the ATP consuming processes can be increased considerably during the quasi-steady state period. The reason for the enhanced constancy of ATP is that an increased k_{ATPase} value leads to a lowered rate of the 2,3-P_2G mutase while the 2,3-P_2G phosphatase is not affected initially. Therefore during the quasi-steady state period more 2,3-P_2G is degraded than formed and this additional flux yields ATP at the pyruvate kinase step. This role of 2,3-P_2G to act as energy source is employed for blood preservation conditions [12].
The control strengths of the enzymes in the quasi-steady state are given in Table 1 (right column). The hexokinase-phosphofructokinase system has the most pronounced influence on the glycolytic flux in agreement with the previous models [1, 2].

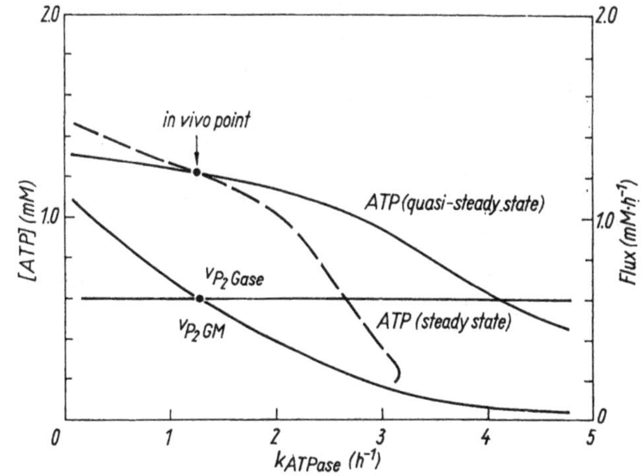

Fig. 5. ATP concentration and fluxes through the 2,3-P_2G bypass in the quasi-steady state as functions of the rate constant of the ATP-consuming processes (k_{ATPase}).

The continious lines are those for the quasi-steady state at unrelaxed 2,3-P_2G concentrations, the broken line corresponds to the true steady state (cf. Fig. 2). The quasi-steady state curves were computed with $[2,3\text{-}P_2G] = 4.49$ mM. At the in vivo point $v_{P2GM} = v_{P2Gase}$ and $[ATP]_{steady\ state} = [ATP_{quasi\text{-}steady\ state}$.

Discussion

The present paper concentrates on the consequences of the introduction into the model of new rate laws for the hexokinase-phosphofructokinase system and for the 2,3-P_2G phosphatase. The basic difference from the previous deductions is the mechanism of ATP regulation. Whereas an important role as an energy buffer has previously been ascribed to the 2,3-P_2G-bypass, the phosphofructokinase has now turned out to be of overwhelming importance. Whereas previously a non-ATP-producing shunt was shut off in case of ATP overconsumption, the ATP production is now increased. It is possible, however, that the 2,3-P_2G bypass is significant for the ATP regulation during *p*H changes [4]. The present model explains the inverse changes of ATP and the glycolytic flux in a limited range [5—7].

Most deductions from the previous model are confirmed by the present one. This refers to the time hierarchy, to the existence of quasi-steady states, to the role of the conservation sum of the oxidation equivalents and to the behaviour of the glycolytic system in vitro [2]. Furthermore, the present model is able to describe all data which could be fitted with the previous model with the exception of the data on pyruvate kinase deficiency. The present model predicts an increase of the 2,3-P_2G concentration which is too great while the decrease of the ATP concentration is too small.

The present model is in qualitative agreement with a recent simulation of the upper part of the erythrocyte glycolysis [17]. The authors obtained an ATP dependence of the glycolytic flux similar to that shown in Fig. 1. Quantitative differences appear to be due to the use of a smaller inhibition constant for G6P of the hexokinase.

References

[1] Rapoport, T. A., R. Heinrich, G. Jacobasch, and S. M. Rapoport: Eur. J. Biochem. **42**, 107 (1974)
[2] Rapoport, T. A., R. Heinrich, and S. M. Rapoport: Biochem. J. **154**, 449 (1976)
[3] Schröter, W.: Stoffwechsel und Membranpermeabilität von Erythrozyten und Thrombozyten. I. International Symposium, Wien 1968, p. 50
[4] Rapoport, I., H. Berger, R. Elsner, and S. M. Rapoport: Eur. J. Biochem. (in press)
[5] Murphy, J. R.: J. Lab. clin. Med. **61**, 567 (1963)
[6] Whittam, R., and M. E. Ager: Biochem. J. **97**, 214 (1965)
[7] Feig, S. A., S. B. Shohet, and D. E. Nathan: J. clin. Invest. **50**, 1731 (1971)
[8] Otto, M., G. Jacobasch, and S. Svetina: Acta biol. med. germ. **36**, 581 (1977)
[9] Gerber, G., H. Preissler, R. Heinrich, and S. M. Rapoport: Eur. J. Biochem. **45**, 39 (1974)
[10] Otto, M., R. Heinrich, B. Kühn, and G. Jacobasch: Eur. J. Biochem. **49**, 169 (1974)
[11] Sel'kov, E. E.: Eur. J. Biochem. **59**, 151 (1975)
[12] Rapoport, T. A., and R. Heinrich: Bio-Systems **7**, 120 (1975)
[13] Ginsburg, S., J. G. Smith, F. M. Ginsburg, J. Z. Reardon, and J. K. Aikawa: Blood **20**, 722 (1962)
[14] Syllm-Rapoport, I., D. Schmidt, and E. Hoffmann: Folia haemat. **78**, 161 (1962)
[15] Syllm-Rapoport, I., G. Jacobasch, S. Prehn, and S. M. Rapoport: Blood **33**, 617 (1969)
[16] Jacobasch, G.: Ph.D. Thesis, Humboldt-Universität zu Berlin, Math.-naturwiss. Fak. 1970
[17] Ataullakhanov, F. I., V. M. Vivitsky, A. M. Zhabotinky, B. N. Kholodenko, and L. I. Erlich: Bioficika (in press)
[18] Gerber, G., H. Berger, G.-R. Jänig, and S. M. Rafofort: Eur. J. Biochem. **38**, 563 (1973)

The author's addresses: Dr. T. A. Rapoport, Zentralinstitut für Molekularbiologie der Akademie der Wissenschaften der DDR, 1115 Berlin-Buch, Lindenberger Weg 70; Dr. M. Otto, and Dr. R. Heinrich, Institut für Physiologische und Biologische Chemie, 104 Berlin, DDR, Hessische Str. 3—4

Acta biol. med. germ., Band 36, Seite 469—473 (1977)

Zentralinstitut für Molekularbiologie der Akademie der Wissenschaften der DDR,
Bereich Methodik und Theorie, 1115 Berlin-Buch, DDR

Elementare Eigenschaften des Energiestoffwechsels der roten Blutzelle

M. GLENDE, TH. GEIER und J. G. REICH

Es erscheint uns wichtig, den Bewegungsablauf und die prinzipiell möglichen Dauerzustände im Energiestoffwechsel der roten Blutzelle zu verstehen. Als theoretisches Modell liegt dabei das stöchiometrische Schema des anaeroben Kohlehydratstoffwechsels zugrunde, wie es allgemein aus Übersichtsarbeiten und Stoffwechselkarten bekannt ist.

Die Bewegungen der Metabolitkonzentrationen werden durch einen Satz von gewöhnlichen Differentialgleichungen beschrieben. Die Koeffizienten dieser Differentialgleichungen sind die kinetischen Parameter der betrachteten Reaktionen. Ihre numerischen Werte haben wir als Größenordnung aus Literaturdaten ermittelt.

Um das Modell zu vereinfachen, haben wir für die einzelnen Reaktionen unkomplizierte kinetische Formen (meist bilineare Ansätze [1]) zugrunde gelegt und außerdem die Zeithierarchie der Metabolite und Reaktionen für die Vereinfachung berücksichtigt [2].

Als wichtigste Variable im Energiestoffwechsel wird das ATP betrachtet. Seine Umsatzzeit liegt im Erythrozyten bei ca. $^1/_2$ Std. Nach 1—2 Std. hat sich im Energiestoffwechsel eine angenähert stationäre Bewegung eingestellt. Andere Größen, wie 2,3-DPG, Summe der Adenin- und Pyridinnukleotide, Summe der intrazellulären Phosphate usw., die sich um eine oder mehrere Größenordnungen langsamer bewegen, bleiben während dieser Zeit nahezu unverändert. Sie wirken auf die Bewegung des ATP als Driftparameter und können im Modell zunächst konstant gesetzt werden.

Es hat sich als vorteilhaft erwiesen, die Zahl und die Lage der stationären Zustände im Energiestoffwechsel nicht nur für einzelne Parametersätze, sondern über ganze Parameterbereiche zu untersuchen. Die Ergebnisse werden in Parameterdiagrammen dargestellt. Wir verstehen darunter eine graphische Darstellung, in der die stationäre ATP-Konzentration (Ordinate) gegen einen variierten Parameter aufgetragen wird. Als Parameter kann auch eine langsame Stoffwechselvariable fungieren, z. B. 2,3-DPG, dessen Konzentration sich in der charakteristischen Zeit des Energiestoffwechsels nur geringfügig ändert.

Der Einfachheit halber werden wir hier nur den stationären Wert der ATP-Konzentration in Abhängigkeit von einem Belastungsparameter (Geschwindigkeitskonstante einer verallgemeinerten ATPase) untersuchen. Für den Energiestoffwechsel der roten Blutzelle haben wir folgende Parameterabhängigkeiten gefunden (s. Abb. 1):

— Es existiert ein stabiler stationärer Zustand. Dieser Zustand wird durch Änderung der Belastung ausschließlich monoton beeinflußt (Abb. 1 a).

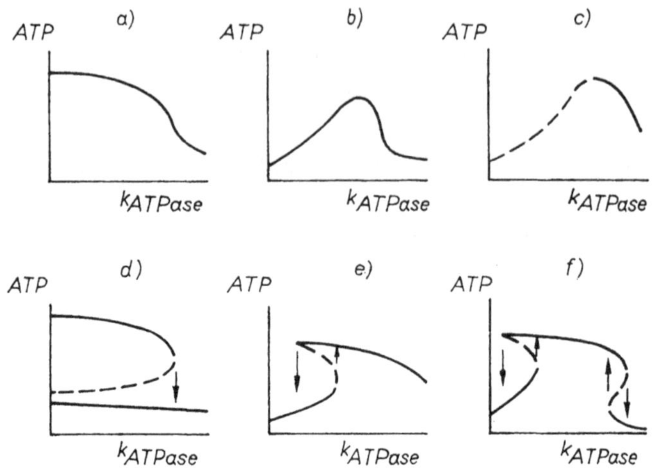

Abb. 1. Typen der stationären Belastungscharakteristik im Energiestoffwechsel der roten Blutzelle.

In den Belastungsdiagrammen ist die stationäre ATP-Konzentration (ATP), die sich nach Ablauf der Bewegung im Energiestoffwechsel (in Zeiten < 100 min) einstellt, gegen die Geschwindigkeitskonstante einer verallgemeinerten ATPase (Energiebedarf der Zelle) aufgetragen. Die einzelnen Belastungscharakteristiken sind typisch für bestimmte Reaktionen und Stoffwechselwege, die sämtlich in der roten Blutzelle aktiv sein können. Stationäre Zustände auf den durchgehend gezeichneten Kurven (———) sind stabil, die auf den gebrochenen Kurven (— — —) instabil. Um instabile stationäre Zustände (c) sowie auf den verschiedenen Hystereseformen (e, f) können schnelle bzw. langsame Relaxationsschwingungen enstehen

— Es existiert ebenfalls nur ein stabiler stationärer Zustand. Dieser Zustand ist jedoch von Parameteränderungen nicht monoton abhängig (Abb. 1 b).

In beiden Fällen ist der vom ATP angestrebte stationäre Zustand für einen gegebenen Parametersatz eindeutig. Er ist nicht von den Ausgangswerten der Metabolitkonzentrationen abhängig.

— Der einzig mögliche (von Null verschiedene) stationäre Zustand im Energiestoffwechsel wird in bestimmten Parameterbereichen instabil (Abb. 1 c). Befindet sich das Stoffwechselsystem in diesem Parameterbereich, dann geht es in einen bewegten stationären Zustand über. Wir haben in solchen Fällen autonome Schwingungen des ATP im Stundenbereich gefunden [3].

— In bestimmten Parameterbereichen existieren mehrere alternative stabile stationäre Zustände; dabei können verschiedene Hystereseformen auftreten (Abb. 1 d, e, f). Welchen der stabilen stationären Zustände das ATP anstrebt, ist in diesen Fällen von seinem Ausgangswert, also von der Vorgeschichte des Stoffwechselsystems abhängig.

Zwischen verschiedenen stabilen stationären Zuständen sind Übergänge möglich. Diese Übergänge können über eine Veränderung der Variablen oder der Parameter ausgelöst werden: Wird die ATP-Konzentration von außen ausgelenkt, so kann das Stoffwechselsystem in den Einzugsbereich des alternativen stabilen stationären Zustands geraten und kehrt dann nicht mehr in seinen Ausgangszustand zurück. In der roten Blutzelle treten sprunghafte Änderungen der ATP-Konzentration durch den Wechsel der Bindungseigenschaften des Hämoglobins im Oxygenierungszyklus ständig auf. Bei kritischen Parameterwerten ändert sich die Zahl und die Stabilität der stationären Zustände. Werden diese kritischen Parameterwerte in bestimmter Richtung überschritten, treten Schalteffekte auf (Pfeile in Abb. 1 d, e, f). Der bisherige stationäre Zustand hört auf zu existieren. Das Stoffwechselsystem geht in den alternativen stationären Zustand über. Diese Umschaltungen können irreversibel sein (Abb. 1 d); eine Rückführung des Stoffwechsels in den Zustand vor der Umschaltung ist in solchen Fällen durch den Parameter, der die Umschaltung ausgelöst hat, nicht mehr möglich. An den Hysteresen (Abb. 1 e, f) können reversible Umschaltungen in beide Richtungen erfolgen.

Die Parameterbewegung, die die Umschaltungen auslöst, ist meist von außen induziert. Einige Parameter werden jedoch vom Energiestoffwechsel selbst gesteuert. Die verschiedenen stationären ATP-Konzentrationen können dann eine langsame Parameterdrift in unterschiedliche Richtungen auslösen. An den Hysteresen können auf diese Weise durch Hin- und Rückschaltungen langsame Relaxationsschwingungen entstehen. In den Modellen des Erythrozytenstoffwechsels haben wir mehrfach solche Langzeitschwingungen gefunden. Ihre Periode ist stark parameterabhängig. In den von uns durchgerechneten Beispielen lag sie zwischen 10 und 500 Stunden [4].

Die theoretische Analyse des Energiestoffwechsels der roten Blutzellen ergibt also folgende Grundtatsachen:

1. Der erreichte stationäre Zustand im Energiestoffwechsel ist nicht notwendig eindeutig. Je nach den Ausgangsbedingungen sind mehrere stationäre Bewegungsformen möglich.

2. Zwischen den verschiedenen Zuständen finden Übergänge statt. Sie werden ausgelöst durch aufgezwungene Metabolitschwankungen, bei denen der aktuelle Einzugsbereich des Stoffwechselsystems sich ändert, oder durch die Überschreitung kritischer Parameterwerte, an denen Schalteffekte entstehen.

3. Um instabile stationäre Zustände sowie auf verschiedenen Hystereseformen können autonome Schwingungen entstehen. In diesen Fällen trifft das steady-state-Konzept nicht zu.

Welche Konsequenzen ergeben sich nun für die Theorie des Erythrozytenstoffwechsels? Diese Frage kann nicht allgemein, sondern nur für bestimmte Stoffwechselsituationen beantwortet werden. Abschließend soll dies für den Fall der Konservierungsbedingungen skizziert werden.

Die Gefahr bei der Blutkonservierung besteht bekanntlich im Zusammenbruch des Energiestoffwechsels. Die einfachste Vorstellung wäre, daß sich in der Konserve Parameterwerte ändern, die einen eindeutigen stationären Zustand verschieben. Bei Reinfusion würden die alten Parameterwerte und damit der Normalzustand der roten Blutzellen wiederhergestellt. Diese Vorstellung ist sinnvoll nur für den Fall einer eindeutigen Abhängigkeit des stationären Zustandes von den Milieufaktoren (vgl. Abb. 1 a, b). Bei der vorliegenden stöchiometrischen Struktur des Erythrozytenstoffwechsels ist der Hysteresefall (vgl. Abb. 1 e, f) viel wahrscheinlicher. Wir haben dann im einfachsten Fall zwei stationäre Ruhezustände, einen mit hohem und einen mit niedrigem ATP-Gehalt. Durch die Lagerung im nichtphysiologischen Milieu geht dann die rote Blutzelle in den schlechten Zustand über, und der Zellzusammenbruch wird über die bekannten Zwischenstufen wie Angriff auf die Adeninnukleotide, Membranschädigungen usw. ausgelöst. Neben den bekannten Parameteränderungen wie *p*H-Verschiebung usw. spielen u. E. Änderungen des Oxygenierungszustandes des Hämoglobins eine bedeutende Rolle. Über den Oxygenierungszustand in der Blutkonserve haben wir keine verläßlichen Daten gefunden. Wir wissen aber aus Modelluntersuchungen, daß sowohl eine zu hohe als auch eine zu niedrige Oxygenierungsrate bei vorliegender Hysterese die metabolische Katastrophe auslösen kann [5].

Ein unerwartete Erklärungsmöglichkeit für den Konservierungsschaden bieten die Kurzzeitschwingungen des Oxygenierungszustandes. Die Hauptfunktion der roten Blutzelle ist der Sauerstofftransport, und der erfordert ein ständiges Umschalten des Oxygenierungszustandes mit einer Periode von einigen Minuten, je nach Kreislaufzeit. Im gleichen Rhythmus wechselt auch die Bindungskapazität des Hämoglobins für ATP und DPG. Mit jedem Oxygenierungszyklus erfolgt also über die Metabolitbindung an Hämoglobin ein heftiger Stoß in die Konzentration wesentlicher Variablen des Energiestoffwechsels. Wir haben nun im Mo-

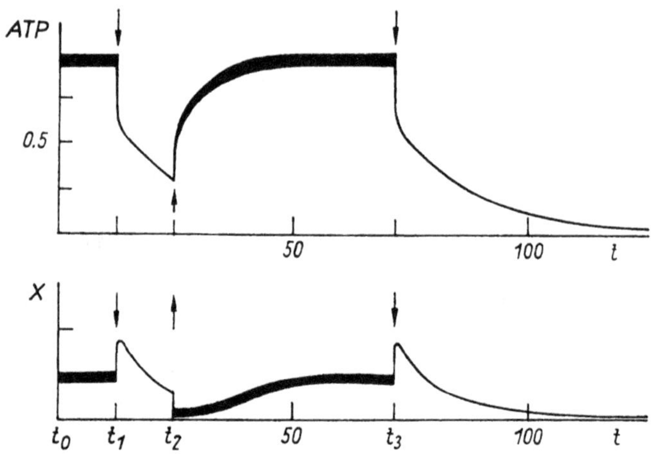

Abb. 2. Stabilisierung der ATP-Konzentration durch Schwingung der Pufferkapazität für intermediäre Metabolite.

In einem einfachen Modell des Energiestoffwechsels wurde die Pufferkapazität für Intermediat X (analog zur peroidischen Änderung der Bindungseigenschaften von Hämoglobin im Oxygenierungs-Desoxygenierungszyklus) einer schnellen exogen veranlaßten Schwingung ausgesetzt.

Die Abbildung zeigt, daß durch die Vibration der Pufferkapazität (t_0 bis t_1) ein stabiler Dauerzustand mit hohen ATP-Konzentrationen eingestellt wird, der sofort zusammenbricht, wenn die Parameterschwingungen ausgeschaltet werden („Zyklus Aus" ↓ bei t_1). Bei t_2 („Zyklus Ein" ↑) ist die mögliche Erholung durch Wiedereinschalten der Parameterschwingung gezeigt; erneutes Ausschalten („Zyklus Aus" ↓ bei t_3) löst wiederum den Zusammenbruch aus.

Ein ähnliches Verhalten läßt sich auch für den Fall zeigen, daß die Pufferkapazität für freies ATP (analog zur wechselnden ATP-Bindung an Hämoglobin im Oxygenierungszyklus) schnelle Schwingungen ausführt [6].

dell gefunden, daß diese aufgezwungenen Schwingungen einen mittleren Dauerzustand des Energiestoffwechsels stabilisieren können, der sofort zusammenbricht, wenn diese äußere Bewegung aufhört, wobei der Energiestoffwechsel sich sich aber auch erholen kann, wenn diese Bewegung rechtzeitig wieder einsetzt (vgl. Abb. 2) [6]. Analoga dieser Stabilisierung durch Vibration findet man auch in der Physik, z. B. bei geeignet bewegten Pendeln [7].
Trifft diese Hypothese der Vibrationsstabilisierung zu, dann ist es die Funktion der roten Blutzelle, die die Intaktheit des Energiestoffwechsels garantiert. Alle Bedingungen, die den Sauerstoffaustausch unterbrechen (in vitro in der Konserve, in vivo bei Stase oder bei Schädigung der Zelle) lösen letztendlich den Zusammenbruch des Energiestoffwechsels aus.

Literatur

[1] Reich, J. G., E. E. Sel'kov, Th. Geier u. V. Dronova: Studia biophys. **54**, 57 (1976)
[2] Reich, J. G., u. E. E. Sel'kov: BioSystems **7**, 39 (1975)
[3] Geier, Th., M. Glende u. J. G. Reich: In Vorbereitung
[4] Glende, M., Th. Geier u. J. G. Reich: In Vorbereitung
[5] Geier, Th., M. Glenede u. J. G. Reich: In Vorbereitung
[6] Glende, M., Th. Geier, W. Meiske u. J. G. Reich: In Vorbereitung
[7] Bogoljubow, N. N., u. J. A. Mitropolski: Asymptotische Methoden in der Theorie nichtlinearer Schwingungen. Nauka, Moskau 1974 (in Russisch)

Acta biol. med. germ., Band 36, Seite 475 (1977)

Physiologisch-Chemisches Institut der Friedrich-Schiller Universität, 69 Jena, DDR

Regulation of red cell glycolysis by calcium ions

H. FRUNDER

There is no measurable $^{45}Ca^{2+}$ uptake by human red cells during 2 hrs incubation in saline with 30 μM $^{45}Ca^{2+}$. The ionophore A 23187 causes a $^{45}Ca^{2+}$ influx under conditions of no net movements of Mg^{2+} from the intra- to the extracellular space and vice versa. Radioactive equilibrium between extra- and intracellular $^{45}Ca^{2+}$ is reached within 1 hr. Immediately after ionophore addition glucose consumption and lactate formation are doubled and constant for 2 hrs. ATP decreases continuously, K^+ leaks rapidly in exchange with Na^+ and cells shrink. All effects of the ionophore are completely counterbalanced by removal of Ca^{2+} with EGTA, or here, with the exception of an increased glycolytic flux, by high extracellular Mg^{2+} concentrations with net movement of Mg^{2+} into the intracellular space.
Inhibition of the Ca^{2+}-Mg^{2+}-dependent ATPase by ruthenium red or La^{3+} markedly reduces glycolysis and speeds up the $^{45}Ca^{2+}$ influx and the potassium efflux. However, the ATP level decreases now faster and much more than in the absence of inhibitors. Apparently, both inhibitors cause a redistribution of Ca^{2+}, so that other ATP-consuming processes than the Ca^{2+}-Mg^{2+}-ATPase are activated and glycolysis inhibited. Next, „thick'' hemolyzates were examined, prepared by different means from red cell sediments with hematocrits $> 95\%$ and fortified with NAD.
Glycolysis is doubled in these hemolyzates compared with intact cells without ionophore. Removal of Ca^{2+} by EGTA-addition reduces the lactate formation markedly, concomitantly with a remarkable increase of the levels of triose-phosphates and fructosebiphosphate. Other metabolite levels were without greater deviations. Addition of Ca^{2+} in excess of EGTA abolished these changes.
Then, $^{45}Ca^{2+}$ preincubated cells were hemolyzed and these "thick" hemolyzates separated in fractions with high and low membrane contents by high speed centrifugation. In these fractions acetylcholine esterase and ^{45}Ca were measured in parallel with the rate of glycolysis and metabolite changes during incubation at 37 °C.
Under certain conditions lactate formatio decreases in parallel with the removal of acetylcholine esterase activity. ATP, which vanishes during longer incubations of complete hemolysates, remains constant or increases in all hemolysate fractions with low acetylcholine esterase. Other conditions of hemolysate preparation and fractionation, however, result in fractions without parallelism between lactate formation and acetylcholine esterase.
Under all conditions there is a fairly good parallelism between lactate formation and ^{45}Ca content of the fractions. This Ca-containing hemolysate fraction is characterized and its action on hemolysate glycolysis discussed.

(Original papers see U. TILL, H. PETERMANN, I. WENZ and H. FRUNDER (this symposium pp. 597) and D. BROX, B. PETERMANN and H. FRUNDER (this symposium pp. 611)

Acta biol. med. germ., Band 36, Seite 477—480 (1977)

Academisch Ziekenhuis, Sectie Medische Enzymologie, Utrecht, Nederland

Kinetics of human erythrocyte hexokinase: influence of temperature, ATP^{4-} and Mg^{2+}

G. RIJKSEN and G. E. J. STAAL

Purified human erythrocyte hexokinase (EC 2.7.1.1) is inhibited competitively with respect to $MgATP^{2-}$ by uncomplexed Mg^{2+} ($K_i = 16—18$ mM) and ATP^{4+} ($K_i = 1.60$ mM) at 37 °C. No real activitation at low concentrations of Mg^{2+} can be detected and no allosteric behaviour was observed under the conditions tested. The apparent activation at small excess of Mg^{2+} can be fully explained by the raised concentration of substrate and simultaneously lowered concentration of ATP^{4-} when introducing a small excess of free magnesium. This can be seen in Fig. 1 in which is plotted the initial velocity of the hexokinase reaction versus excess of magnesium at fixed $[ATP]_{total}$ of 0.65 mM. The closed dots represent the experimental values, the open dots are calculated according to the following rate equation describing the initial velocity of the hexokinase reaction in the presence of the competitive inhibitors Mg^{2+} and ATP^{4-}.

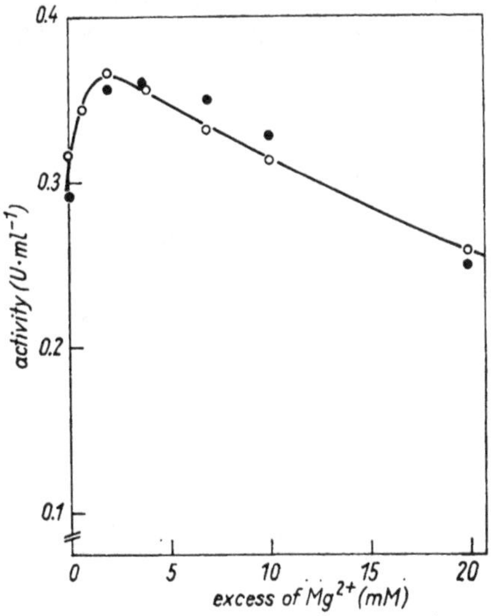

Fig. 1. Initial velocity of the hexokinase reaction at 37 °C versus excess of magnesium

$$\frac{V}{\frac{\text{MgATP}^{2-}}{\text{gATP}^{2-}]}\left(1+\frac{[\text{Mg}^{2+}]}{K_i\,\text{Mg}^{2+}}+\frac{[\text{ATP}^{4-}]}{K_i\,\text{ATP}^{4-}}\right)+\frac{K_m\,\text{glucose}}{[\text{glucose}]}+\frac{K_m\,\text{glucose}\;K_m\text{MgATP}^{2-}}{[\text{glucose}]\;[\text{MgATP}^{2-}]}\left(1+\frac{[\text{Mg}^{2+}]}{K_i\,\text{Mg}^{2+}}+\frac{[\text{ATP}^{4-}]}{K_i\,\text{ATP}^{4-}}\right)}$$

This rate equation is valid assuming a rapid equilibrium random mechanism in which the binding of either substrate does not influence the affinity of the complex for the other substrate. The values of the kinetic constants in this rate equation obtained from kinetic studies are summarized in Table 1. A stability constant of 20000 mole^{-1} was used to calculate the levels of free and complex ions. It can be seen from Fig. 1 that this rate equation gives a rather good fit with the experimental results. These results on the inhibition by free magnesium and ATP^{4-} are consistent with the reports on hexokinase from other tissues [1, 2] and confirm the conclusion of PURICH and FROMM [3] about the non-allosteric behaviour of the

Table 1
Kinetic constants, describing the hexokinase reaction

Parameter	Value	Parameter	Value
V	0.70 U · ml^{-1}	K_i Mg^{2+}	16–18 mM
K_m $MgATP^{2-}$	0.50 mM	K_i ATP^{4-}	1.60 mM
K_m glucose	0.064 mM		

enzyme. However they are in contradiction to the results of GERBER et al. [4] on human erythrocyte hexokinase. These authors reported the activation by uncomplexed Mg^{2+} up to 4 mM. At higher concentrations an inhibition was observed. Two binding sites for magnesium were suggested, one for its activating and one for its inhibitory effect. Furthermore no inhibition by uncomplexed ATP was detected.

The temperature dependence of the enzyme was studied in relation to the presence of Mg^{2+} and ATP^{4-}. Although GERBER et al. [4] found a straight-line relationship between log V and 1/temperature (°K) in the range 4—37 °C, we observed a break in the Arrhenius plot dependent on the presence of Mg^{2+} or ATP^{4-} (Fig. 2).

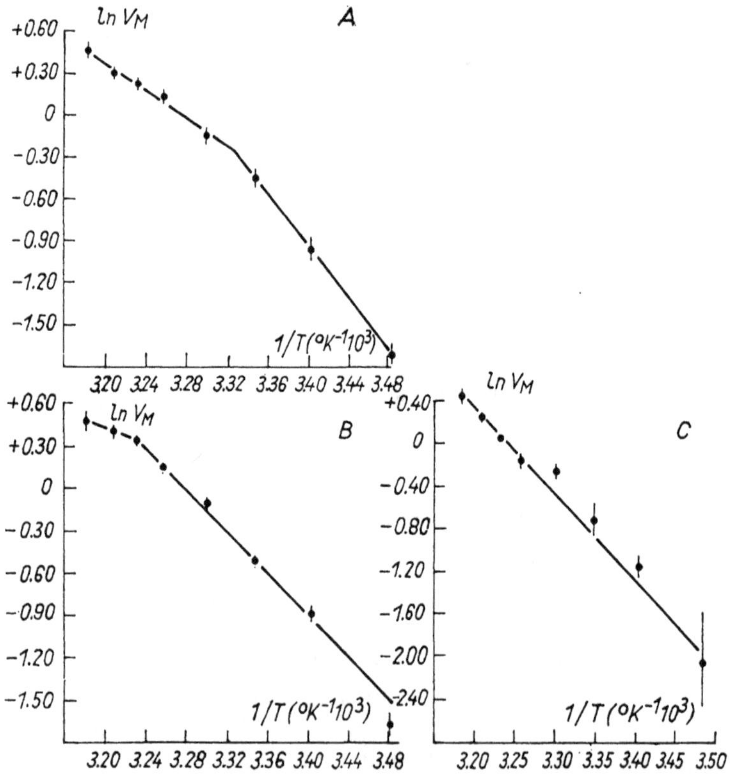

Fig. 2. Arrhenius plots of hexokinase activity versus the temperature. These plots are drawn by computer using a program derived for the statistical treatment of Arrhenius plots using 1/variance of the maximal velocity as weighting factors (brackets)

At equal amounts of Mg^{2+} and ATP^{4-} a breakpoint is observed at 27.5 °C (Fig. 2A), which is shifted to 36 °C in the presence of an excess of 5 mM Mg^{2+} (Fig. 2B). No break-point is observed in the presence of an excess of ATP^{4-} (Fig. 2C).

The parameters describing the Arrhenius plots are summarized in Table 2.

Table 2

Parameters describing the Arrhenius plots

Ln V_{max} estimated in the presence of	T_{break} °C	$T < T_{br}$ E_α(kcal mole^{-1})	$T > T_{br}$ E_α(kcal mole^{-1})
$[Mg^{2+}]_{total} = [ATP^{4-}]_{total}$	27.5 ± 0.7	18.7 ± 0.1	9.7 ± 0.8
excess of $[Mg^{2+}] = 5.0$ mM	36.0 ± 0.4	14.7 ± 0.6	5.9 ± 0.3
excess of $[ATP^{4-}] = 2.5$ mM	—	17.9 ± 0.8	

The possibility that the breaks in the Arrhenius plots reflect a conformational transition between two different forms of the enzyme or a structural alteration in the protein involving the catalytic site was investigated by studying the dependence of the K_m on temperature. A constant value for the affinity constant for $MgATP^{2-}$ was found up to about 36 °C (Fig. 3). At higher temperatures a gradual increase of K_m was found. This phenomenon was independent of the presence of Mg^{2+} or ATP^{4-}. So we may conclude that this phenomenon is not likely to be related to the effects shown in the Arrhenius plots, but rather reflects either a gradual conformational change at temperatures over about 36 °C to a different form of the enzyme or a denaturation effect at the catalytic site which does not express itself in the maximal velocity.

As our purified human erythrocyte hexokinase preparation contains a certain amount of lipid [5], the possibility that the breaks in the Arrhenius plots represent a phase transition of these lipids must be considered.

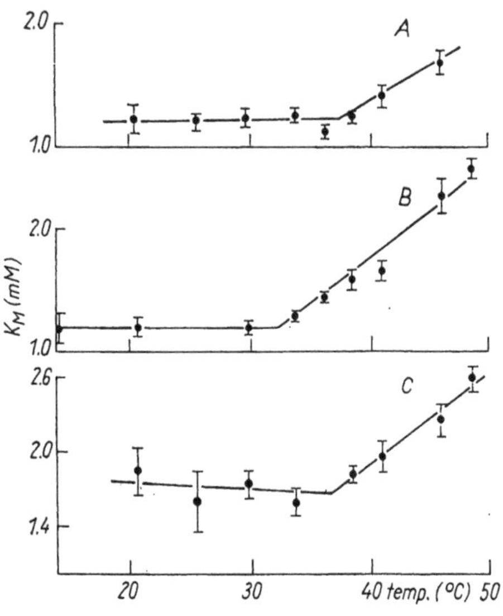

Fig. 3. Plots of K_m for $MgATP^{2-}$ versus temperature in the presence of equal amounts of Mg^{2+} and ATP^{4-} (A), excess of 5 mM of Mg^{2+} (B) and excess of 2.5 mM of ATP^{4-} (C). The values for K_m and the standard error of K_m (brackets) are calculated from Lineweaver-Burk plots using the method of the least squares with weighting factors proportional to 1/(velocity)^4.

References

[1] NING, J., D. L. PURICH, and H. J. FROMM: J. biol. Chem. **244**, 3840 (1969)
[2] PURICH, D. L., and H. J. FROMM: J. Biol. Chem. **246**, 3456 (1971)
[3] PURICH, D. L., and H. J. FROMM: Biochem. J. **130**, 63 (1972)
[4] GERBER, H., H. PREISSLER, R. HEINRICH, and S. M. RAPOPORT: Eur. J. Biochem. **45**, 39 (1974)
[5] RIJKSEN, G., and G. E. J. STAAL: Biochem. biophys. Acta (in press)

Acta biol. med. germ., Band 36, Seite 481–490 (1977)

Department of Biological Chemistry, University of Sassari, Sassari (Italy)
Department of Biological Chemistry, University of Torino, Torino (Italy)
Institute of Physiological Chemistry, University of Jena, 69 Jena (GDR)

The role of red cell membrane in the regulation of glycolysis and the 2,3-bisphosphoglycerate-cycle

P. Arese, A. Bosia, G. P. Pescarmona, and U. Till

Summary

Pyruvate and *K*-ferricyanide stimulation of net ATP and 2,3-bisphosphoglycerate synthesis is very probably due to enhancement of glyceraldehyde 3-phosphate dehydrogenase activity. Significant peculiarities in the *K*-ferricyanide effect and its depression by non-penetrating-SH inhibitors at low concentrations were noted and suggested that membrane-bound enzymes play a substantial part in the synthesis of ATP and 2,3-bisphosphoglycerate. Experiments with isolated ghosts showed their ATP- and 2,3-bis-phosphogylcerate-building capacity. Pulse-labeling with ^{32}P-P_i and determination of specific radioactivities in intracellular inorganic phosphate and ATP-γ-P demonstrated that the ferricyanide-stimulated compartment utilizes only intracellular inorganic phosphate for ATP (and 2,3-bisphosphoglycerate) synthesis, and does so only when extracellular inorganic phosphate is present.

Since most ATP-consuming processes are localized in the membrane, this may clearly be seen as the site of an ATP-synthesizing ability. 2,3-bisphosphoglycerate (DPG) is a typical cytoplasm-oriented molecule; however, membrane involvement in DPG-metabolism would help to explain some discrepancies between the in vitro and in vivo behaviour of the DPG-cycle enzymes [1]. To assess the role of the membrane in ATP and DPG metabolism, intact human red cells, isolated ghosts and ghost fragments were utilized in the present study.

Intact red cells confronted with a membrane-confined agent acting on redox couples — *K*-ferricyanide (FCN) — were preincubated at slightly alkaline *p*H to increase the NADH/NAD-ratio and unbalance phosphofructokinase, which is alkali-activated, and glyceraldehyde 3-phosphate dehydrogenase (GAPDH), which is inhibited by the increased NADH level [2]. The typical metabolite pattern is characterized by high and increasing fructose, 1,6-bisphosphate (FDP), bishydroxyacetone phosphate (DAP), glyceraldehyde 3-phosphate (GAP) levels and by low 1,3-bisphosphoglycerate (1,3-DPG), 3- and 2-phosphoglycerate (3-PG, 2-PG), phosphoenolpyruvate (PEP) and pyruvate levels. Addition of easily penetrating substrates (such as pyruvate or oxaloacetate), which rapidly oxidize NADH by using endogenous lactate- or malate dehydrogenase, eliminates the GAPDH-inhibition and causes a cascade-like activation of 3-phosphoglycerate kinase, DPG-mutase and pyruvate kinase: FDP, DAP, and GAP fall: 1,3-DPG, DPG, 3-PG, 2-PG, PEP increase: a net synthesis of ATP is observed [2]. FCN is a nonpenetrating strong electron acceptor [3, 4]; there is no reason why its effect should be different from that of pyruvate or oxaloacetate. FCN should drain electrons out through the membrane, oxidize NADH and activate GAPDH.

Addition of 2 mM FCN to alkali-preincubated red cells, in fact, brought about metabolite changes quite similar to those noted after pyruvate or oxaloacetate: however, very significant distinctions were noted, since addition of FCN only:

1) was accompanied by proton outflow and medium acidification,

2) caused no oxidation of NADH,

3) had no effect in red cells preincubated without glucose,

4) had little effect on concentrated hemolysates obtained by sonication. These peculiarities can be accounted for by the following tentative hypothesis:

a) ATP and DPG synthesis is at least partly conducted on the red cell membrane

b) The red cell ATP and DPG levels are regulated by nonpenetrating redox reagents which interact with membrane-localized enzymes, primarily with GAPDH [5].

If this hypothesis is correct, it should be possible to abolish the FCN effect by treating the red cells with non-penetrating-SH reagents, since these may be expected to inhibit the membrane GAPDH, which is known to be most sensitive to low concentrations of SH-blockers [6, 7]. The use of slowly or non-penetrating SH-reagents, such as N-ethylmaleimide, p-chloromercuribenzoate, p-chloromercuribenzenesulfonate abolishes the FCN effect either completely or drastically in nearly all cases, while the pyruvate effect is not inhibited (Fig. 1).

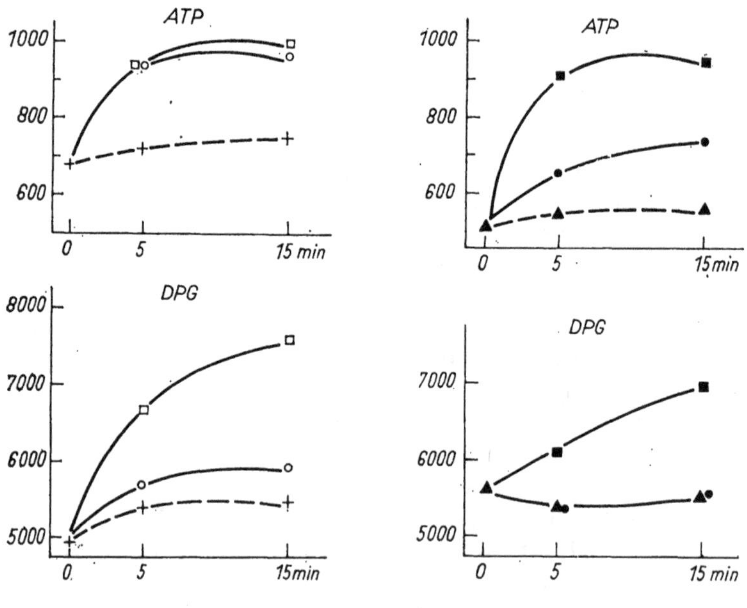

Fig. 1. Effect of pyruvate (Pyr) and ferricyanide (FCN) on ATP and DPG levels in red cells preincubated with p-chloromercuribenzenesulfonate (PCMBS).

Red cells were preincubated with PCMBS (final concentration 1 mM) for 3 min and washed once at 4 °C. Pyr and FCN (final concentration 2 mM) were added immediately after withdrawal of time-0 samples. T: 38 °C; pH 8.0; hematocrit: 50%. Metabolite levels in nmoles/ml packed red cells. For further experimental details, see [5].

+ — — + controls; □——□ pyruvate; ○——○ ferricyanide; ▲ — — ▲ controls + PCMBS; ■——■ pyruvate + PCMBS; ●——● ferricyanide + PCMBS

The activity of GAPDH depends on the presence of inorganic phosphate (P_i). Red cells preincubated at alkaline *p*H and washed free of intra- and extracellular P_i show no pyruvate or FCN effect; however, the metabolite levels immediately move upon addition of extracellular P_i. As shown by Fig. 2, preincubation with p-chloromercuribenzenesulfonate only removes the FCN effect.

Clearer evidence that the membrane is directly involved in ATP and DPG synthesis is given by experiments performed with isolated white ghosts. Table 1 shows that intact as well as sonicated ghosts actively transform FDP to DPG, ATP and 3-PG. The build-up rates depend on the *p*H and ionic composition of the incubation fluid. At nearly intracellular conditions, 1.6 μmoles DPG, 0.5 μmoles ATP, and 2.5 μmoles 3-PG were produced by 10^{10} ghosts per hour at 37°C. The amount of DPG produced by the isolated ghosts equals or even exceeds the total DPG-building capacity of the intact human red cell incubated under comparable conditions, assuming a lactate production of 3.3 μmoles/10^{10} red cells/hr and 35–50% of the glycolytic flux going through the DPG cycle [8]. The ghost metabolism is still subject to control; in the absence of ADP, no ATP and little DPG are formed, due to block of the 3-phosphoglycerate kinase reac-

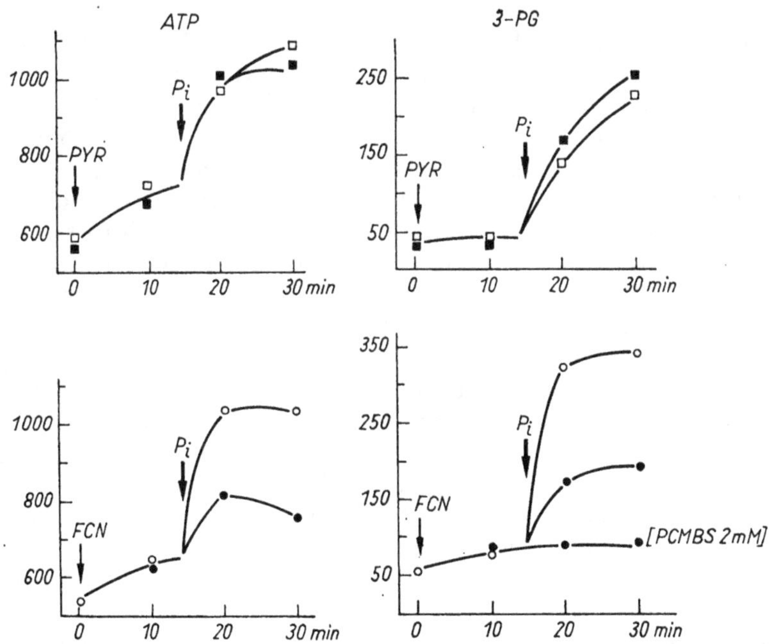

Fig. 2. Effect of sequential addition of pyruvate (Pyr) or ferricyanide (FCN) and inorganic phosphate (P_i) on ATP and 3-PG levels in red cells preincubated with p-chloromercuribenzenesulfonate (PCMBS).

Red cells were preincubated with PCMBS (final concentration 1 mM) for 20 min and washed once at +4 °C. Pyr or FCN (final concentration 2 mM) and P_i (final concentration 20 mM) were added immediately and 15 min after withdrawal of time-0 sample. T: 38 °C; *p*H: 8.0; hematocrit: 50%. Metabolite levels in nmoles/ml packed red cells. For further experimental details, see [5].

□ pyruvate; ○ ferricyanide; ■ pyruvate + PCMBS; ● ferricyanide + PCMBS

Table 1
Production of DPG, ATP and 3-PG by sonicated white ghosts under different incubation conditions

Incubation[1] medium	pH	DPG	ATP	3-PG
I	7.5	1.6	0.5	2.5
II	7.5	2.2	1.4	2.8
III	7.5	3.5	6.7	6.5
III	8.5	4.8	7.4	8.2

[1] Composition of the media (mmoles/l); I: Na^+ 25, K^+ 142, Mg^{++} 2.5, inorg. phosphate 0.8, HCO_3^- 93, Cl^- 84, NAD^+ 0.5, ADP 0.25, AMP 0.25, FDP 2, 3-PG 0.1, GSH 3, TRA-HCl 50, pH 7.5.
II: Na^+ 100, K^+ 20, Mg^{++} 2.5, inorg. phosphate 10, SO_4^{2-} 10, Cl^- 90, NAD^+ 0.5, ADP 0.25, AMP 0.25, FDP 2, 3-PG 0.1, GSH 3, TRA-HCl 50, pH 7.5.
III: Na^+ 100, K^+ 20, Mg^{++} 2.5, inorg. phosphate 10, SO_4^{2-} 10, Cl^- 90, NAD^+ 2, ADP 1, FDP 2, TRA-HCl 50, pH 7.5 or 8.5. Mean production in µmoles/10^{10} ghosts/hr at 37 °C.

tion, and to the limited availability of 3-PG, which is the co-substrate for the DPG-mutase reaction. DPG synthesis is particularly sensitive to very small changes in 3-PG within the physiological range; a drop of 20 nmoles/ml is followed by a 50% diminution in DPG synthesis [8]. The major possibility of artifacts derives from the fact that the enzymes stick to the ghost in an unspecific manner. There are reports showing that ATP or NAD in millimolar amounts partially dissociates GAPDH from the membrane [9, 10]. To rule out the possibility that DPG and ATP formation was due to the presence of GAPDH as an artifact on the membrane, 1.5 mM ATP was added to leaky ghosts and washed out after 13 hrs incubation at +4 °C. This treatment caused a 70—80% drop in ATP and DPG synthesis. However, if the ghosts were filled with equimolar amounts of ATP and stripped hemoglobin, the drop was much less evident. This shows that addition of only one effector without hemoglobin is in a sense artifactual as well [8]. Moreover, in recent reports the non-artifactual association of the key glycolytic enzymes to red cell membranes has been assessed [11—13]. Tillmann et al. [14] show that a number of glycolytic enzymes can be extrated by mechanical disruption of white ghosts, after two-fold extraction with cold toluene. These "cryptic" activities would be amply able to cope with the steady-state glycolytic flux of the whole red cell at physiological pH. As to the membrane association of GAPDH, Kant and Steck [12] and McDaniel et al. [13] have shown that red cell and muscle, but not yeast, GAPDH specifically bind to high affinity sites of the inner surface of the membrane. The enzyme is associated in 1:1 ratio with the band 3 polypeptide [15]; this is the predominant polypeptide of red cell membrane and is involved in anion translocation and other transport processes [16—18].

If we assume that the results of our ghost experiments are not artifactual and accept them as confirmations of the redox stimulation experiments, we can turn to the whole cell model for a better differentiation of the "membrane" and the cytoplasmic compartments. One interesting question is whether intracellular or extracellular P_i is preferentially used in the course of redox stimulation. As shown in Fig. 3, red cells filled with different amounts of P_i were stripped free of extracellular P_i by passage through Sephadex G75. At 20 mM intracellular P_i, net synthesis of ATP and DPG occurs after pyruvate stimulation, but not after FCN stimulation. The same was observed at lower intracellular P_i (about 2 mM) (Fig. 4). However, when about 60 mM extracellular P_i was added, no further stimulation with pyruvate but a net synthesis with FCN was observed. The most obvious interpretation of these results is that intracellular P_i is not accessible to the membrane system, or that a certain amount of extracellular P_i is required for the membrane system to work. Follow-up of ^{32}P-P_i incorporation into phosphorylated compounds is a suitable way of assessing this problem. Fig. 5 shows the experimental scheme employed to evaluate the source and relative amounts of P_i utilized by GAPDH after FCN or pyruvate stimulation. Erythrocytes loaded with ^{32}P-P_i behave as an entry-limited system, so that the radioactivity exchange across the membrane is small compared with the turnover among phosphate esters. A short time after the addition of the isotope there is a difference of many orders of magnitude in the specific radioactivity of intra- und extracellular P_i. GAPDH transfers the ^{32}P-P_i to the 1-position of 1,3-DPG, which is present in amounts too low for assay. The label will be then transferred by the 3-phosphoglycerate kinase to the γ-position of ATP, which is suitable for analysis. Specific intra-

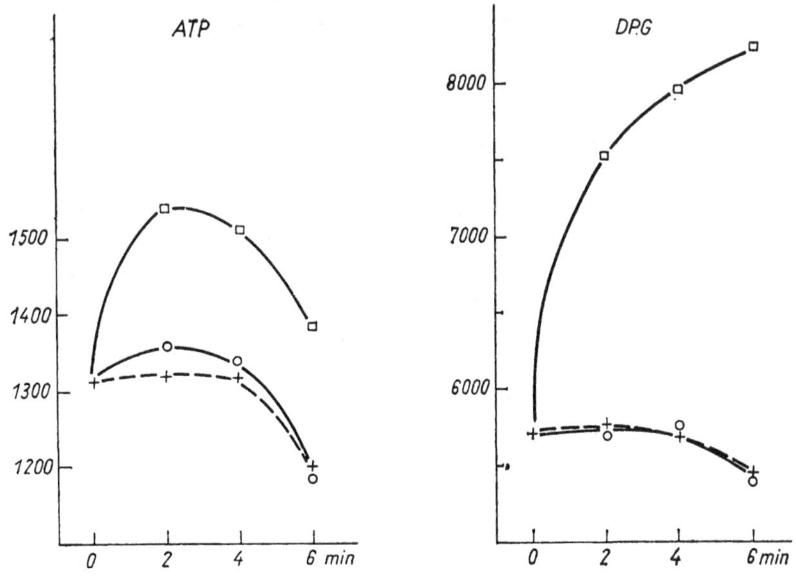

Fig. 3. Effect of pyruvate (Pyr) or ferricyanide (FCN) on ATP and DPG levels in P_i-enriched red cells at very low extracellular P_i (extracellular $P_i \approx 20\ \mu M$; intracellular $P_i = 20$ mM).

Pyr and FCN (final concentration 2 mM) were added immediately after withdrawal of time-0 samples. T: 38 °C; pH: 8.0; hematocrit: 50%. Metabolite levels in nmoles/ml packed red ceels.

+ — — + controls; □——□ pyruvate; ○——○ ferricyanide.

cellular P_i and ATP-γ-P radioactivities were therefore measured and related to that of the precursor. The experiment was performed in control red cells, and after stimulation of GAPDH by pyruvate and FCN. If cytoplasmic GAPDH is stimulated, it may be assumed that the intracellular P_i will be used preferentially. In that case, intracellular labelled ^{32}P–P_i should be the preferential pre-

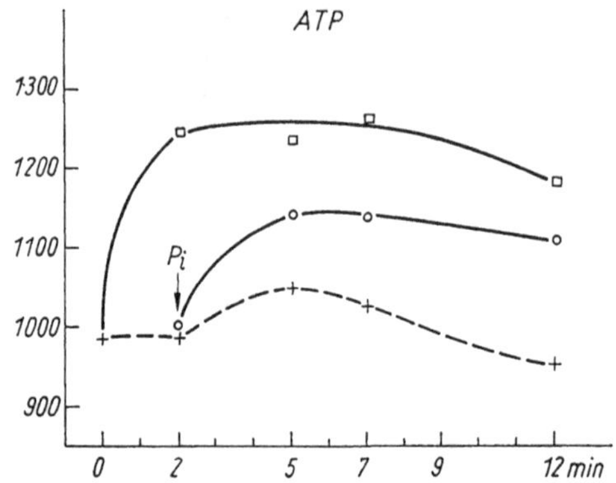

Fig. 4. Effect of sequential addition of pyruvate (Pyr) or ferricyanide (FCN) and inorganic phosphate (P_i) on ATP levels at very low extracellular P_i (extracellular $P_i \approx 20$ µM; intracellular $P_i = 2$ mM). Pyr or FCN (final concentration 2 mM) and P_i (final concentration 60 mM) were added immediately and 2 min after withdrawal of time-0 sample. T: 38 °C; pH: 8.0; hematocrit: 50%. Metabolite levels in nmoles/ml packed red cells.
+ — — + controls; □——□ pyruvate; ○——○ ferricyanide

cursor for the labelling of ATP-γ-P, whose specific radioactivity should thus not exceed that of intracellular P_i. On the other hand, if the GAPDH of the membrane compartment is stimulated, preferential utilization of extracellular P_i is to be expected. Extracellular labelled P_i should be the preferential precursor for ATP-γ-P, and its specific radioactivity should be much higher than that of intracellular P_i. The tracer flow pattern shown in Fig. 5 is a simplification, since it does not take into account label dilution due to the DPG-bypass. This simplification is justified by our results, which show that after a short-time labelling, ATP-γ-P was much more highly labelled than DPG. Due to the much higher labelling of extracellular compared with intracellular phosphate, exact assessment of intracellular specific radioactivity is dependent on rapid and complete separation of the cells from highly labelled extracellular precursor.

Fig. 6 shows the experimental arrangement. The red cell sediment and the incubation fluid were mixed in a small chamber (volume 0.15 ml) and incubated for 80 s at 37 °C. The incubation was stopped by passing the mixture through a cooled Sephadex G75 column. The cells were separated from the radioactive precursor and immediately quenched by perchloric acid. Elution was performed with a cold phosphate-free medium. A typical separation is shown in Fig. 6 (inset). The red cells and the extracellular radioactive phosphate are clearly separated, however, red cells were only sampled up to the elution peak. Assessment of the specific radioactivity of intracellular P_i and ATP-γ-P used Till et al.'s combination of chromatographic and precipitation methods [19].

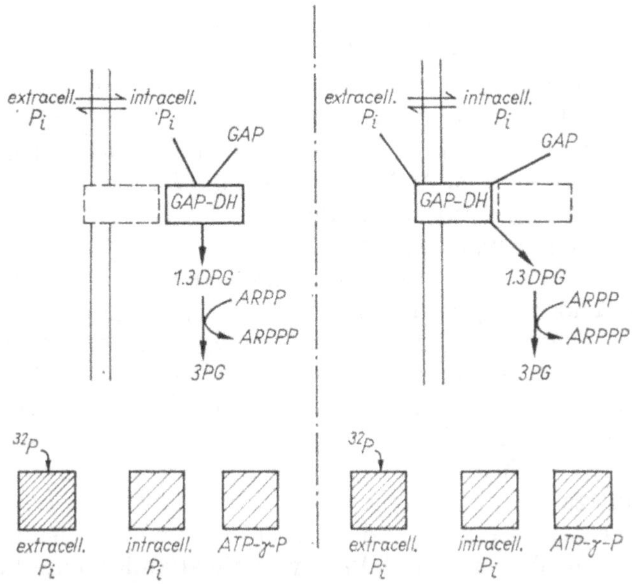

Fig. 5. Scheme for the follow-up of ^{32}P-P_i incorporation in ATP-γ-P after GAPDH stimulation by pyruvate of ferricyanide. The intensity of shadowing is an index of the specific radioactivity. For further explanations, see text

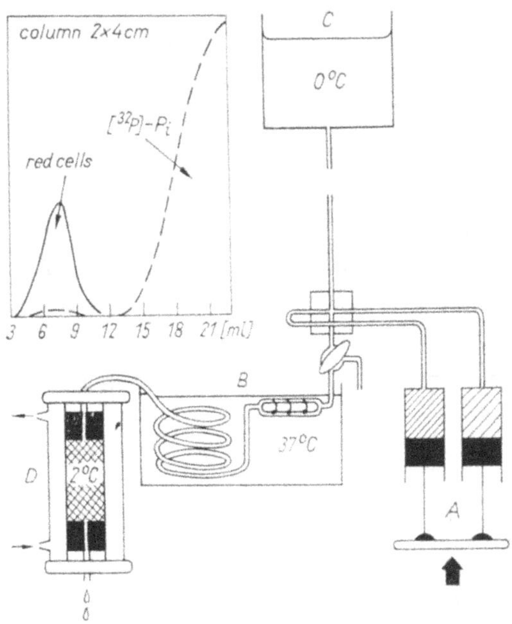

Fig. 6. Experimental set-up for short-term labelling of intracellular P_i and ATP-γ-P by highly labelled ^{32}P-P_i.
A) Continuously driven syringes containing packed red cells (left) and ^{32}P-P_i-buffer (right); B) mixing chamber; C) elution buffer without P_i; D) cooled G75 Sephadex column for separation of red cells from highly labelled extracellular precursor (see inset and text).

Fig. 7 shows the results of a typical experiment. Stimulation of red cell GAPDH was performed at very low extracellular P_i (approximately 40 μM) (left) and at very high extracellular P_i (approximately 60 mM) (right). In both cases, the sum of the intracellular ATP and P_i-pools was identical. The lower part of the figure represents the movements of ATP (broken line) and P_i (continuous line) after pyruvate or FCN-stimulation. At low extracellular P_i, pyruvate elicited the expected ATP increase and a stoichiometric P_i decrease: after FCN, ATP was unchanged. At high extracellular P_i, both pyruvate and FCN elicited net ATP increases. In the unstimulated red cells at low extracellular P_i, the specific radioactivity of P_i and ATP-γ-P are almost identical; this points at fast exchange between them by means of GAPDH-3-phosphoglycerate kinase. At high extracellular P_i, a net influx of P_i occurs, while the exchange at the GAPDH-3-phosphoglycerate kinase is expected to remain constant. This accounts for the higher specific radioactivity of P_i with respect to ATP-γ-P. After stimulation with pyruvate and FCN at high extracellular P_i, both specific radioactivities are very similar and reach higher levels. This points to a stimulation of the radioactivity exchange across the membrane. Such a stimulation is observed at low extracellular P_i also. We note that after pyruvate stimulation, ATP-γ-P is more highly labelled than P_i; this means that GAPDH possibly turns over a small portion of extracellular P_i.

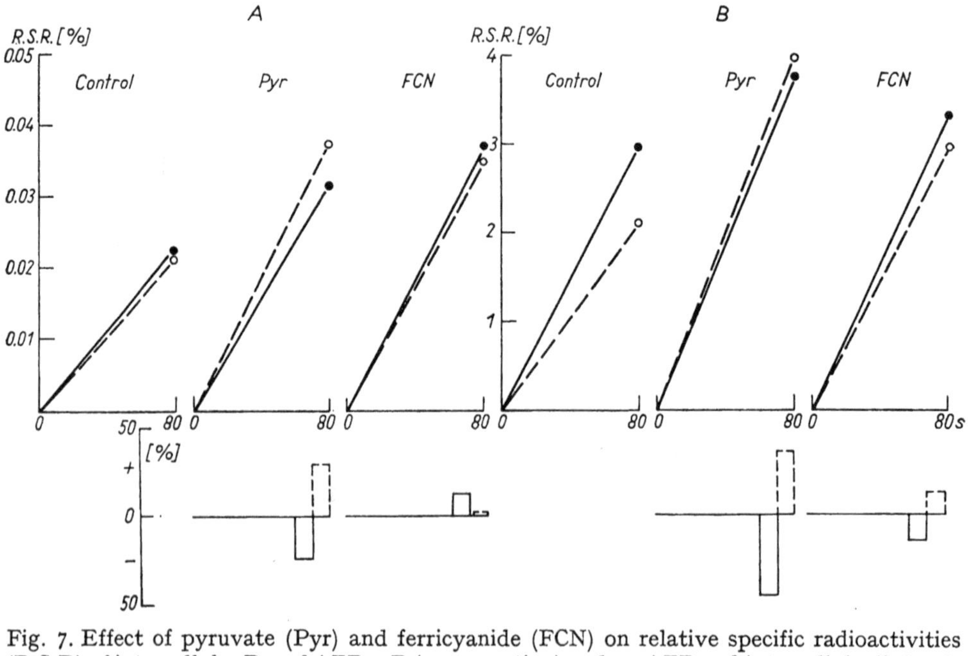

Fig. 7. Effect of pyruvate (Pyr) and ferricyanide (FCN) on relative specific radioactivities (R.S.R) of intracellular P_i and ATP-γ-P (uppersection) and on ATP and intracellular P_i levels (lower section).

A) extracellular $P_i \approx 40$ μM; intracellular P_i (at time 0) ≈ 2 mM; B) extracellular P_i (at time 0) ≈ 60 mM; intracellular P_i (at time 0) ≈ 2 mM. Pyr and FCN: final concentration 2 mM. T: 37 °C; *p*H 8.0; hematocrit: 50% For further explanations, see Fig. 6 and text.

●——● intracellular P_i, ○ — — ○ ATP-γ-P

The tracer experiments did not confirm our expectations; we assumed that in case of stimulation of the membrane by FCN, GAPDH would preferentially utilize extracellular P_i. This is clearly not the case, since it was calculated that utilization of extracellular P_i would increase the relative specific radioactivity of ATP-γ-P two or three times that of intracellular P_i. It can be concluded that membrane GAPDH utilizes only intracelluar P_i and does so only in the presence of extracellular phosphate.
An explanation may, perhaps, be found on the relationship which exists between band 3 protein and GAPDH [15]. It is currently assumed that band 3 protein is a class of related transmembrane polypeptides which encompasses many transport functions, such as anion transport and Na^+/K^+-ATPase [16—18, 20]. It is conceivable that membrane GAPDH is kept in an inactive form and unable to utilize intracellular P_i, unless band 3 channels are in a "transport conformation" and there is an active inward flux of inorganic phosphate. The speculative nature of such explanation is evident, but it is tempting to combine the different indications obtained in vitro and in vivo by such different experimental model and techniques and so build up a functional picture of membrane metabolism, with a close association of ATP and DPG building reactions with the phosphate transport system.

Acknowledgements

Thanks are due to Drs. Köhler, Petermann, Wenz, Ruschke and Mrs. Schulz for their valuable help in some of the experiments. Travel grants to P. Arese and U. Till were provided by FEBS-Jugendfonds and C.N.R.. The work was supported by the Consiglio Nazionale delle Ricerche (Rome, Italy).

References

[1] Rose, Z. B.: J. biol. Chem. **243**, 4810 (1968)

[2] Arese, P., A. Bosia, and G. P. Pescarmona: VI. Internationales Symposium über Struktur und Funktion der Erythrozyten, Berlin 1970. S. Rapoport and F. Jung (Eds.). Akademie-Verlag, Berlin 1972, p. 325

[3] Szekely, M., S. Manyai, and F. B. Straub: Acta physiol. hung. **3**, 371 (1952)

[4] Mishra, R. K., and H. Passow: J. Membrane Biol. **1**, 214 (1969)

[5] Arese, P., A. Bosia, and G. P. Pescarmona: VII. Internationales Symposium über Struktur und Funktion der Erythrozyten, Berlin 1973. S. Rapoport und F. Jung (Eds.). Akademie-Verlag, Berlin 1975, p. 91

[6] Givol, D.: FEBS Lett. **5**, 153 (1969)

[7] Harrigan, P. J., and D. R. Trentham: Biochem. J. **124**, 573, (1971)

[8] Arese, P., A. Bosia, and G. P. Pescarmona: FEBS Lett. **49**, 33 (1974)

[9] Shin, B. C., and K. L. Carraway: J. biol. Chem. **248**, 1436 (1973)

[10] Letko, G., and R. Bohnensack: FEBS Lett. **39**, 313 (1974)

[11] Duchon, G., and H. B. Collier: J. Membrane Biol. **6**, 138 (1971)

[12] Kant, J. A., and T. L. Steck: J. biol. Chem. **248**, 8457 (1973)

[13] Mc Daniel, C. F., M. E. Kirtley, and M. J. A. Tanner: J. biol. Chem. **249**, 6478 (1974)

[14] Tillmann, W., A. Cordua, and W. Schroter: Biochim. biophys. Acta **382**, 157 (1975)

[15] Yu, J. and T. L. Steck: J. biol. Chem. **250**, 9176 (1975)

[16] Cabantchik, Z. I., and A. Rothstein: J. Membrane Biol. **10**, 311 (1972)

[17] Passow, H., H. Fasold, L. Zaki, B. Schumann, and S. Lepke: Proceedings of the 9th Meeting of the Federation of European Biochemical Societies, Budapest 1974, Vol. **35**, p. 197. G. Gardos u. I. Szasz (Eds.) Publishing House of the Hungarian Academy of Sciences, Budapest 1975
[18] Avruch, J., and G. Fairbanks: Proc. natn. Acad. Sci. U.S.A. **69**, 1216 (1972)
[19] Till, U., W. Kohler, I. Ruschke, A. Kohler, and W. Losche: Eur. J. Biochem. **35**, 167 (1973)
[20] Rothstein, A., Z. I. Cabantchik, and P. Knauf: Fedn Proc. Fedn Am. Socs exp. Biol. **35**, 3 (1976)

Acta biol. med. germ., Band 36, Seite 491—505 (1977)

Department of Food Science and Technology, Faculty of Agriculture, Kyoto University, Kyoto, Japan

Regulation of 2,3-bisphosphoglycerate metabolism in erythrocytes by a multifunctional enzyme[1]

H. Chiba, K. Ikura, H. Narita and R. Sasaki

Summary

Bisphosphoglyceromutase, 2,3-bisphosphoglycerate phosphatase and phosphoglyceromutase have been purified from human or pig erythrocytes. Three enzymes were copurified throughout all purification steps. Three fractions (peaks I, II, and III) which were chromatographically separable and had three activities in different ratios were obtained. Peak III which contained the main bisphosphoglyceromutase and 2,3-bisphosphoglycerate phosphatase activities was purified to homogeneity by electrophoretic and ultracentrifugal analysis. The homogeneous preparation had the phosphoglyceromutase activity. The three activities were lost at the same rate during thermal inactivation. Thus, bisphosphoglyceromutase'and 2,3-bisphosphoglycerate phosphatase activities, which are responsible for 2,3-bisphosphoglycerate metabolism in erythrocytes, are displayed by the same enzyme protein which had phosphoglyceromutase activity. The peak III enzyme from human erythrocytes was subjected to chemical modification by trinitrobenzenesulfonate. The three enzyme activities were inactivated by trinitrobenzenesulfonate at the same rate. The sulfhydryl content of the enzyme was unchanged during trinitrophenylation, indicating that derivatization was through the amino group. Trinitrophenylation of about one amino group per mol of the enzyme resulted in complete loss of the three activities. Both 2,3-bisphophoglycerate and 1,3-bisphosphoglycerate inhibited trinitrophenylation and effectively protected the enzyme from inactivation. Although monophosphoglycerates did not show any protective effect at concentrations which should be adequate based upon their kinetic constant, they were protective at higher concentrations. Inactivation by trinitrophenylation was an apparent first-order reaction. The dissociation constant of the enzyme 2,3-bisphosphoglycerate complex was determined by analysing the first order reaction on the assumption that the protective effect of 2,3-bisphosphoglycerate was due to competition with trinitrobenzenesulfonate. The dissociation constant was in good agreement with kinetic constants of 2,3-bisphosphoglycerate in the enzyme reactions, which indicated that 2,3-bisphosphoglycerate did indeed exert its protective effect through competition with trinitrobenzenesulfonate for an amino group of the enzyme. The protective effect of monophosphoglycerates could be rationalized with kinetic evidence that 2-phosphoglycerate at high concentrations interacts with the 2,3-bisphosphoglycerate binding site. These results indicate that the enzyme exhibits the three enzyme activities at a common active site at which one amino group essential for binding of bisphosphoglycerates is located. Based on the multifunctional properties of this enzyme, a possible mechanism was discussed for regulation of 2,3-bisphosphoglycerate metabolism in human erythrocytes.

[1] A part of this paper has been published (Ref. [9]) and the remainigs have been submitted for publication (Ikura, K., et al., Eur. J. Biochem., and Sasaki, R., et al., Agric Biol. Chem. in press).

Abbreviations: 2,3-DPG, 2,3-bisphosphosglycerate; 1,3-DPG, 1,3-bisphosphoglycerate; 3-PGA, 3-phosphoglycerate; 2-PGA, 2-phosphoglycerate; MPG, monophosphoglycerate; DPG, bisphosphoglycerate; TNBS, trinitrobenzenesulfonate; NbS_2, 5,5'-dithiobis (2-nitrobenzoic acid); SDS, sodium dodecyl sulfate; DPG mutase, bisphosphoglyceromutase (EC 2.7.5.4); 2,3-DPG phosphatase, 2,3-bisphosphoglycerate phosphatase (EC 3.1.3.13); PGA mutase, phosphoglyceromutase (EC 7.5.3.).

Introduction

2,3-bisphosphoglycerate (2,3-DPG) is an essential reactant in the enzymatic interconversion between 3 PGA and 2 PGA, which is catalysed by PGA mutase. In the erythrocytes of human and most mammals, 2,3-DPG is a major component of the organic acid-soluble phosphate and its concentration is much greater as it would be expected of its requirement in the PGA mutase reaction. Since several biological functions besides the primary function described above has been assigned to 2,3-DPG in human erythrocytes [1, 2], much attention have been focused on the regulation of levels for this compound. There are three enzyme catalysing reactions in which 2,3-DPG is directly involved: DPG mutase (Eq. 1), 2,3-DPG phosphatase (Eq. 2) and PGA mutase (Eq. 3).

$$\text{1,3-DPG} + \text{3-PGA} \xrightarrow[\text{DPG mutase}]{} \text{2,3-DPG} + \text{3-PGA} \quad (1)$$

$$\text{2,3-DPG} \xrightarrow[\text{2,3-DPG phosphatase}]{} \text{3-PGA} + P_i \quad (2)$$

$$\text{3-PGA} + \text{2,3-DPG} \underset{\text{PGA mutase}}{\rightleftharpoons} \text{2-PGA} + \text{2,3-DPG} \quad (3)$$

These three enzyme activities were believed to be manifested by different protein molecules. In a continuing study of yeast PGA mutase, we have found that this enzyme can catalyse 2,3-DPG synthesis (Eq. 1) and its breakdown (Eq. 2) and that these three enzyme activities are exhibited at a common active site [3—5]. In the meantime, the papers suggesting resemblances of these three enzymes in human erythrocytes appeared [6—8]. In view of the physiological significance of 2,3-DPG in the erythrocytes, we have undertaken to investigate the enzymes responsible for the 2,3-DPG metabolism with the aim of revealing whether their activities are attributable to one protein or not. We report here that 2,3-DPG in human erythrocytes is metabolized by one multifunctional protein carrying the PGA mutase activity and that the three enzyme activities are manifested at a common active site. Based on these results, a new regulatory mechanism will be proposed for the 2,3-DPG metabolism in human erythrocytes. We also discuss universal occurrence of such multifunctional protein showing the experimental results on pig erythrocytes.

Material and methods

Enzyme assay

The three enzyme activities, DPG mutase, 2,3-DPG phosphatase and PGA mutase, were measured as described previously [9]. One unit of enzyme activity is defined as the amount of enzyme which catalyzes the conversion of 1 μmol of the substrate to the product per min. Specific activity is expressed as units per mg of protein.

Isoelectric focusing, polyacrylamide gel electrophoresis and ultracentrifugation were carried out as described previously [9].

Protein concentration

Protein was determined by measuring the absorbance at 280 nm. Its concentration was calculated from the absorbance at 280 nm based on the value $A^{1\%}_{280\,nm} = 10$. This value was used throughout all purification steps. A value for $A^{1\%}_{280\,nm}$ of the purified peak III enzyme from human erythrocytes was determined to be 16.0 ± 0.4 by using a micro-Kjeldahl method.

Inactivation by TNBS

Trinitrophenylation of the enzyme was performed in borate buffer containing KCl and NaCl. These salts were added to simulate intracellular conditions.
Details of experimental conditions are described in the legend of figures and tables. The extent of trinitrophenylation was monitored directly in the inactivation mixture at 367 nm, assuming a molar absorption coefficient of 10,800 M^{-1} cm^{-1} for the trinitrophenyl group [10, 11].

Sulfhydryl determination

By using, Nbs_2, the sulfhydryl groups were determined according to the method of Ernest and Kim [12,13]. At various times during trinitrophenylation, a 0.3 ml sample was diluted into 0.9 ml of 40 mM borate buffer; *p*H 8.0, containing 2 mM EDTA and 5 mM lysine. Lysine instantly stopped further trinitrophenylation of the enzyme by consuming excess TNBS. After extensive dialysis against 40 mM borate buffer; *p*H 8.0, containing 2 mM EDTA, the sample was adjusted to contain 2% SDS and then an excess of Nbs_2 was added. The absorbance at 412 nm was determined against a protein blank and a molecular absorption coefficient of 13,600 M^{-1} cm^{-1} was used to calculate the mols of sulfhydryl group present [14].

Results

Purification of enzymes

As the purification procedure of enzyme from human erythrocytes has been published [9] and that from pig erythrocytes is described in the accompanying paper [15], the purification procedures of enzymes from both erythrocytes are briefly summarized in Fig. 1. Simultaneous purification of DPG mutase, 2,3-DPG phosphatase and PGA mutase from hemolysate of human or pig erythrocytes indicated that these three enzymes were not distinguishable in all purification procedures tested. Three fractions (peaks I, II and III) having the three enzyme activities in the different ratios were obtained by column chromatography using DEAE-Sephadex, Sephadex G-100, phosphocellulose and hydroxyapatite. As shown in Table 1, most of the two enzyme activities for metabolizing 2,3-DPG were found in peak III for both human and pig erythrocytes, while the PGA mutase activity was extremely rich in peak I.
Peak I probably functions as PGA mutase in the glycolysis and peak III is responsible for the 2,3-DPG metabolism. Peak II does not seem to play directly an

Table 1

Distribution of three enzyme activities in the hemolysate
These values were calculated by correcting the recovery percent of the enzyme activities in each purification step

Erythrocyte	Peak	Activity (%)		
		DPG mutase	2,3-DPG phosphatase	PGA mutase
Human	I	0.3	0.5	55
	II	9.0	7.0	27
	III	90.7	92.5	18
Pig	I	1.2	1.5	92
	II	3.6	3.9	1.6
	III	95.2	94.6	6.4

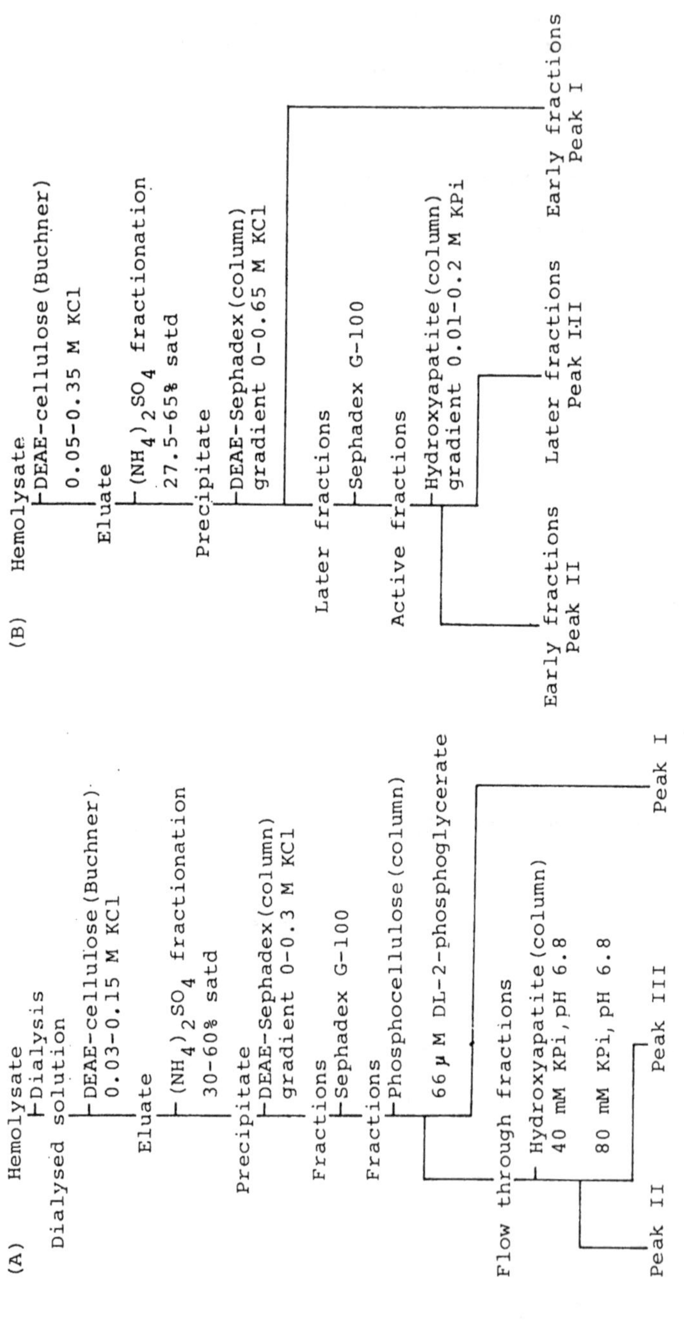

Fig. 1. Scheme for the purification of DPG mutase, 2,3-DPG phosphatase and PGA mutase (A) from human erythrocytes; (B) from pig erythrocytes

important role in the 2,3-DPG metabolism, although little is known of this peak. As shown in Table 2, specific activities of the three enzyme activities from both erythrocytes were very similar, although the specific activity of PGA mutase from human erythrocytes was somewhat higher.

Table 2
Specific activities of peak III

Enzyme sources	Specific activity		
	DPG mutase	2,3-DPG phosphatase	PGA mutase
Human erythrocyte	9.8	2.0	9.8
Pig erythrocyte	11.4	1.7	5.3

Homogeneity and molecular properties of peak III material

Peak III enzymes from human and pig erythrocytes were homogeneous on polyacrylamide gel electrophoresis (Fig. 2). Each enzyme preparation showed one band on SDS polyacrylamide gel electrophoresis and a single and symmetrical boundary on analytical ultracentrifugation (data not shown). Table 3 shows some molecular properties of peak III enzymes from human and pig erythrocytes. As far as we tested, we found no remarkable difference between the enzyme from human and that from pig. They appear to be composed of two subunits with a similar molecular weight.

Table 3
Molecular properties of peak III enzyme.
Molecular weights in the absence of SDS were determined by the sedimentation equilibrium and those in the presence of SDS were determined by polyacrylamide gel electrophoresis. pI = isoelectric point

Enzyme sources	$s_{20,w}$	Molecular weight		*pI*
		− SDS	+ SDS	
Human erythrocyte	4.1 S	61000	29000	5.1
Pig erythrocyte	4.0 S	59000	28000	5.1

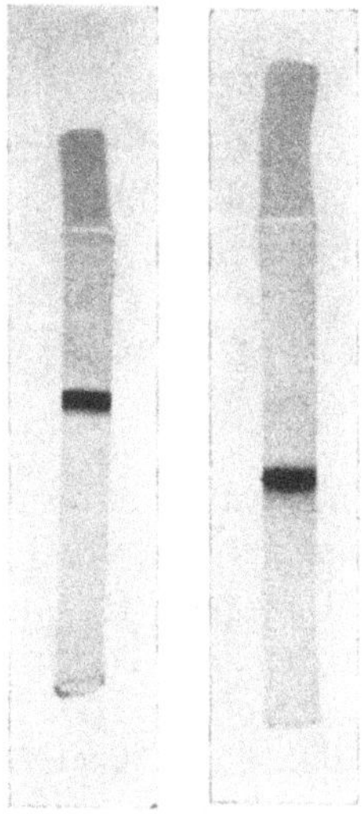

Fig. 2. Electrophoretic homogeneity of the purified enzymes in polyacrylamide gel electrophoresis.

(A) peak III from human erythrocytes; (B) peak III from pig erythrocytes

Thermal inactivation of peak III enzyme

Fig. 3 shows the thermal inactivation curves of peak III material from human and pig erythrocytes. The three enzyme activities of the peak III enzyme from human as well as pig were lost at the same rate, indicating that the three enzyme activities are exhibited by the same enzyme protein.

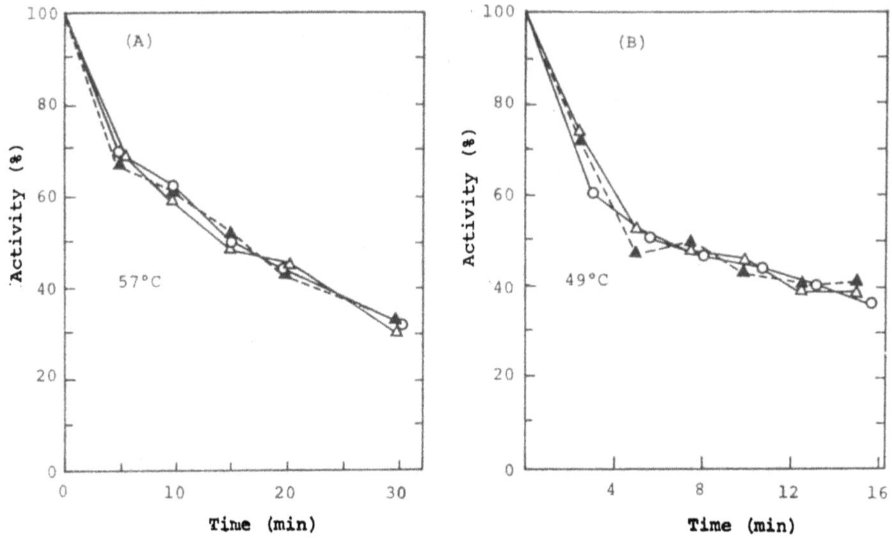

Fig. 3. Thermal inactivation curves of the three activities of the purified enzyme. (A) peak III from human erythrocytes (125 µg/ml in 10 mM Tris-HCl buffer, *p*H 7.5, containing 0.1 mM EDTA and 0.1 mM dithiothreitol); (B) peak III from pig erythrocytes (216 µg/ml in the same buffer). △——△ DPG mutase; ▲– – –▲ 2,3-DPG phosphatase; ○——○ PGA mutase. The sample solutions were heated at the indicated temperatures and proteins were removed at intervals to assay the three enzyme activities

Chemical modification of peak III enzyme from human erythrocytes

For the better understanding of not only the regulatory mechanism of 2,3-DPG metabolism but also of the mechanism by which one enzyme catalyzes three kinds of reactions, it is necessary to reveal whether the three activities of peak III enzyme are exhibited at a common active site or different sites. The peak III enzyme from human erthrocytes was subjected to chemical modification by TNBS.

Inactivation of the three enzyme activities by trinitrophenylation

Fig. 4 shows the results of the chemical modification experiments of the peak III protein. Rapid trinitrophenylation of the enzyme and parallel inactivation of the three enzyme activities were observed on treatment with TNBS. The presence of 2,3-DPG strongly repressed the trinitrophenylation and protected against the inactivation of the three activities efficiently and to the same extent (Fig. 4A). It should be noted that 2,3-DPG is a substrate for both 2,3-DPG phosphatase and PGA mutase, and acts as an inhibitor which is competitive with 1,3-DPG in the DPG mutase reaction [16]. In the absence of 2,3-DPG, inactivation and trinitrophenylation shows a linear relationship up to 60% inactivation of the enzyme (Fig. 4B). Extrapolation of the linear portion of this curve to zero activity reveals that incorporation of about 0.82 mol of trinitrophenyl group per mol of enzyme inactivates the enzyme completely. Deviation from unity probably originates in an error in calculating $A^{1\%}_{280\,nm}$ or in an erroneous molecular

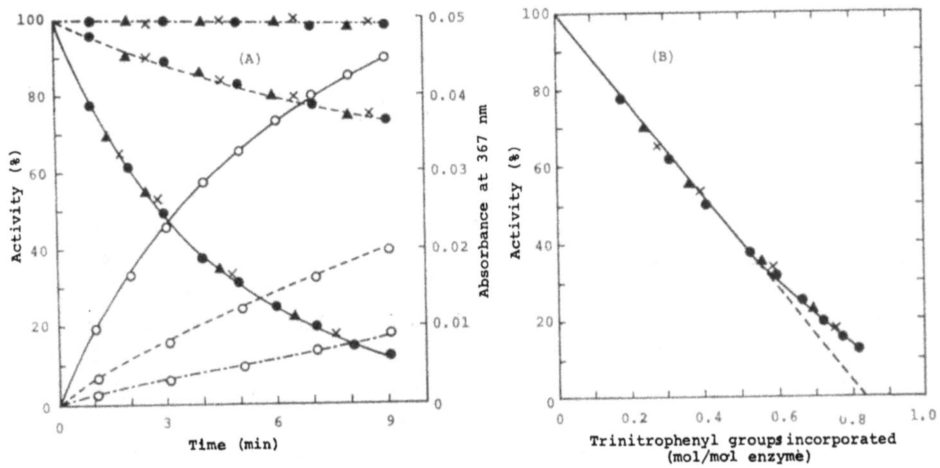

Fig. 4. Inactivation of the three enzyme activities of peak III protein from human erythrocytes by trinitrophenylation.

● PGA mutase; ▲ DPG mutase; × 2,3-DPG phosphatase; o absorbance at 367 nm. The concentrations of 2,3-DPG were: 0 mM (———), 1 mM (--------) and 5 mM (-·-·-·-·-).

(A) Time course of trinitrophenylation and loss of the three enzyme activities. The reaction mixture contained 5.1 μM enzyme, 0.12 mM TNBS and various concentrations of 2,3-DPG in buffer A (25 mM borate buffer, *p*H 7.65, containing 0.1 mM EDTA, 0.13 M KCl and 0.02 M NaCl). The temperature was 25 °C. The absorbance at 367 nm was monitored continuously against an appropriate protein blank to determine the extent of trinitrophenylation. At various times, a 0.05 ml portion was removed and diluted with buffer A for assaying three enzyme activities.

Percent activitiy was based on the activity obtained in the absence of TNBS. (B) Residual activities versus mol of trinitrophenyl group incorporated per mol of enzyme in the absence of 2,3-DPG. A plot was done by using data obtained in the absence of 2,3-DPG. Mols of the trinitrophenyl group incorporated per mol of enzyme were calculated from (A) as described in the text

weight or both. Although TNBS is generally considered to be an amino-specific reagent, at high concentrations it reacts with thiol groups [17]. In order to ascertain that the inactivation resulted in modification of amino group, the sulfhydryl content was determined for the native and for the trinitrophenylated enzymes. The sulfhydryl content remained at the value of 5.7 for the native enzyme through 85% inactivation (Table 4). It is concluded from these results that the inactivation by TNBS is attributable to trinitrophenylation of one amino group which is essential for the three enzyme activities.

Table 4

Sulfhydryl content of peak III enzyme during trinitrophenylation.

The reaction mixture for trinitrophenylation was the same as in the experiment of Fig. 4. At various times during incubation, a 0.05 ml portion was removed for the PGA mutase assay and a 0.3-ml portion for the determination of sulfhydryl content. Sulfhydryl determinations were carried out as described in "Material and methods". Percent activity was based on the activity obtained in the absence of TNBS

Incubation time	Percent activity	Sulfhydryl group/ mol enzyme
min		
0	100	5.7 ± 0.3
2.5	55	5.7
4.0	38	5.6
8.0	15	5.8

Effects of phosphate compounds on inactivation of the enzyme by TNBS

Table 5 summarizes effects of phosphate compounds on inactivation of the PGA mutase activity by TNBS and their known functions in the three reactions. It was confirmed that the DPG mutase and 2,3-DPG phosphatase activities showed the same pattern as the PGA mutase activity regarding inactivation with TNBS in the presence and absence of the compounds tested. Of the three activities, the PGA mutase activity could be measured accurately, since it was relatively insensitive to the phosphate compounds brought from the reaction mixtures for trinitrophenylation. For this reason, the PGA mutase activity is expressed as a representative of the three activities. A significant protective effect was observed in the presence of 2,3- or 1,3-DPG. MPG did not protect the enzyme from inactivation at the concentrations which would be adequate based upon K_m of DPG mutase [18] and K_i of 2,3-DPG phosphatase [6]. At high concentrations, MPG exerted a protective action. Inorganic phosphate and phosphoglycolate were not protective. Both compounds were stimulatory toward 2,3-DPG phosphatase [6,9] and inhibitory toward DPG mutase [9, 16] and PGA mutase. Since enzymatic hydrolysis of 2,3-DPG is negligible under the conditions used for chemical modification, it is evident that the protective effect in the presence of 2,3-DPG results from the binding of this compound with the enzyme. The 1,3-DPG preparation

Table 5

Effects of various compounds on TNBS-inactivation and their known functions in the three enzyme reactions.

S = substrate; I = inhibitor; A = activator. Functions of phosphate compounds were cited from references [6, 9, 16 and 18]. The reaction mixture contained 2.0 μM enzyme, 0.1 mM TNBS and the compounds at indicated concentrations in buffer A (see legend for Fig. 4). After incubation for 8 min at 25 °C, a 0.2 ml portion was removed for the PGA mutase assay. 100% activity was the enzyme activity obtained in the absence of TNBS. N; not known.

Compound added	Conc. (mM)	Activity (%)	Functions		
			PGA mutase	DPG mutase	2,3 DPG phosphatase
None	—	24	—	—	—
2,3-DPG	0.27	53	S	I	S
	1.00	80			
	5.00	97			
1,3-DPG	0.02	40	N	S	N
	0.08	47			
	0.23	57			
3-PGA	0.10	24	S	S	I
	2.50	50			
	10.00	74			
2-PGA	0.10	25	S, I	S, A	I
	2.70	46			
	5.00	59			
Pi	6.25	25	I	I	A
P-glycolate	1.25	30	I	I	A

always contains 3-PGA and inorganic phosphate because of the lability of the acyl-phosphate group in 1,3-DPG. Therefore, the DPG mutase activity of the enzyme can produce 2,3-DPG from the 1,3-DPG preparation without addition of 3-PGA. Since conversion of most of the added 1,3-DPG to 2,3-DPG was observed under the conditions in Table 5, the protective effect in the presence of 1,3-DPG might be due to the 2,3-DPG produced. In order to examine the effect of 1,3-DPG on inactivation of the enzyme, experimental conditions were selected so as to repress the conversion of 1,3-DPG to 2,3-DPG as much as possible. Repression was partially fulfilled by carrying out the modification at low temperature and alkaline *p*H in the presence of a high concentration of TNBS. TNBS reacts with amino groups at alkaline *p*H more rapidly than at an acidic one [19]. Results under these conditions are summarized in Table 6. The enzyme was inactivated to 17% of the original activity by incubation with TNBS for 3.75 min. The inactivation was greatly reduced by increasing the 1,3-DPG concentration in the reaction mixture. It was found from separate experiments that 15—25% of the 1,3-DPG initially added was converted to 2,3-DPG during the incubation for 3.75 min in the absence of TNBS. When 2,3-DPG was added to the reaction mixture at the concentrations corresponding to 25% of 1,3-DPG, the protective effect was much weaker than that of 1,3-DPG. From these results, it is clear that 1,3-DPG is also capable of protecting the enzyme from inactivation by TNBS. These results suggest that one amino group essential to the three activities plays an important role in the binding of bisphosphoglycerates to the enzyme.

Table 6

Protective effect of 1,3-DPG on TNBS-inactivation.

The reaction mixture contained 0.25 μM enzyme, 0.7 mM TNBS, 0.1 mM EDTA, 0.13 M KCl, 0.02 M NaCl and 1,3-DPG or 2,3-DPG at the indicated concentrations in 25 mM borate buffer, *p*H 8.0. After incubation for 3.75 min at 0 °C, the PGA mutase activity was assayed by using a 0.2 ml portion. Percent activity was based on the activity obtained in the absence of TNBS

1,3-DPG (mM)	Activity (%)	2,3-DPG (mM)	Activity (%)
0	17	0	17
0.043	47	0.010	22
0.140	53	0.035	28
0.250	74	0.063	34

Protection by 2,3-DPG

A dissociation constant of the binary complex of the enzyme and 2,3-DPG was determined by analysing the protective effect of the phosphoglycerate. The value obtained was compared with kinetic constants of 2,3-DPG in the enzyme reactions. Fig. 5A shows that the inactivation of the enzyme with TNBS is an apparent first order reaction in the presence of 2,3-DPG at various concentrations. Based on the assumption that the protective effect of 2,3-DPG is due to competi-

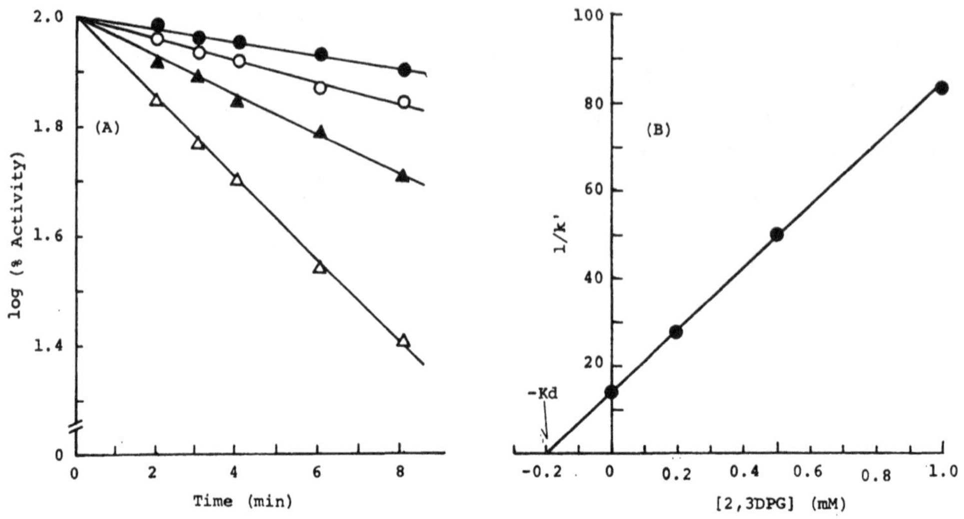

Fig. 5. Determination of the dissociation constant (Kd) of the 2,3-DPG · enzyme complex from the protective effect.

The concentrations of 2,3-DPG were: 0 mM (△), 0.2 mM (▲), 0.5 mM (○) and 1.0 mM (●). (A) First order plot of the inactivation of the PGA mutase activity on peak III enzyme. The incubation mixture contained 7.0 μM enzyme, 0.1 mM TNBS and 2,3-DPG at indicated concentrations in buffer A (see legend for Fig. 4). The temperature was 25 °C. At various times, a 0.05 ml-portion was removed for PGA mutase assay. Percent activity was based on the activity obtained in the absence of TNBS. (B) A plot of $1/k'$ versus the concentrations of 2,3-DPG. Values of k' were calculated from the slopes of lines in (A) (see text)

tion with TNBS, the inactivation of the enzyme in the presence of 2,3-DPG may be expressed with the following equation,

$$\frac{d[EI]}{dt} = k([E]_0 - [EI] - [E \cdot 2,3\text{-DPG}]) \tag{4}$$

where k, E, $[E]_0$ and [EI] are an apparent first order rate constant, the enzyme, the initial concentration of the enzyme and the concentration of the trinitrophenylated inactive enzyme, respectively. The dissociation constant (K_d) of the enzyme · 2,3-DPG complex can be related to the residual activity of the enzyme by following equations:

$$\log \frac{[E]_0 - [EI]}{[E]_0} = -k't \tag{5}$$

$$k' = \frac{k}{2.303} \cdot \frac{K_d}{K_d + [2,3\text{-DPG}]} \tag{6}$$

Values of k' could be calculated from the slopes of the straight lines in Fig. 5A. A plot of l/k' versus the concentration of 2,3-DPG was linear (Fig. 5B) and yielded a dissociation constant (K_d) of 180 μM. These experiments were carried out in the presence of 0.13 M KCl and 0.02 M NaCl to simulate intracellular conditions of ionic strength and ion species [20]. The same kind of experiments was performed in the presence of 0.1 M KCl, because basic properties of DPG mutase and 2,3-DPG phosphatase have been studied at this ionic strength [6, 18]. The pattern of enzyme inactivation was similar to that of Fig. 5 (data not shown). Values of k' were 0.085, 0.050, 0.038 and 0.026 min^{-1} for 0, 0.05, 0.10 and 0.20 mM of 2,3-DPG. A value for the dissociation constant (Kd) was calculated to be 82 μM. Kinetic studies on the PGA mutase activity of peak III enzyme were performed in order to determine a K_m value for 2,3-DPG. Fig. 6A shows that the PGA mutase is inhibited by substrate (2-PGA) competitively with respect to 2,3-DPG. The apparent K_m values for 2,3-DPG, obtained from the intercepts on the abscissa of Fig. 6A, are equal to $K_m(1 + [2PGA]/Ki)$ [9]. Plotting the apparent K_m values versus the concentration of 2-PGA gave a straight line (Fig. 6B). From the intercept on the ordinate, the K_m for 2,3-DPG was calculated to be 200 μM in the presence of 0.13 M KCl and 0.02 M NaCl. In the presence of 0.1 M KCl, the K_m value decreased to 80 μM (data not shown). The constants for 2,3-DPG are summarized in Table 7. Good agreement of the kinetic constants with the dissociation constants leads to the conclusion that 2,3-DPG protects the enzyme through competition with TNBS for the one amino group located at the 2,3-DPG-binding site. Kinetic studies on the 2,3-DPG phosphatase of peak III enzyme were performed to determine K_m value for 2,3-DPG in the presence of 0.13 M KCl and 0.02 M NaCl. When the concentration of 2,3-DPG was varied from 0.15 to 0.50 mM, the enzyme activity increased (data not shown). Lineweaver-Burk plot gave a K_m value of 250 ± 50 μM which was much greater than that reported by Rose and Liebowitz [6].

At present, the source of this discrepancy is not clear. It has been reported that 2,3-DPG inhibits DPG mutase competitively with 1,3-DPG with a K_i of 53 μM

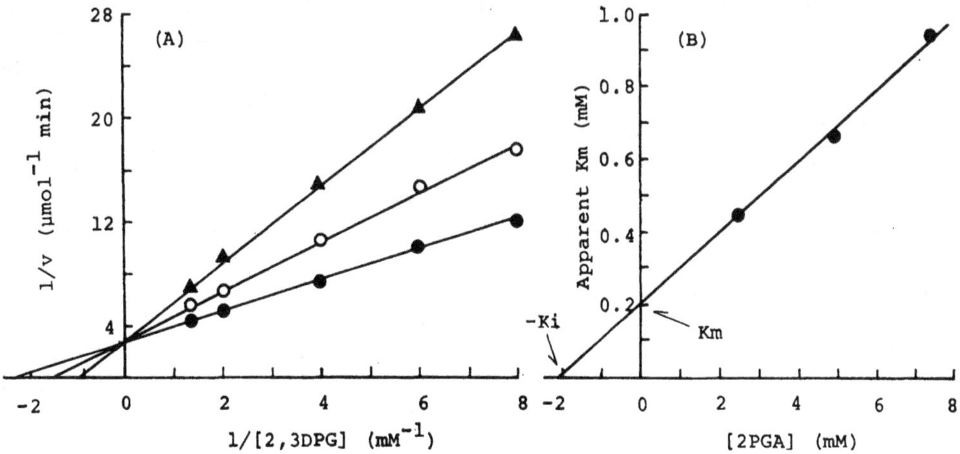

Fig. 6. Determination of K_m value of 2,3-DPG from the kinetics of PGA mutase reaction. The concentrations of 2-PGA were: 2.5 mM (●), 4.9 mM (○) and 7.5 mM (▲). The PGA mutase activity (v) is expressed as μmol of 3-PGA produced per min.

(A) Lineweaver-Burk plot at different 2-PGA concentrations. The reaction mixture for assaying the PGA mutase activity on peak III enzyme contained 7.3 μM enzyme, 2-PGA and 2,3-DPG at indicated concentrations in 2.0 ml of buffer A (see legend for Fig. 4). The temperature was 25 °C. (B) A plot of the apparent Km value versus the concentration of 2-PGA. Apparent Km values were calculated from the intercepts on the ordinate in (A) (see text)

Table 7

Constants for 2,3-DPG and MPG.
TNP-lation = trinitrophenylation. The right column shows the method by which each constant was determined

Compound	Constant	Condition: 0.13 M KCl 0.02 M NaCl	Condition: 0.1 M KCl	Method
		μM		
2,3-DPG	K_d	180	82	TNP-lation
	K_m	200	80	PGA mutase
	K_m	250 ± 50	—	2,3-DPG phosphatase
	K_i	—	53	DPG mutase (Ref. [18])
		mM		
2-PGA	K_d	2.5	—	TNP-lation
2-PGA	K_i	2.0	—	PGA mutase
3-PGA	K_d	2.6	—	TNP-lation

in the presence of 0.1 M KCl [18]. These values are comparable to the constants on 2,3-DPG as seen in Table 7.

Protection by MPG

Kinetic studies on the PGA mutase activity in the previous section showed that 2-PGA could interact with the 2,3-DPG-binding site.

The protective effect of MPG observed at high concentrations (see Table 5) may be attributable to the binding of the MPG to the DPG-binding site. With the same argument as in 2,3-DPG, an inhibition constant of 2-PGA in the PGA mutase reaction was compared with a dissociation constant of the enzyme · 2-PGA complex. The dissociation constant (K_d) was determined by analysing the protective effect of 2-PGA in the same way as for 2,3-DPG. Inactivation of the enzyme by TNBS in the presence of 2-PGA was an apparent first order reaction (data not shown). Values of k' were calculated to be 0.080, 0.058, 0.042 and 0.029 min^{-1} for 0, 1.1, 2.7 and 5.0 mM of 2-PGA. A dissociation constant of 2.5 mM was calculated by plotting $1/k'$ versus the concentration of 2-PGA. The inhibition constant of 2-PGA (K_i) was calculated to be 2.0 mM from the intercept on the abscissa of Fig. 6B. Good agreement of the inhibition constant with the dissociation constant indicates that the protective effect of 2-PGA is through the direct interaction of this compound with the DPG-binding site (Table 7). The dissociation constant of 3-PGA was 2.6 mM (Table 7), although the inhibition constant was not determined.

Discussion

Until now, the purification of DPG mutase and 2,3-DPG phosphatase from human erythrocytes has been independently attempted with little consideration for the possibility that the two activities are due to a common protein. Much attention has been paid to whether 2,3-DPG phosphatase is separable from PGA mutase or not, because the PGA mutases from various sources have 2,3-DPG phosphatase activity. This paper has described the simultaneous purification of DPG mutase,

2,3-DPG phosphatase and PGA mutase from human and pig erythrocytes. The most notable of these results is that the main DPG mutase and 2,3-DPG phosphatase activities (peak III) responsible for the 2,3-DPG metabolism are displayed by the same enzyme protein. In addition, the enzyme molecule had PGA mutase activity. Our conclusion is based on the following observations. (1) The DPG mutase/2,3-DPG phosphatase activity ratio was nearly constant throughout the purification steps. (2) Both activities on chromatography and isoelectric focusing behaved as if they were due to the same protein (3) The final preparation displayed a single band on polyacrylamide gel electrophoresis in the presence and absence of SDS, and was homogeneous on ultracentrifugal analysis.

(4) The DPG mutase, 2,3-DPG phosphatase and PGA mutase activities of peak III were lost at the same rate on thermal inactivation.

Peak III enzyme from human erythrocytes is composed of two subunits of molecular weight 29,000 daltons [9]. Rose [7] has shown with the DPG mutase from human erythrocytes that one phosphoryl group per subunit is covalently bound upon incubation of the enzyme with either 1,3-DPG or 2,3-DPG. This enzyme probably is identical to peak III enzyme. It is of interest that complete inactivation of the enzyme is achieved by trinitrophenylation of one amino group per dimer. These observations might be explained by the model of Levitzki et al. [21] in which binding of a ligand on one subunit hinders subsequent ligand binding of the neighbouring subunit. This type of interaction has been termed "half-of the sites reactivity" and is reported for several enzymes having subunit structure [21]. A conformational change of one subunit induced by the binding of TNBS may be transmitted to the neighbouring subunit so that neither TNBS nor bisphosphoglycerates can bind to it. Although further studies on the three enzyme activities are required to show clearly the mechanism by which they are exhibited at a common active site, a tentative model (Fig. 7) is presented based upon the results in this paper and upon kinetic studies on 2,3-DPG phosphatase [6] and DPG mutase [7, 16, 18]. Peak III enzyme seems to possess two separable sites on the enzyme molecule. One is exclusively available for MPG and another for DPG. MPG can bind to the DPG-site with lower affinity. When two sites are occupied by MPG and 1,3-DPG, the DPG mutase reaction proceeds (enzyme species I). With DPG mutase from human erythrocytes, an ordered sequential pathway has been presented in which 1,3-DPG binds with the free enzyme, followed by the combination with 3-PGA to form a ternary complex [16]. 2,3-DPG is a potent inhibitor competitive with 1,3-DPG [16, 18]. If the two sites are occupied by 2 or 3-PGA and 2,3-DPG, the enzyme becomes PGA mutase (enzyme species II). When the DPG site is filled by 2,3-DPG, the enzyme is converted into 2,3-DPG phosphatase (enzyme species III). It has been shown that the filling of the MPG-site with 2,3-DPG is essential for the functioning of the 2,3-DPG phosphatase of yeast PGA mutase [4]. It is not known whether the MPG site is filled by 2,3-DPG when peak III protein functions as 2,3-DPG phosphatase. The inhibitory effect of MPG on 2,3-DPG phosphatase is due to the conversion of the enzyme into PGA mutase. Inorganic phosphate and phosphoglycolate probably interact with MPG site to stimulate 2,3-DPG phosphatase (enzyme species III). These compounds are inhibitors for the other two enzyme activities. With this model, known func-

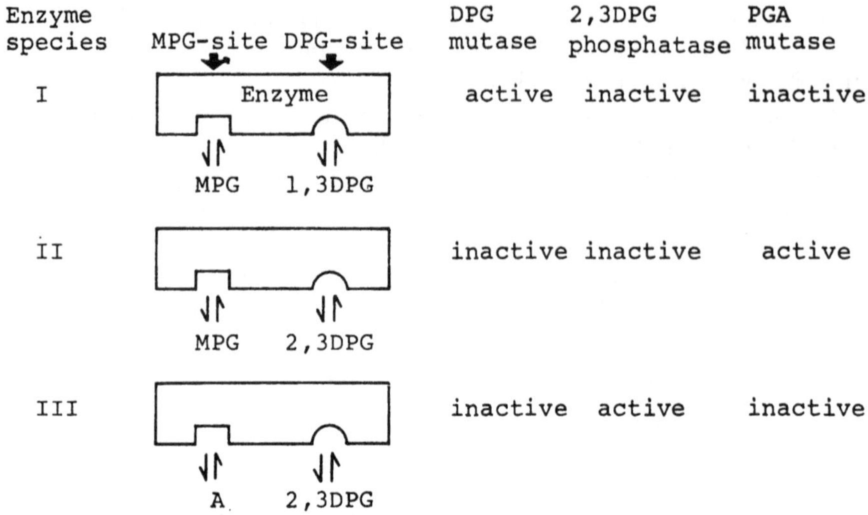

Fig. 7. A tentative model of the mechanism by which the three enzyme activities are displayed at one active site.
A = activator

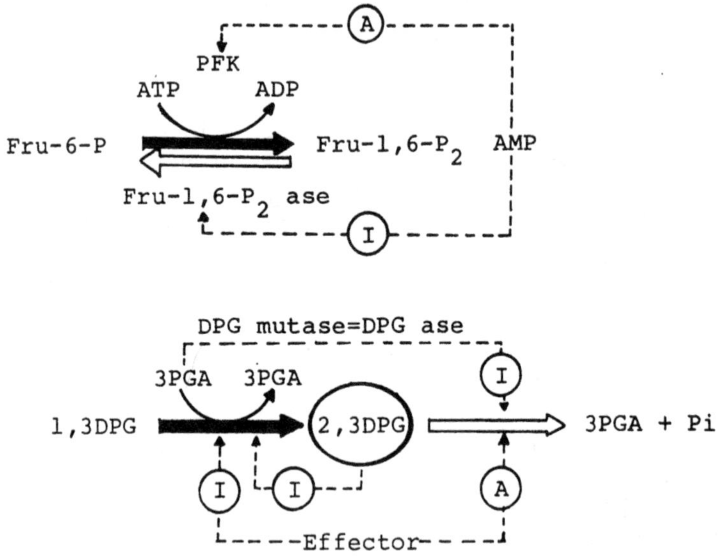

Fig. 8. A possible regulatory mechanism for 2,3-DPG metabolism.
A = activator; I = inhibitor

tions of phosphate compounds in Table 5 seem to be interpretable to a great extent, although not completely.

Physiologically it is of importance that the two enzyme activities responsible for 2,3-DPG metabolism are displayed at a common active site on one protein. The kind of reaction which occurs depends on the affinities of substrates with the enzyme and on substrate concentrations in the cells. In addition, as seen in the case of inorganic phosphate, the effect of a ligand on one of the two enzyme activities can be magnified by exerting the opposite effect on the other (Fig. 8). Normally an irreversible reaction accompanying the consumption of a high energy bond and its reverse reaction are catalyzed by different enzymes. Allosteric transition is a prevailing mechanism for regulation of these enzymes. Regulation of the reactions is effectively achieved by an effector exerting one effect on one enzyme and the opposite effect on the other. Interconversion between fructose-1,6-diphosphate and fructose-6-phosphate is a typical case in which AMP activates phosphofructokinase and inhibits fructose-1,6-diphosphatase (Fig. 8). From this point of view, the regulatory mechanism proposed for 2,3-DPG metabolism in human erythrocytes would be a new type of metabolic regulation achieved with one multifunctional enzyme capable of catalyzing two irreversible reactions.

Occurrence of such multifunctional enzyme in pig erythrocytes as well as human indicates that 2,3-DPG levels in mammalian erythrocytes rich in this compound may be regulated by one multifunctional protein.

References

[1] Duhm, J., and E. Gerlach in: The human red cell in vitro. T. J. Greewalt and G. A. Jamieson (Ed.). Grune and Stratton, New York 1974, p. 111
[2] Sasaki, R., K. Ikura, S. Katsura, and H. Chiba: Agric. biol. Biochem. (in press)
[3] Sasaki, R., E. Sugimoto, and H. Chiba: Biochim. biophys. Acta **227**, 584 (1971)
[4] Sasaki, R., M. Hirose, E. Sugimoto, and H. Chiba: Biochim. biophys. Acta **227**, 595(1971)
[5] Sasaki, R., S. Utsumi, E. Sugimoto, and. H. Chiba: Eur. J. Biochem. (in press)
[6] Rose, Z. B., and J. Liebowitz: J. biol. Chem. **245**, 3232 (1970)
[7] Rose, Z. B., and R. G. Whalen: J. biol. Chem. **248**, 1513 (1973)
[8] Rosa, R., J. Gaillardon, and J. Rosa: Biochem. biophys. Res. Commun. **51**, 536 (1973)
[9] Sasaki, R., K. Ikura, E. Sugimoto, and H. Chiba: Eur. J. Biochem. **50**, 581 (1975)
[10] Goldfarb, A. R.: Biochemistry **5**, 2570 (1966)
[11] Coffee, C. J., R. A. Bradshew, B. R. Goldin, and C. Frieden: Biochemistry **10**, 3516 (1971)
[12] Ernest, M. J., and K.-H. Kim: J. biol. Chem. **249**, 5011 (1974)
[13] Ernest, M. J., and K.-H. Kim: J. biol. Chem. **249**, 6770 (1970)
[14] Ellman, G. L.: Archs Biochem. Biophys. **82**, 70 (1959)
[15] Sasaki, R., K. Ikrua, H. Narita, and H. Chiba: Acta biol. med. germ. **36**, (1977)
[16] Rose, Z. B.: J. biol. Chem. **243**, 4810 (1968)
[17] Okuyama, T., and K. Satake: J. Biochem., Tokyo **47**, 454 (1960)
[18] Rose, Z. B.: Archs Biochem. Biophys. **158**, 903 (1973)
[19] Satake, K., T. Okuyama, M. Ohashi, and T. Shinoda: J. Biochem., Tokyo **47**, 654 (1960)
[20] Berger, H., G.-R. Jänig, G. Gerber, K. Ruckpaul, and S. M. Rapoport: Eur. J. Biochem. **38**, 553 (1973)
[21] Levitzki, A., W. B. Stallcup, and D. E. Koshland, Jr.: Biochemistry **10**, 3371 (1971)

Acta biol. med. germ., Band 36, Seite 507–513 (1977)

Institut für Physiologische Chemie, Lehrstuhl II
der Universität Erlangen-Nürnberg, BRD

Interrelationship between energy metabolism from various substrates and the 2,3-bisphosphoglycerate bypass in human erythrocytes

K. Brand, and K.-H. Quadflieg

Summary

Metabolism of the substrates D-ribose, xylitol, D-xylulose, D-fructose, D-glucose and mixtures of these compounds were studied in human erythrocytes. The metabolic rates obtained with the various substrates strongly affected the intracellular levels of ATP and 2,3-bisphosphoglycerate. Small amounts of substrate utilization resulted in a decrease of the ATP and more pronounced of the 2,3-bisphosphoglycerate concentration while carbon utilization rates beyond 14 µg atom C/ml packed cells/120 min yielded constant levels of ATP and 2,3-bisphosphoglycerate. From these results it can be concluded that a carbon utilization rate of 14 µg atom C/ml cells/120 min is able to cover the ATP requirement of the red cells under steady state conditions. Based on the carbon utilization rates obtained with the various substrates and the rates of 2,3-bisphosphoglycerate decomposition an attempt is made to calculate the contribution of the 2,3-bisphosphoglycerate bypass to substrate metabolism. In case of xylitol as substrate the decrease in the 2,3-bisphosphoglycerate content provides the regeneration of NAD thus facilitating uptake and metabolism of xylitol.

Introduction

Glucose, fructose and xylitol separately or in mixture are widely used as an energy source in parenteral nutrition and therefore knowledge about their complete intracellular metabolism in red blood cells which are exposed to these compounds is of importance. In contrast to glucose metabolism [1], (for review see [2]) no extensive studies on the rates of fructose and xylitol [3, 4] metabolism via the different pathways (Fig. 1) in red cells and their relationship to energy metabolism and the 2,3-bisphosphoglycerate pathway have been reported.

Material and methods

Enzymes and coenzymes were obtained from Boehringer, Mannheim; D-xylulose from Koch-Light Laboratories Ltd., Colmbrook, England; D-glucose, D-fructose, D-ribose and xylitol from E. Merck, Darmstadt. Uniformly ^{14}C-labelled D-glucose, D-fructose, D-ribose and xylitol were purchased from Amersham, Buchler GmbH, and CoKG, Braunschweig. All other chemicals were of analytical reagent grade.

Freshly isolated erythrocytes from healthy human subjects (2.5 ml of packed cells) were suspended in Krebs-Ringer phosphate buffer, *p*H 7.4 or 7.6 (free of calcium ions), and incubated at 37 °C for 120 min. The incubation was started by the addition of the following U-^{14}C-labelled substrates: xylitol, D-glucose, D-fructose, D-ribose and combinations of these substrates as well as unlabelled D-xylulose. At time 0 and after 2 hrs the incubation was terminated by the addition of perchlorid acid. $^{14}CO_2$ was trapped by 0.5 ml of 2-phenylethylamine. In the neutralized perchloric acid extracts, metabolic intermediates and products and not utilized substrate(s) were determined enzymatically or by isotopic techniques. 2,3-bisphosphoglycerate was assayed according to Ericson and De Verdier [5]. From these meas-

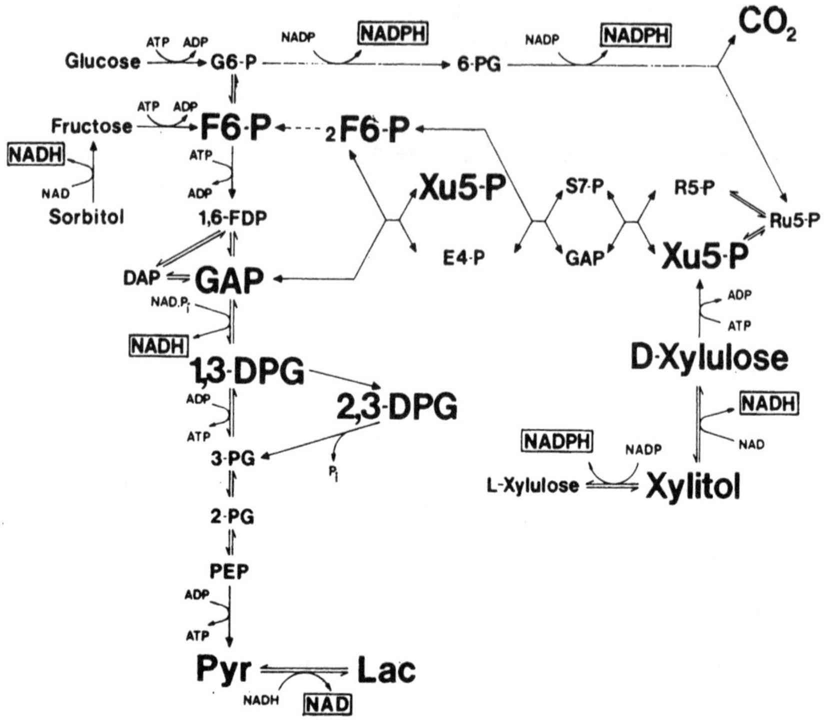

Fig. 1. Pathways of glucose, fructose and xylitol metabolism in mammalian erythrocytes

urements at time 0 and after 120 min the amount of substrate uptake, the change in the concentrations of intermediates and the amount of formation of products could be calculated in order to construct balances of carbon metabolism of the different substrates.

Results and discussion

The various studied substrates were utilized by different rates: the lowest uptake was obtained with D-ribose followed by xylitol, D-fructose, D-xylulose and D-glucose whereas with mixtures of these substrates carbon uptake was more pronounced. In order to get a wide range of carbon utilization rates, incubations were carried out at *p*H 7.4 as well as at *p*H 7.6 in order to partially release the hexokinase-phosphofructokinase system from inhibition [6].

As shown in Fig. 2 the concentration of 2,3-bisphosphoglycerate in erythrocytes is strongly influenced by the rates of substrate metabolism. When erythrocytes were incubated without substrate 2,3-bisphosphoglycerate was found to decompose at a rate of 1034 nmole/ml cells/120 min regardless if incubations were carried out at *p*H 7.4 or 7.6. This decomposition rate of 517 nmole/ml cells/60 min is in perfect agreement with that reported by DUHM et al. [7] and RAPOPORT et al. [8] and corresponds to the rate of 2,3-biphosphoglycerate breakdown in the glycolyzing erythrocyte under steady state conditions. With substrates like D-ribose, xylitol, D-fructose or D-xylulose different low rates of carbon utilization were obtained and concomitantly different levels of 2,3-bisphosphoglycerate have been determined.

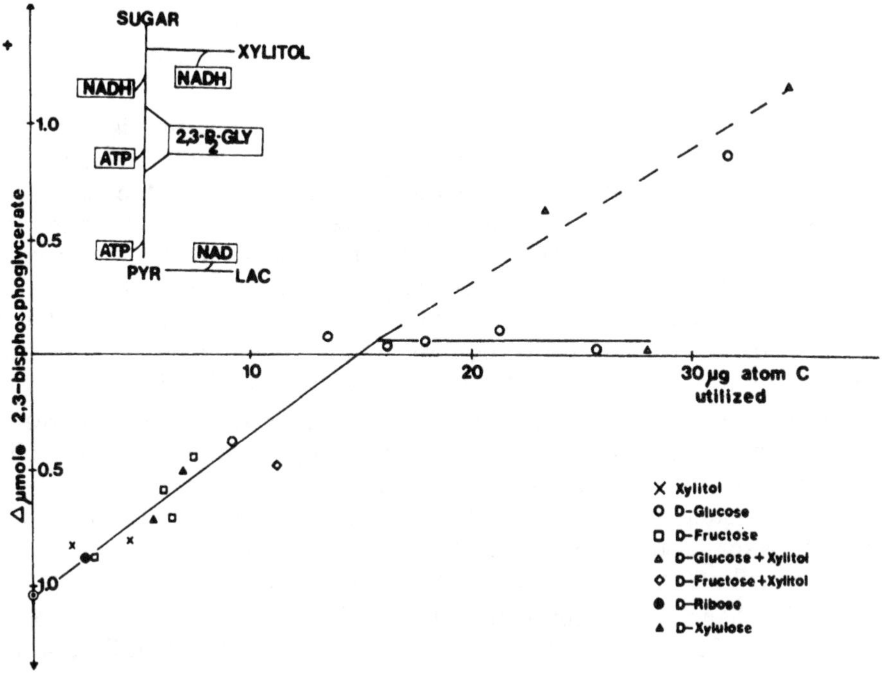

Fig. 2. Relationship between carbon utilization rates from various substrates and changes of the 2,3-bisphosphoglycerate level in human erythrocytes during the incubation period. Mean values of 1—4 experiments with duplicates each are given as μg atom C or μmole/ml packed erythrocytes/120 min of incubation at *p*H 7.4 or 7.6. Carbon utilization is expressed as μg atom C which have been utilized from a given substrate under the experimental conditions for the formation of ATP and 2,3-bisphosphoglycerate. This amount corresponds to those carbon atoms which have been degraded beyond 1,3-bisphosphoglycerate. Accumulation of intermediates above this compound as well as CO_2 production in the pentose cycle therefore have been subtracted from the amount of carbon uptake determined. Metabolic rates above 20 μg atom C/ml cells/120 min have been obtained by incubating 2.5 ml of red cells with D-glucose or D-glucose + xylitol in a final volume of 12.5 ml at *p*H 7.6

From these results it appears that up to a carbon utilization rate of 14 μg atom C/ml cells/120 min the net-decomposition rate of 2,3-bisphosphoglycerate is positively correlated to the rate of substrate utilization. When substrates were utilized at higher rates, e. g. at *p*H 7.6, the 2,3-bisphosphoglycerate content did not decrease but remained constant (solid line). Only in the presence of electron acceptors like phenazine methosulphate or pyruvate accumulation of 2,3-bisphosphoglycerate did occur at *p*H 7.6 (dashed line).

From these results the following conclusions can be drawn:

1. The net-decomposition rate of 2,3-bisphosphoglycerate at various carbon utilization rates can be calculated by subtracting the rate of 2,3-bisphosphoglycerate breakdown with a given substrate from that determined without substrate (see calculation).

2. At carbon utilization rates between 14—28 μg atom C/ml cells/120 min the concentration of 2,3-bisphosphoglycerate was found to be constant indicating

that steady state conditions for this intermediate have been achieved. Within this range the rate of formation and degradation of 2,3-bisphosphoglycerate is equal but it could be that this rate exceeds that of 1034 nmole/ml cells/120 min.

3. As long as 2,3-bisphosphoglycerate does not accumulate NAD is regenerated by the degradation of 2,3-bisphosphoglycerate to lactate and available for the glyceraldehyde-3-phosphate dehydrogenase reaction. Only in the presence of other NADH reoxidizing systems such as phenazine methosulphate or pyruvate the 2,3-bisphosphoglycerate level was found to be increased without reducing the glycolytic rate. This result further suggests that at high metabolic rates the NAD/NADH ratio controls the glycolytic flow rate since intermediates preceeding the glyceraldehyde-3-phosphate dehydrogenase reaction were found to be increased while after addition of the oxidant these intermediates declined.

4. 2,3-bisphosphoglycerate obviously does not inhibit the hexokinase reaction under the experimental conditions since the rate of glucose metabolism could be further enhanced by the addition of phenazine methosulphate in spite of the fact that 2,3-bisphosphoglycerate did accumulate. It therefore appears that the inhibition of hexokinase by 2,3-bisphosphoglycerate reported by Gerber et al. [9] does not contribute significantly to the control of the glycolytic flux.

In Fig. 3 the change in the concentration of both ATP and 2,3-bisphosphoglycerate during the incubation period is plotted versus the carbon utilization rates from various substrates. Since in erythrocytes the degradation of one mole of 2,3-bisphosphoglycerate gives rise to the formation of one mole of ATP both compounds can be regarded as a single pool of energy sources. When the red cells were incubated without substrate the change in the concentration of [ATP + 2,3-bisphosphoglycerate] amounted to 1286 nmole/ml cells/120 min on average regardless if the incubations were carried out at *p*H 7.4 or at 7.6. This amount can be regarded as the minimum ATP requirement under the experimental conditions in vitro. Similar to the graph in Fig. 2 also the change in the concentrations of [ATP + 2,3-bisphosphoglycerate] are positively correlated to the rates of substrate utilization up to 14 μg atom C/ml cells/120 min. Beyond this rate the levels of both compounds remained constant. From these results the following conclusions can be drawn:

1. At a carbon utilization rate of 14 μg atom C/ml cells/120 min metabolic steady state has been achieved as indicated by the fact that the levels of ATP and 2,3-bisphosphoglycerate remain constant. Since the rates of formation and consumption of ATP are balanced it can be concluded that 14 μg atom C/ml cells/120 min are able to cover the ATP requirement of erythrocytes at steady state conditions.

2. At steady state conditions 1.034 μmole 1,3-bisphosphoglycerate/ml cells/120 min can be assumed to have entered the 2,3-bisphosphoglycerate bypass compensating for the decomposition rate. The actual amount of carbon atoms which is metabolized via the glycolytic pathway thus forming ATP therefore can be calculated to be $14 - 3 \times 1{,}034 = 10.9$ μg atom C/ml cells/120 min. Since 3 μg atom C metabolized via the glycolytic pathway give rise to the formation of 1 μmole ATP the ATP requirement for the glycolizing red blood cells under steady state conditions can be calculated to be $10.9 : 3 = 3.6$ μmole ATP/ml cells/120 min of incubation.

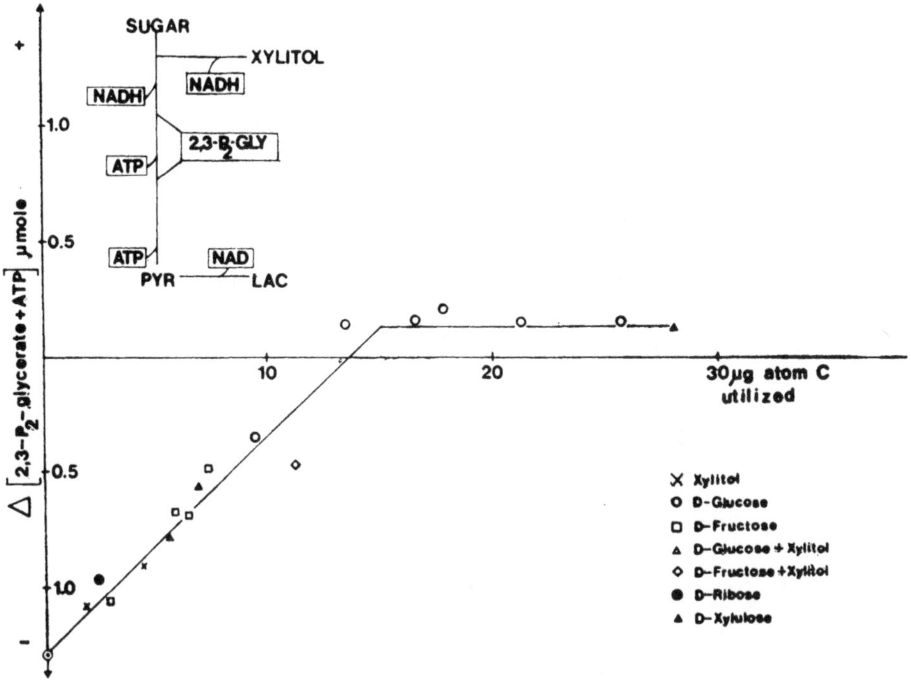

Fig. 3. Relationship between carbon utilization rates from various substrates and changes of the levels of [ATP + 2,3-bisphosphoglycerate] in human erythrocytes during the incubation period.

Mean values of 1—4 experiments with duplicates each are given as µg atom C or µmole/ml packed cells/120 min of incubation at *p*H 7.4 or 7.6. For details see legend to Fig. 2

3. If the ATP consumption rate remains constant also under enhanced metabolic rates all carbon atoms utilized during the incubation exceeding the amount of 10.9 µg atom C/ml cells/120 min must have been metabolized via the bypass because the concentration of ATP remained constant. In this case the share of the bypass to carbon metabolism increases with increasing metabolic rates. At a rate of 21.8 µg atom C/ml cells/120 min e. g. the share of the bypass to carbon metabolism can be calculated to be 50%.

4. After releasing the hexokinase-phosphofructokinase system from inhibition it appears that there is no further control of carbon metabolism within a wide range of metabolic rates (14—28 µg atom C/ml cells/120 min). This could be explained by assuming that carbon metabolism proceeds mainly via the 2,3-bisphosphoglycerate bypass without net formation of ATP and NADH. Our results therefore support the concept that the overall flux of glycolysis is mainly determined by hexokinase and phosphofructokinase as has been proposed by T. A. Rapoport et al. [10].

Calculations

As shown on Table 1 the rates of carbon utilization via the 2,3-bisphosphoglycerate bypass and the 3-phosphoglycerate kinase reaction can be calculated from the experimental data. The procedure of calculation is explained in the legend to

Table 1

Calculation of flow rates via the 2,3-bisphosphoglycerate bypass and the 3-phosphoglycerate kinase reaction in human erythrocytes Mean values of 2—6 experiments are ng atom or nmole/ml packed cells/120 min of incubation at *p*H 7.4. The amount of 1,3-bisphosphoglycerate which entered the 2,3-bisphosphoglycerate bypass has been calculated by subtracting the decrease of 2,3-bisphosphoglycerate in incubations with various substrates from that determined without substrate (1034 nmole/ml cells/120 min). The difference between the carbon utilization rate (column 2, defined in the legend to Fig. 2) and the rate of C atoms entering the bypass (column 5) yields the rate of carbon atoms which are metabolized via the 3-phosphoglycerate kinase reaction (column 6)

Substrate	Carbon atoms utilized for ATP and 2,3-P_2-glycerate formation	Δ[2,3-P_2-glycerate]	1,3-P_2-glycerate entering the bypass	C atoms entering the bypass	C atoms metabolized via the 3-phosphoglycerate kinase reaction	Glycolytic ATP formation	Share of the bypass to carbon metabolism
	—	−1034	—	—	—	—	—
D-Ribose	2434	−884	150	450	1984	661	23%
Xylitol	4444	−808	226	678	3766	1255	18%
D-Fructose	6497	−700	334	1002	5495	1832	18%
D-Xylulose	6935	−513	521	1563	5372	1791	29%
D-Glucose	9431	−373	661	1983	7448	2483	27%

the Table. This calculation can be applied only to metabolic rates in which steady state is not yet achieved. Under these conditions however it is interesting to note that the share of bypass to carbon metabolism is rather constant at the different metabolic rates ranging between 18 to 29%. Beyond a carbon utilization rate of 14 µg atom C/ml cells/120 min two possibilities can be assumed to occur:

1. The ATP consumption rate remains constant and corresponds to the ATP requirement determined to be 3.6 µmole/ml cells/120 min. In this case, as stated below, the share of the bypass would increase with increasing metabolic rates. Since no net formation of ATP does occur in this pathway a waste of ATP would be the consequence.

2. The ATP consumption rate increases with higher metabolic rates e. g. by increased ATPase activities. In this case the share of the bypass could be assumed to remain in the range between 15—30%.

Which one of the two possibilities is realized in the red blood cells remains still open. Our results, however, allow to conclude that at all metabolic rates a significant amount of carbon atoms are metabolized via the 2,3-bisphosphoglycerate bypass.

References

[1] Brand, K., P. Arese, and M. Rivera: Hoppe-Seyler's Z. physiol. Chem. **351**, 501 (1970)
[2] Jacobasch, G., S. Minakami, and S. M. Rapoport in: Cellular and Molecular Biology of Erythrocytes. H. Yoshikawa and S. Rapoport (ed.). Urban und Schwarzenberg, München, Berlin, Wien 1974, p. 55
[3] Asakura, T., K. Adachi, S. Minakami, and S. Yoshikawa: J. Biochem., Tokyo **62**, 184 (1967)
[4] Quadflieg, K.-H., and K. Brand: Z. ErnährWiss. **14**, 268 (1975)
[5] Ericson, A., and C.-H. De Verdier: Scand. J. clin. Lab. Invest. **29**, 85 (1972)
[6] Minakami, S., and H. Yoshikawa: J. Biochem., Tokyo **59**, 145 (1966)
[7] Duhm, J., B. Deuticke, and E. Gerlach: Pflügers Arch. ges. Physiol. **306**, 329 (1969)
[8] Rapoport, I., H. Berger, R. Elsner, and S. M. Rapoport in: Abstracts of the VIIIth International Berlin Symposium on Structure and Function of Erythrocytes, Berlin 1976 p. 98
[9] Gerber, G., H. Preissler, R. Heinrich, and S. Rapoport: Eur. J. Biochem. **45**, 39 (1974)
[10] Rapoport, T. A., R. Heinrich, G. Jacobasch, and S. Rapoport: Eur. J. Biochem. **42**, 107 (1974)

Acta biol. med. germ., Band 36, Seite 515—521 (1977)

Institut für Physiologische und Biologische Chemie der Humboldt-Universität zu Berlin, 104 Berlin, DDR

*p*H-Dependent changes of 2,3-bisphosphoglycerate

I. Rapoport, H. Berger, R. Elsner, and S. M. Rapoport

Summary

A systematic study of the *p*H dependent changes in the range 6.6—7.4 of 2,3-bisphosphoglycerate (2,3-DPG) was performed in the presence and absence of glucose during transitional and steady states.

The results indicate that 2,3-DPGase breaks down 2,3-DPG nearly independent of *p*H at a rate of 480 μmoles 2,3-DPG/l cells · h. The 2,3-DPG mutase is practically completely inhibited below *p*H 6.9. The 2,3-DPG level in the presence of glucose reaches a *p*H dependent steady state after about 18 h.

The share of the 2,3-DPG bypass in the steady state decreases from 24% at *p*H 7.4 to 12% at *p*H 7.0. The formation of pyruvate corresponds to the breakdown of 2,3-DPG after consumption of an unknown reducing substance.

With a concentration of 4—5 mmoles/l cells 2,3-bisphosphoglycerate (2,3-DPG) is the phosphate ester with the highest concentration in the human red blood cell. A variety of effectors are assumed to change the 2,3-DPG level under physiological and pathological conditions:

1. The effect of deoxygenation on the level of 2,3-DPG has attracted considerable attention [1—5].
2. Some inorganic ions influence indirectly the 2,3-DPG concentration by affecting the hexokinase-phosphofructokinase-system. Others do so directly by affecting the bypass [6—16].
3. There is an inverse ratio between the hemoglobin content of the cell and the 2,3-DPG-concentration [17—19].
4. Some hormones [20—23] and vitamins [24, 25] are supposed to influence the 2,3-DPG level.
5. Several drugs [26—28] affect the 2,3-DPG concentration, possibly by affecting ATP consuming processes of the membrane.
6. Genetic defects of enzymes can change the 2,3-DPG concentration of the cell [29—34].
7. Under physiological conditions the most important effector is the concentration of hydrogen ions. 2,3-DPG increases during alcalosis, while it decreases during acidosis. The original observations of a breakdown of 2,3-DPG [35—38] at low *p*H values, both in vivo and in vitro have been confirmed repeatedly [39—47]. Nevertheless, no systematic study of this phenomenon has been performed yet, nor has its mechanism been elucidated. At present there is a lack of understanding by what mechanism steady state concentrations of 2,3-DPG are established under various physiological and pathological conditions. The in vitro experiments carried out so far do not permit valid extrapolations to the conditions in vivo. The purpose of the present study was an investigation of the *p*H-dependent kinetics of

the changes of 2,3-DPG both during transitional and steady states. It included an assessment of the flux through the 2,3-DPG bypass.

Fig. 1 shows the rate of 2,3-DPG breakdown as a function of *p*H with and without glucose. The rates of breakdown are derived from short-term experiments lasting for 6 hrs. During this period the 2,3-DPG concentration decreased in a linear fashion, which means that initial velocities were measured down to a concentration of 1 mM 2,3-DPG. Without glucose a linear rate of breakdown of 480 μmoles/l cells · h was observed irrespective of *p*H. In the presence of glucose at *p*H values of 6.9 or less the same rate of 2,3-DPG breakdown occurred. Above *p*H 6.9 the net rate of 2,3-DPG degradation decreased and became zero at a *p*H exactly maintained at 7.4. The constant and linear rate of breakdown of 2,3-DPG in intact erythrocytes in the absence of glucose which is independent of *p*H and the changing pattern of metabolites suggest two conclusions: first, that the 2,3-DPG ase is saturated above 1 mM 2,3-DPG. Secondly, that the activity of the saturated enzyme is not significantly affected by *p*H or metabolites.

Since below *p*H 6.9 2,3-DPG breaks down at the same rate with or without glucose the 2,3-DPG mutase must be almost completely inhibited, while the flux through the DPGase remains constant. The inhibition of the mutase cannot be due to a decrease in the concentration of its substrate 1,3-DPG both for experimental and theoretical reasons. It rather represents a *p*H dependence of the intrinsic properties of the mutase.

The narrow span of 0.3 intracellular *p*H units between complete inhibition of the mutase and its normal functioning suggests a cooperative behaviour of the enzyme.

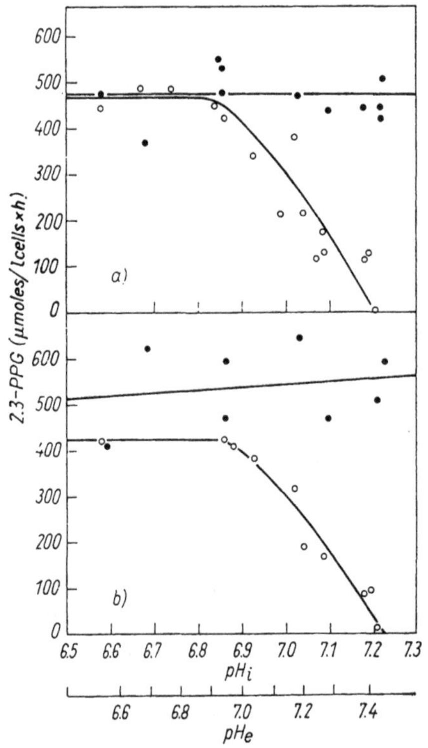

Fig. 1. Initial rates of breakdown of 2,3-DPG.

a) Calculated from 2,3-DPG degradation;
b) calculated from the formation of inorganic phosphate according to the formula:

$$\frac{\Delta\text{inorganic phosphate} - \Delta\text{nucleotide phosphate}}{2}\ [\mu\text{moles/l cells} \cdot \text{h}]$$

to correct for the breakdown of the adenine nucleotides

○ with glucose, ● without glucose

This marked effect of *p*H on the activity of the 2,3-DPG mutase has not been observed so far on purified enzyme preparations [48].
Fig. 2 shows the 2,3-DPG breakdown without glucose when a 2,3-DPG concentration of less than 1 mM is reached. Under these conditions a slow-down of the rate of degradation is apparent. Moreover a *p*H dependence of the activity of the 2,3-DPGase becomes noticeable. The K_m decreases from 384 ± 25 μM at pH_e 7.41 to 286 ± 20 μM at pH_e 7.04. The K_m values thus differ from those determined on purified enzyme preparations by HARKNESS and ROTH [49] and by ROSE and LIEBOWITZ [48] but are in good agreement with those of CHIBA et al. [50].
In the presence of glucose a new steady state of 2,3-DPG is attained after 17—20 hrs. The time requirement for that is independent of the final concentration of 2,3-DPG (Fig. 3).

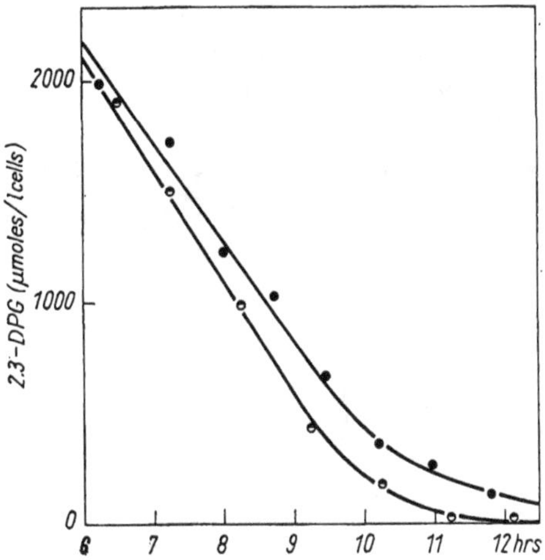

Fig. 2. *p*H dependence of 2,3-DPG breakdown without glucose at
● *p*H 7.405, ◓ *p*H 7.04

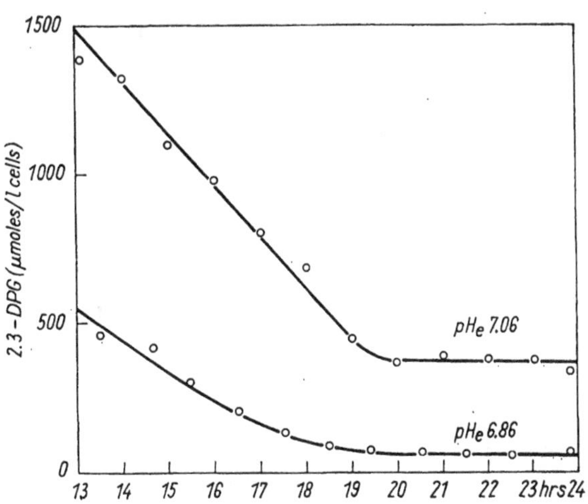

Fig. 3. Changes of 2,3-DPG in the presence of glucose in long term experiments at different *p*H-values

Since DPG is an effector of a number of glycolytic enzymes the glycolytic system changes from a quasi steady state to a true new steady state after about 19 hrs.

The steady state levels of 2,3-DPG are of course strongly dependent on *p*H, as may be seen from Fig. 4. Because of the non-linearity of the *p*H-dependence the ratio (Δmmoles 2,3-DPG/l cells)/0.1 *p*H is not constant. In contrast to the behaviour of 2,3-DPG, ATP changes by less than 10% over the entire *p*H range.

The DPG broken down appears in the sum of pyruvate and lactate. Fig. 5 shows the formation of pyruvate with and without glucose at different *p*H-values. Without glucose the pyruvate formation is independent of *p*H. There is hardly any accumulation for the first two hours. After the third hour pyruvate increases at a rate equal to the 2,3-DPG breakdown. Since no pyruvate appears during the first two hours, one has to postulate that about 1.5 mmoles of reducing equivalents are present to reduce pyruvate to lactate. Glutathione, a possible candidate, did not change its reduction state within the critical period. The character of the reducing substance is not yet known. In the presence of glucose the pyruvate formation becomes *p*H dependent. It is very low at pH_e 7.4 and increases at lower *p*H values without ever reaching the value occurring in the absence of glucose.

A consideration of the share of the 2,3-DPG bypass on the overall glycolytic flux under various conditions is of interest (Table 1). Since under conditions of the

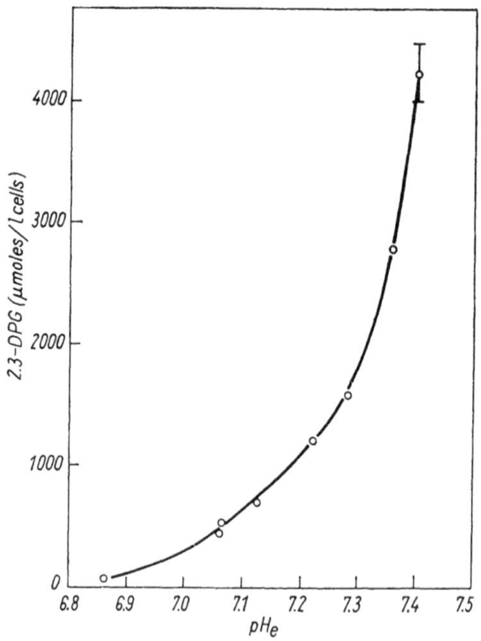

Fig. 4

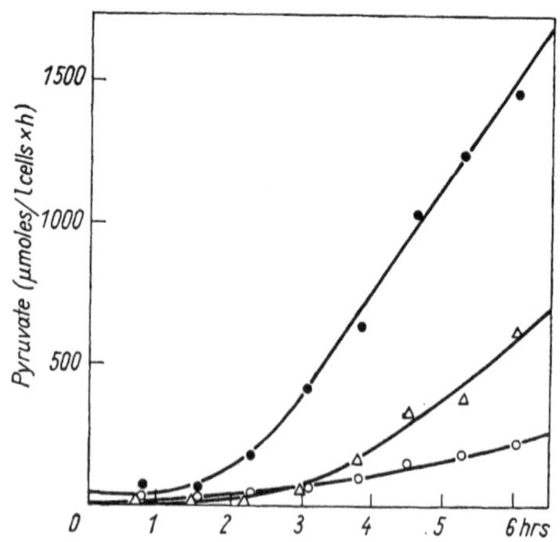

Fig. 5

Fig. 4. The influence of *p*H on the steady state concentration of 2,3-DPG in the presence of glucose

Fig. 5. The formation of pyruvate at different *p*H values with and without glucose
○ with glucose, pH_e 7,415; ● without glucose, pH_e 7,415; △ with glucose, pH_e 6,90.

transitional state glucose influx and formation of the sum of lactate and pyruvate do not balance, and the influx in the bypass is practically nil, meaningful estimates have to be related to the output. As shown in Table 1 with decreasing pH the share of the 2,3-DPG bypass on the glycolytic flux rises to 35% at pH_e 7.0. In the steady state, glucose consumption and lactate formation are equal within the limits of analytical errors. The glycolytic rate is higher than in the transient state during the first 6 hours. The reason is probably the diminished inhibition of the hexokinase-phosphofructokinase-system because of the low level of the inhibitor 2,3-DPG. The flux through the 2,3-DPG bypass is constant over the range of pH_e 7.4 to 7.2 and amounted to 24%. This is in good agreement with other data from the literature [51, 52]. At pH_e 7.0 the steady state concentration of 2,3-DPG corresponds to the K_m of the 2,3-DPGase and therefore the flux through the bypass is about half that at higher pH-values. The share of the bypass in the steady state is lower than that during the transient period owing to the changes in the glycolytic flux and in the flux through the bypass. The share amounted to 12% at pH_e 7.0.

Table 1
The share of the 2,3-DPG-bypass on the glycolytic flux in the transient and the steady state

pH_e	Glycolytic flux (µmoles/l cells)		Flux through the 2,3-DPG-bypass (µmoles/l cells)		Share of the bypass	
	transient state (pyruvate + lactate)	steady state (lactate)	transient state	steady state	transient state	steady state
7.4	—	2000	—	480	—	24%
7.0	1410	2000	480	240	35%	12%

If one compares these in vitro results in the 2,3-DPG metabolism with those under in vivo conditions one can say that both the marked pH dependence of the steady state 2,3-DPG-concentration as well as the time course of the changes of 2,3-DPG are in very good agreement with the observations in acidotic and hypoxic conditions in vivo [43, 46, 47, 53—56].
DPG attains a new steady state later than all other metabolites of the glycolytic chain. Therefore these results are an experimental proof for the existence of a time hierarchy in an enzymatic system [51].

References

[1] Hamasaki, N., and S. Minakami: J. Biochem., Tokyo **72**, 981 (1972)
[2] Oski, F. A., A. J. Gottlieb, W. W. Miller, and M. Delivoria-Papadopoulos: J. clin Invest. **49**, 400 (1970)
[3] Rörth, M.: Scand. J. clin. Lab. Invest. **26**, 43 (1970)
[4] Gerlach, E., and J. Duhm: Oxygen Affinity of Haemoglobin and Red Cell Acid Base Status. M. Rørth and P. Astrup (Eds). Munksgaard, Copenhagen 1972, p. 552
[5] Rapoport, I., H. Berger, S. Rapoport, R. Elsner, and G. Gerber: Biochim. biophys. Acta **428**, 193 (1976)
[6] Kühn, B., G. Jacobasch, and S. Rapoport: FEBS Lett. **38**, 354 (1974)

[7] ROSE, I. A., and J. V. B. WARMS: J. biol. Chem. 241, 4848 (1966)
[8] TSUBOI, K. K., and K. FUKUNAGA: J. biol. Chem. 240, 2806 (1965)
[9] MINAKAMI, S., and H. YOSHIKAWA: J. biol. Chem. 59, 145 (1966)
[10] RIZZO, S. C., and F. E. ECKEL: Am. J. Physiol. 211, 429 (1966)
[11] JACOBASCH, G.: Folia haemat. 89, 376 (1968)
[12] RAPOPORT, S.: Essays Biochem. 4, 69 (1968)
[13] RØRTH, M., and J. ASTRUP: Scand. J. clin. Lab. Invest., Suppl. 126 (1972)
[14] RAPOPORT, S., D. MARETZKI, Ch. SCHEWE, and G. JACOBASCH: Oxygen Affinity of Haemoglobin and Red Cell Acid Base Status. M. RØRTH and P. ASTRUP (Eds). Munksgaard, Copenhagen 1972, p. 528
[15] LICHTMAN, M. A., and D. A. MILLER: J. Lab. clin. Med. 76, 267 (1970)
[16] DITZEL, J.: Horm. Metab. Res. 5, 471 (1973)
[17] EATON, J. W., and G. J. BREWER: Proc. natn. Acad. Sci. U.S.A. 61, 756 (1968)
[18] HJELM, M.: Försvarsmedicin 5, 195 (1969)
[19] VALERI, R. C., C. G. ZAROULIS, and N. L. FORTIER: Oxygen Affinity of Haemoglobin and Red Cell Acid Base Status. M. RØRTH and P. ASTRUP (Eds). Munksgaard, Copenhagen 1972, p. 650
[20] GAHLENBECK, H., and H. BARTELS: Klin. Wschr. 46, 547 (1968)
[21] VERSMOLD, H., K. HORN, H. WINDTHORST, and K. P. RIEGEL: Respiration Physiol. 18, 26 (1973)
[22] RØRTH, M., and S. F. NYGAARD: Oxygen Affinity of Haemoglobin and Red Cell Acid Base Status. M. RØRTH and P. ASTRUP (Eds). Munksgaard, Copenhagen 1972, p. 599
[23] GORSHEIM, D., M. DELIVORIA-PAPADOPOULOS, F. A. OSKI, and F. H. GARDNER: Clin. Res. 20, 487 (1972)
[24] WOOD, L. A., and A. BEUTLER: Br. J. Haemat. 25, 611 (1973)
[25] KUROKAWA, I., T. KIMURA, T. KIDA, T. NAGAI, and M. KAMIMURA: J. Vitam. 18, 237 (1972)
[26] GERLACH, F., J. DUHM, and E. DEUTICKE: Adv. exp. Med. Biol. 6, 155 (1970)
[27] KRAVATH, R. E., G. ABEL, A. COLLI, H. McNAMARA, and M. I. COHEN: Biochem. Pharmac. 21, 2656 (1972)
[28] GILLIES, I. D. S., B. D. BIRD, J. NORMAN, E. C. GORDON-SMITH, and J. G. WHITWAM: Br. J. Anaest. 42, 561 (1970)
[29] BOWDLER, A. J., and T. A. J. PRANKERD: Acta haemat. 31, 65 (1964)
[30] DELIVORIA-PAPADOPOULOS, M., F. A. OSKI, and A. Y. GOTTLIEB: Science, N. Y. 165, 601 (1969)
[31] LABIE, D., M. COHEN SOLAL, and V. BYČKOVÁ: Bull. Soc. Chim. biol. 52, 1271 (1970)
[32] VERSMOLD, H., E. KOHNE, E. KLEIHAUER, and K. RIEGEL: Mschr. Kinderheilk. 121, 397 (1973)
[33] LABIE, D., J. P. LEROUSE, A. MAJMAN, and C. REYROLLE: FEBS Lett. 9, 37 (1970)
[34] CARTIER, P., D. LABIE, J. P. LEROUSE, A. NAJMAN, and F. DEMAUGRE: Nouv. Rev. fr. Hématol. 12, 269 (1972)
[35] RAPOPORT, S.: Biochem. Z. 289, 411 (1936)
[36] GUEST, G. M., and S. RAPOPORT: Am. J. Dis. Child. 58, 1072 (1939)
[37] RAPOPORT, S., and G. M. GUEST: J. biol. Chem. 129, 781 (1939)
[38] GUEST, G. M., and S. RAPOPORT: J. Lab. clin. Med. 26, 190 (1940)
[39] MINAKAMI, S., T. SAITO, C. SUZUKI, and N. YOSHIKAWA: Biochem. biophys. Res. Commun. 17, 748 (1964)
[40] ASAKURA, T., Y. SATO, S. MINAKAMI, and H. YOSKIKAWA: Clinica chim. Acta 14, 840 (1966)
[41] RØRTH, M.: Adv. exp. Med. Biol. 6, 57 (1970)
[42] DE LECUW, R. J. M., M. L. HAMELINK, and B. F. VISSER: Pflügers Arch. ges. Physiol. 328, 238 (1971)
[43] ASTRUP, P.: Försvarsmedicin 5, 199 (1969)
[44] ASTRUP, P.: Adv. exp. Med. Biol. 6, 67 (1970)
[45] BELLINGHAM, A. J., J. C. DETTER, and C. LENFANT: J. clin. Invest. 50, 700 (1970)
[46] ALBERTI, K. G. M. M., J. M. DARLEY, F. M. EMERSON and T. D. E. HOCKADAY: Lancet 1972/I, 391

[47] Raich, P. C., J. M. Rodriguez, J. N. Desai, and N. T. Shakidi: Am. J. med. Sci. **265**, 147 (1973)
[48] Rose, Z. B., and J. Liebowitz: J. biol. Chem. **245**, 3232 (1970)
[49] Harkness, D. R., and S. Roth: Biochem. biophys. Res. Commun. **34**, 849 (1969)
[50] Chiba, H., K. Ikura, M. Narita, and R. Sasaki: Acta biol. med. germ. **36**, 491 (1977)
[51] Rapoport, T. A., R. Heinrich and S. M. Rapoport: Biochem. J. **154**, 449 (1976)
[52] W. Schröter in: Stoffwechsel und Membranpermeabilität von Erythrocyten und Thrombocyten. I. Internationales Symposium, Wien 1968. E. Deutsch, E. Gerlach and K. Moser (Eds). Thieme, Stuttgart 1968, p. 50
[53] Lenfant, C., J. Torrance, E. Englisch, C. A. Finch, C. Reynafarje, J. Ramos, and J. Faura: J. clin. Invest. **47**, 2652 (1968)
[54] Astrup, P., M. Rørth, and C. Thorshauge: Scand. J. clin. Lab. Invest. **26**, 48 (1970)
[55] Duhm, J., and E. Gerlach: Pflügers Arch. ges. Physiol. **326**, 254 (1971)
[56] Rørth, M., S. F. Nygaard, H.-H. Parving, V. Hansen, and T. Kalsig: Scand. J. clin. Lab. Invest. **31**, 447 (1973)

Acta biol. med. germ., Band 36, Seite 523—529 (1977)

Uppsala Datacentral, Uppsala Universitet, Kliniskt-kemiska centrallaboratoriet, Akademiska sjukhuset, Uppsala, Sweden
and Fysiologisk Institut, Odense Universitet, Odense, Denmark

The molecular function of hemoglobin as reflected in ligand binding data: Analysis of data on erythrocytes

T. GROTH, C.-H. DE VERDIER, and L. GARBY

Summary

Hemoglobin oxygen binding data on erythrocytes at different pH, pCO_2 and bisphosphoglycerate concentrations have been analyzed in terms of an extended version of the Herzfeld-Stanley model of 1972.
The binding of oxygen to subunits when the tetramer is in the quaternary oxy conformation was found to be insensitive to moderate changes in pH and pCO_2 (0.71 ± 0.05 mm Hg^{-1}). Utilizing this circumstance it has been possible to obtain, for the first time, unique estimates of energy parameters related to hemoglobin cooperativity and effector action.
At 37 °C, pH 7.2 and pCO_2 22 mm Hg the following parameter values were obtained:
The allosteric constant: $(1.5 \pm 0.4) \cdot 10^4$;
the oxygen binding constant of the deoxy state: $(5.4 \pm 0.3) \cdot 10^{-3}$ mm Hg^{-1};
the 2,3-bisphosphoglycerate binding constants: $(3.3 \pm 1.3) \cdot 10^3 \, l \cdot mol^{-1}$ (deoxy), $(1.3 \pm 0.5) \cdot 10^2 \, l \cdot mol^{-1}$ (oxy).
Quaternary transition most likely takes place after binding of the second O_2 molecule.
Following the concepts of Perutz the results suggest that
(1) protons and carbon dioxide act as constraint effectors and/or as quaternary effectors;
(2) the difference in total conformational energy between the two quaternary ligand-free states is almost exclusively confined to molecular constraints and very little to the difference in quaternary conformational energy.
The consistency of the results indicate that the model may be regarded as a useful tool for the description of the functional interrelations in the hemoglobin oxygenation process as reflected in oxygen binding data.

Introduction

Interpretation of hemoglobin oxygen binding data in terms of models for hemoglobin function on a molecular level requires critical consideration of the selection of (1) a theoretical model, (2) the experimental design, and (3) the statistical and numerical methods for the analysis.
During the last two decades different theoretical models of varying sophistication have been published. In 1974 HERZFELD and STANLEY [1] proposed a general model for cooperativity of ligand binding to macromolecules. This model describes the particular states of a macromolecule in terms of differences in energy levels associated with the events that are assumed to take place, i.e. transitions between quaternary conformations and between tertiary conformations, binding of ligand to different conformational forms, interactions between quaternary and tertiary

This work was supported by the Danish Medical Research Council and the Swedish Medical Research Council (Project No. 13X—152) conformations, and changes in subunit aggregation. This formalism is attractive from a theoretical point of view. However, in the analysis of hemoglobin oxygen binding data the usefulness of such detailed models is limited by the difficulty of obtaining unique parameter estimates.

With regard to the experimental design of hemoglobin oxygen binding experiments it is important to consider the influence of hemoglobin and effector concentrations. In most previous studies, the hemoglobin concentration has generally been very low, between 0.01 and 0.12 mmol · l^{-1} tetramer. Recently ACKERS et al. [2] have shown that under these circumstances, the dissociation of tetramer molecules into dimers is not negligible and that the presence of even small amounts of dimers may profoundly affect the shape of the binding curve. Previous data on oxygen binding in whole blood are incomplete with respect to, e.g., information on the concentration of 2,3-bisphosphoglycerate (P_2-Gly) and/or red cell hydrogen ion activity, and thus cannot be used for the purposes of testing detailed models for hemoglobin cooperativity and effector action.

A proper numerical and statistical analysis also requires estimation and due consideration of confidence limits of estimated descriptive parameters. This is the primary basis for a critical judgement of model adequacy and of the significance of obtained numerical values.

In the present communication, several sets of oxygen binding data on intact erythrocytes at different values of pH, pCO_2 and [P_2-Gly] have been analysed in terms of an extended version of the Herzfeld-Stanley model from 1972 [3]. A less advanced study on mainly different data was reported by DE VERDIER et al. [4] in 1973. In accordance with the present general opinion we have, contrary to HERZFELD and STANLEY [1], regarded the oxy quaternary conformation to be without quaternary-tertiary constraints. Furthermore we have described the molecular constraints in the deoxy quaternary conformation in a more adequate way. The association constant of oxygen binding in the oxy quaternary state appeared to be insensitive to moderate changes in pH and pCO_2. Utilizing this circumstance, it has been possible to obtain unique estimates of energy parameters related to hemoglobin cooperativity and effector action.

Materials and methods

Experiments

The oxygen binding data were taken from GARBY et al. [5] and ARTURSON et al. [6]. Using a Radiometer DCA-1 oxygen binding curves were obtained for intact erythrocytes at 37 °C for pH (intracellular) of 7.0, 7.1, 7.2, 7.3: pCO_2 of 22, 77 mm Hg and [P_2-Gly]: 1.4, 4.4, 9.3 mmol · l^{-1}.

Computations

The fractional saturation of oxygen binding sites Y_{O_2}, was calculated from the two-state model by HERZFELD and STANLEY [3]. For the case that P_2-Gly is regarded purely as a quaternary effector Y_{O_2} can be written

$$Y_{O_2} = pO_2 \cdot \frac{K_q\,(1+\widetilde{K}_{\text{deoxy}}\cdot[\text{P}_2\text{-Gly}]_{\text{free}})\cdot(1+K_{\text{deoxy}}\cdot pO_2)^3\,K_{\text{deoxy}} + (1+\widetilde{K}_{\text{oxy}}\,[\text{P}_2\text{-Gly}]_{\text{free}})\cdot(1+K_{\text{oxy}}\cdot pO_2)^3\cdot K_{\text{oxy}}}{K_q\,(1+\widetilde{K}_{\text{deoxy}}\cdot[\text{P}_2\text{-Gly}]_{\text{free}})\cdot(1+K_{\text{deoxy}}\cdot pO_2)^4 + (1+\widetilde{K}_{\text{oxy}}\,[\text{P}_2\text{-Gly}]_{\text{free}})\cdot(1+K_{\text{oxy}}\cdot pO_2)^4} \tag{1}$$

where K_q is the equilibrium constant for the two quaternary conformations in the absence of oxygen, $\tilde{K}_{deoxy}$ and $\tilde{K}_{oxy}$ are the binding constants for P_2-Gly to the two quaternary conformations, and K_{deoxy} and K_{oxy} are the binding constants for oxygen to subunits when the quaternary conformation is given as stated.

The free concentration of P_2-Gly is calculated according to the formula

$$[P_2\text{-Gly}]_{free} = [P_2\text{-Gly}]_{tot} - [Hb] \cdot Y_{P_2\text{-Gly}} \quad (2)$$

where $Y_{P_2\text{-Gly}}$ is the fractional saturation of P_2-Gly. For details see [3].

Following the concepts of PERUTZ et al. [7, 8, 9] the free energy change related to quaternary transition (in the absence of oxygen) can be written

$$RT \ln K_q = E_q + E'_{qt} \quad (3)$$

where

E'_{qt} is the maximal energy related to molecular constraints imposed by bonds, e.g. salt-bridges, stabilizing the deoxy quaternary conformation in the ligand-free state.

E_q is the difference in energy between the two quaternary conformations (oxy-deoxy) exclusive of constraint energy.

Binding of oxygen to a subunit induces a change in tertiary conformation, so termed induced-fit binding. During oxygenation energy is released by the reaction of the heme groups with oxygen: RT ln K_{deoxy} and RT ln K_{oxy} for each subunit in the two quaternary conformations, respectively.

In the deoxy quaternary state part of this energy is expended

(1) to break the stabilizing salt-bridges,
(2) to increase the constraint energy due to strains imposed on the oxy tertiary structure when confined in the "wrong" quaternary structure.

Thus, we assume that the energy of the stabilizing salt-bridges goes from $-E'_{qt}$ to 0, and that the constraint energy due to "wrong" conformation increases from 0 to its maximal value E''_{qt} during oxygenation. Consequently the following relation holds

$$K_{deoxy} = K_{oxy} \cdot \exp\left[-(E'_{qt} + E''_{qt})/4RT\right] \quad (4)$$

i.e. K_{deoxy} is the constant for induced-fit oxygen binding of a subunit as constrained by the deoxy quaternary structure. In the unconstrained oxy quaternary conformation (cf. [10]) K_{oxy} is the constant for induced-fit binding.

The parameters (K_q, K_{deoxy}, K_{oxy}, $\tilde{K}_{deoxy}$ and $\tilde{K}_{oxy}$) were estimated by fitting the Y_{O_2}-function in a weighted least-square sense to the experimental data sets.

A computer program for non-linear least-squares curve fitting was used [11]. Weights were given to the data points as calculated from the variance of pO_2 measurements.

Results and discussion

The five-parameter model described by eqns. (1—2) was found to be compatible with all data sets. The deviations between theory and experiment were less than two standard deviations of the experimental error.

Despite strong correlation between K_q and K_{oxy} it turned out that "best" estimates of the oxygen binding constant were rather constant with a small scatter between values for different pH and $p CO_2$: mean value $K_{oxy} = 0.71 \pm 0.05$ mmHg^{-1}. The data were thus reanalysed assuming that K_{oxy} is insensitive to moderate changes in pH and $p CO_2$ (cf. [12], [13]). At $p\text{H} = 7.2$ and $p CO_2 = 22$ mmHg the following estimates were obtained:

$$K_q = (1.5 \pm 0.4) \cdot 10^4;\ K_{deoxy} = (5.4 \pm 0.3) \cdot 10^{-3}\ \text{mmHg}^{-1},$$

$$\tilde{K}_{deoxy} = (3.3 \pm 1.3) \cdot 10^3\ \text{l} \cdot \text{mol}^{-1} \text{ and } \tilde{K}_{oxy} = (1.3 \pm 0.5) \cdot 10^2\ \text{l} \cdot \text{mol}^{-1}).$$

The corresponding free energy changes on quaternary transition and binding of oxygen and P_2-Gly are:

-24.8 kJ · mol^{-1} (K_q), -21.4 kJ · mol^{-1} (K_{deoxy}),

-33.9 kJ · mol^{-1} (K_{oxy}), -20.9 kJ · mol^{-1} ($\tilde{K}_{deoxy}$) and

-12.7 kJ · mol^{-1} ($\tilde{K}_{oxy}$). In Fig. 1 the ten hypothesized states of the hemoglobin molecule are ordered on a relative energy scale in units of kJ · mol^{-1}. The differences in total energy between the two quaternary conformations at different stages of oxygenation are easily calculated from the estimates above. The figure gives a quantitative overview of the thermodynamics of hemoglobin oxygenation. The relative probabilities of the ten states were computed [3] from the estimated parameters and are shown in Fig. 2.

Figs. 1 and 2 indicate that

(1) quaternary transition most likely takes place after oxygenation of the second subunit,

(2) the probabilities of intermediary states are surprisingly low.

Binding of P_2-Gly, preferentially to the deoxy state, will decrease the deoxy energy levels relative to the energy levels of the corresponding oxy states by an amount of 8.2 to 20.9 kJ · mol^{-1}, moving the quaternary switch-over to a later stage of the oxygenation process (Fig. 3).

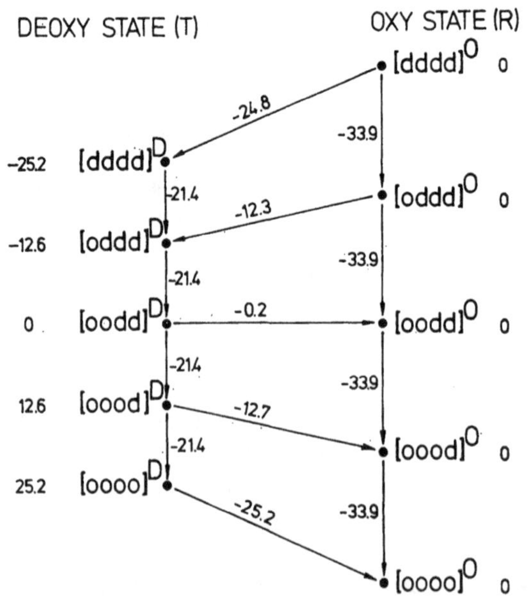

Fig. 1. The free energy levels of the ten different states in the process of hemoglobin oxygenation.

The columns of figures to the left and right give the total constraint energies of the different states (0 for the unconstrained oxy quaternary state) · d and o indicate the tertiary conformation of the subunit (deoxy and oxy respectively): D and O the quaternary conformation

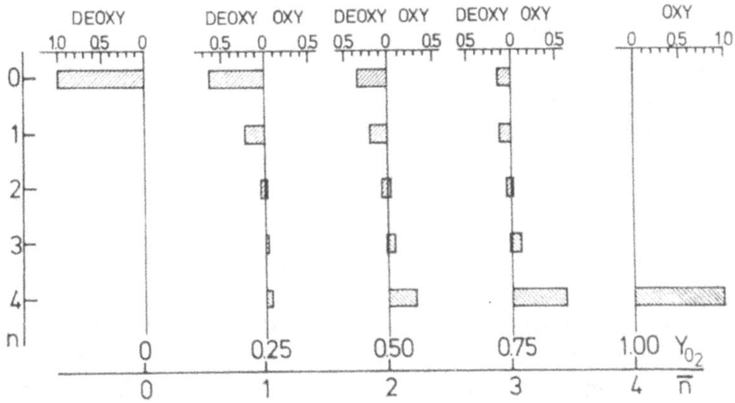

Fig. 2. Probabilities, expressed as mole fraction, for the ten hypothesized states of hemoglobin at five different levels of hemoglobin oxygenation (with an average number of 0, 1, 2, 3 and 4 subunits oxidized)

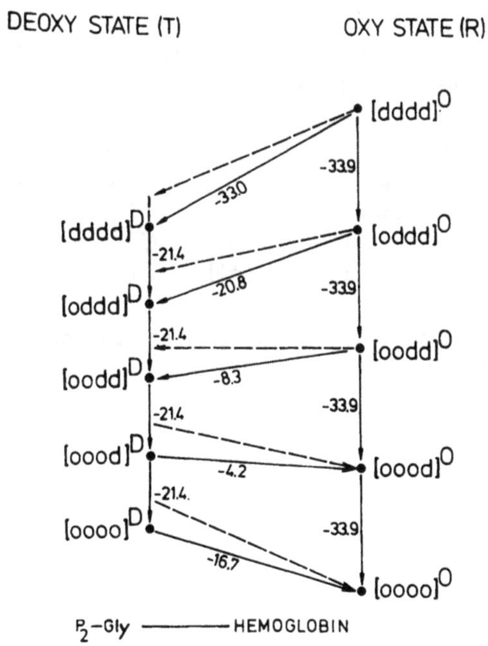

Fig. 3. Bisphosphoglycerate (P_2-Gly) influences the energy levels given in Fig. 1. The preferential binding of P_2-Gly to the deoxy state implies that the energy levels of the different subunit combinations in this state decrease with a value somewhere in the range between the energy level for P_2-Gly binding to the deoxy state and the difference between this value and the energy level for P_2-Gly binding to the oxy state. The diagram refers to the latter alternative

Interpretation of the estimated parameters in terms of molecular mechanisms

From equations (3) and (4) follows that it is only possible to estimate the sums $E_q + E'_{qt}$ and $E''_{qt} + E''_{qt}$, unless specific assumptions are made about the relation between two of the three parameters, or unless one is known from independent investigations.

The estimates of $K_q = 1.5 \cdot 10^4$, $K_{deoxy} = 0.054$ mm Hg^{-1} and $K_{oxy} = 0.71$ mm Hg^{-1} at $pH = 7.2$ and $pCO_2 = 22$ mm Hg give the sums $E_q + E'_{qt} = 24.8$ kJ $\cdot$ $\cdot$ mol^{-1} and $E'_{qt} + E''_{qt} = 50.4$ kJ $\cdot$ mol^{-1}. According to PERUTZ [7] „the bond energy per salt-bridge may reasonably be estimated as 1—2 kcal/mole". Assuming that the deoxy quaternary conformation is stabilized by six salt-bridges E'_{qt} obtains a value of 25.2—50.4 kJ $\cdot$ mol^{-1}. The lower limit of E'_{qt} gives an estimate of E''_{qt} (the maximal constraint energy due to "wrong" conformation) equal to 25.2 $\cdot$ kJ $\cdot$ mol^{-1}, and of E_q (the difference in energy between the two quaternary conformations) equal to -0.4 kJ $\cdot$ mol^{-1}. The upper limit of E'_{qt} gives $E''_{qt} = 0$ and $E_q = -25.6$ kJ $\cdot$ mol^{-1}.

These rough calculations show that our parameter estimates are consistent with PERUTZ' data on bond energy of stabilizing saltbridges. The lower value suggests that the difference in total conformational energy between the deoxy-oxy quaternary states (in the absence of ligand) is almost exclusively confined to molecular constraints and very little to the difference in quaternary conformational energy. This relation between E_q and E'_{qt} seems rather plausible as the bond energy of salt-bridges could be expected to be larger than the types of bonds lumped in E_q. Values of E'_q approaching the upper limit, on the other hand, seem less probable as E''_{qt} by definition should be > 0.

The effect of both protons and carbon dioxide was found to stabilize the quaternary deoxy conformation (Fig. 4). Following the concepts of PERUTZ this figure suggests that both protons and carbon dioxide act as constraint effectors (changing E'_{qt}) and/or quaternary effectors (changing E_q). The effect of carbon dioxide is negligible at pH 7.0 and increases with pH, in accordance with the idea that the changes are linked to formation of carbamate.

The oxygen binding constant, K_{deoxy}, showed a significant fall with pH. This is in agreement with results published by IMAI and YONETANI [12]. Furthermore, ac-

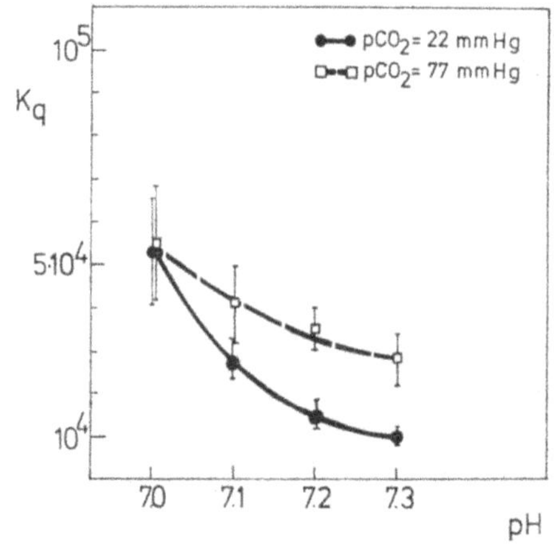

Fig. 4. The effect of pH and carbon dioxide tension on the quaternary equilibrium constant, K_q.

cording to the model concepts applied, protons acting on the salt-bridges (i.e. changing E'_{qt}) should affect not only the quaternary equilibrium (eqn. (3)), but also the oxygen binding affinity of subunits confined in the deoxy quaternary structure (eqn. (4)).

The consistency of the results discussed so far increases the credibility of the model concepts applied. Despite the fact that the model is an over-simplification of the real system, it may at least be regarded as a useful approximation for the description of functional interrelations in the hemoglobin oxygenation process as reflected in oxygen binding data. It does not seem motivated to use more detailed models in this connection, e.g. by introducing parameters related to the difference in oxygen binding of α and β subunits. Such overspecifications will just add to the problem of estimating unique parameter values, unless these new parameters can be established by independent investigations.

References

[1] Herzfeld, J., and H. E. Stanley: J. molec. Biol. **82**, 231 (1974)

[2] Ackers, G. K., M. L. Johnson, F. C. Mills, H. R. Halvorson, and S. Shapiro: Biochemistry **14**, 5128 (1975)

[3] Herzfeld, J., and H. E. Stanley: Biochem. biophys. Res. Commun. **48**, 307 (1972)

[4] de Verdier, C.-H., L. Garby, and T. Groth: Abhandlungen der Akademie der Wissenschaften der DDR, 283 (1973). (Proceedings of the VIIth International Symposium on Structure and Function of Erythrocytes, Berlin, 1973)

[5] Garby, L., M. Robert, and B. Zaar: Acta physiol. scand. **84**, 482 (1972)

[6] Arturson, G., L. Garby, B. Wranne, and B. Zaar: Acta physiol. scand. **92**, 332 (1974)

[7] Perutz, M. F.: Nature, Lond. **228**, 726 (1970)

[8] Perutz, M. F.: Nature, Lond. **237**, 495 (1972)

[9] Perutz, M. F., J. V. Kilmartin, K. Nagai, A. Szabo, and S. R. Simon: Biochemistry **15**, 378 (1976)

[10] Edelstein, S. J.: A. Rev. Biochem. **44**, 209 (1975)

[11] Metzler, C., G. Elfring, and A. McEven: Users Manual, Research Biostatistics. The Upjohn Comp., Kalamazoo, Mich. 1974

[12] Imai, K., and T. Yonetani: J. biol. Chem. **250**, 2227 (1975)

[13] Tyuma, I., K. Shimizu, and K. Imai: Biochem. biophys. Res. Commun. **43**, 423 (1971)

Acta biol. med. germ., Band 36, Seite 531—536 (1977)

[1]Abteilung für Transfusions- und Transplantationswesen des Bereiches Medizin (Charité) der Humboldt-Universität, 104 Berlin, DDR
[2]Bezirksinstitut für Blutspende- und Transfusionswesen, 113 Berlin, DDR

Biochemische Veränderungen von Erythrozyten während einer Lagerung bei 25 °C und Aufrechterhaltung eines konstanten *p*H-Wertes

G. Matthes[1] und D. Strauss[2]

Zusammenfassung

Untersuchungen zum Einfluß einer *p*H-Konstanthaltung auf das Stoffwechselverhalten konservierter Erythrozyten wurden über einen Zeitraum von 7 Tagen bei 25 °C-Lagerung von CD-Konservenblut durchgeführt. ACD-Konservenblut diente als Kontrollwert. Folgende Ergebnisse wurden erhalten:

1. Durch tägliche *p*H-Einstellung im CD-Konservenblut auf *p*H 7,4 wurde der extrazelluläre *p*H-Wert aufrechterhalten, der intrazelluläre *p*H-Wert sank nach 12 Std. bereits auf 7,2 ab.
2. Auf Grund des höheren initialen *p*H-Wertes und der nachfolgenden *p*H-Konstanthaltung im CD-Blut war eine größere Stoffwechselaktivität ersichtlich.
3. Die Erhöhung des Initial-*p*H und nachfolgende *p*H-Konstanz im CD-Blut führte zu einer verbesserten Erhaltung der zellulären Konzentrationen an ATP (bis zum 5. Tag der Lagerung) und 2,3-DPG (innerhalb der ersten 12 Std.).
4. Frühe Veränderungen biochemischer Parameter traten im Verlaufe der Konservierung besonders im Verhalten der Adeninnukleotide auf.

Einleitung

Während der Lagerung von Konservenblut treten Veränderungen im Erythrozytenstoffwechsel auf, die je nach Stabilisatorzusammensetzung ein unterschiedliches Ausmaß besitzen. Durch den Abfall organischer Phosphatester, speziell von zellulärem ATP und 2,3-DPG, wird die Lebensfähigkeit der Erythrozyten im Empfängerorganismus herabgesetzt und die O_2-Transportfunktion des Hämoglobin beeinträchtigt.

Im Verlaufe der Konservierung von Vollblut fällt der *p*H-Wert in der ACD-Konserve von etwa 7,0 auf 6,6 ab. Durch diese *p*H-Veränderung wird besonders der Gehalt an 2,3-DPG beeinflußt. Eine verbesserte Erhaltung von zellulärem 2,3-DPG wurde durch Erhöhung des Ausgangs-*p*H-Wertes in ACD- und CPD-Konservenblut beobachtet [1, 2]. Eine Aufrechterhaltung des *p*H-Wertes in der Konserve kann sowohl durch Anwendung geeigneter Puffersysteme als auch durch ständige Dialyse ermöglicht werden [3].

Das Ziel der vorliegenden Arbeit bestand in der Untersuchung des Einflusses einer *p*H-Konstanthaltung im Konservenblut während der Lagerung bei 25 °C im CD-Medium auf das Verhalten biochemischer Parameter und in der Erfassung frühzeitiger Veränderungen im Stoffwechsel konservierter Erythrozyten. Die Lagertemperatur von 25 °C wurde bewußt gewählt, weil dadurch ein kürzerer Beobachtungszeitraum ermöglicht wird und eine Übertragung von Ergebnissen der 25 °C-Konservierung auf die übliche 4 °C-Konservierung grundsätzlich möglich ist [4—6].

Material und Methoden

Blut wurde aus der Kubitalvene gesunder Spender entnommen und in 37 °C-temperierte Konservenflaschen eingefüllt (Verhältnis Stabilisator: Blut = 1:4). Die Zusammensetzung der verwendeten Stabilisatoren geht aus den Tab. 1 und 2 hervor. Nach Abnahme des Ausgangswertes wurden die Konservenflaschen sofort in ein 25 °C-Wasserbad überführt. Proben wurden zu folgenden Zeitpunkten entnommen: Stunden 0, 1, 2, 4, 8, 12; Tage 1, 2, 3, 4, 5, 6, 7. Täglich wurde der *p*H-Wert der Konserven auf *p*H 7,4 durch tropfenweise Zugabe von 0,2 N NaOH eingestellt. Alle *p*H-Bestimmungen erfolgten bei einer Temperatur von 25 °C, mit Ausnahme des Ausgangswertes (37 °C). Die angegebenen Werte sind jeweils das Mittel aus drei Bestimmungen.

Tabelle 1

Zusammensetzung der verwendeten Stabilisatorlösungen (Angaben in g/l)

Substanz	Mol.-Gewicht	ACD	CD
Glukose · H_2O	198.0	25.0	25.0
Natriumzitrat · 2 H_2O	294.1	13.2	13.2
Zitronensäure · H_2O	210.0	4.8	—

Tabelle 2

Charakteristik der Stabilisatorlösungen

Kriterium	ACD	CD
Zitratgehalt im Konservenblut (mM)	13.0	9.5
Glukosegehalt im Konservenblut (mM)	25.0	25.0
Osmolarität, gemessen (mOsm)	344	330
*p*H	5.1	7.4

Folgende Methoden wurden verwendet:

ATP, ADP, AMP und Laktat wurden enzymatisch unter Verwendung von Testbestecken der Fa. Boehringer, Mannheim, gemessen. Die Glukosebestimmungen wurden im Automatenlabor des Klinikums Berlin-Buch durchgeführt. Die 2,3-DPG-Konzentrationen wurden nach der enzymatischen Mikromethode nach Nygaard und Rørth [7] durchgeführt. Die Bestimmung des extrazellulären *p*H-Wertes (pH_e) wurde am *p*H-Meßgerät der Fa. Clamann und Grahnert, Dresden, durchgeführt. Die Ermittlung der intrazellulären *p*H-Werte (pH_i) erfolgte nach dreimaliger Gefrierhämolyse der konservierten Erythrozyten. Hämoglobin im Vollblut wurde als Zyanmethämoglobin nach DAB 7 bestimmt.

Ergebnisse

Der Ausgangs-*p*H-Wert in der CD-Konserve beträgt bei 37 °C 7,4. Nach Überführung in das 25 °C-Wasserbad wurde mit 0,5 ml 0,1 M Zitronensäure auf einen *p*H-Wert von 7,4 eingestellt. Der extrazelluläre *p*H-Wert fiel innerhalb von 24 Std. auf 7,1 ab und wurde nachfolgend täglich mit 0,2 N NaOH auf 7,4 gehalten Zur Einstellung des *p*H-Wertes waren jeweils folgende Mengen an 0,2 N NaOH notwendig: 1. Tag: 1,6 ml, 2. Tag: 1,6 ml, 3. Tag: 1,4 ml, 4. Tag: 1,2 ml, 5. Tag: 1,0 ml, 6. Tag: 0,4 ml. Der intrazelluläre *p*H-Wert im CD-Blut betrug bei 37 °C 7,38 und fiel im Verlaufe des ersten Lagerungstages auf 7,1 ab. Im Gegensatz zum ACD-Blut wurden hierbei jeweils vor der NaOH-Zugabe identische extrazelluläre und intrazelluläre *p*H-Werte ermittelt (Abb. 1). Im ACD-Blut fiel der *p*H-Wert von 6,95 innerhalb von 7 Tagen auf 6,4 ab. Die Differenz zwischen pH_i und pH_e betrug anfangs 0,1 *p*H-Einheiten und wurde bis zum 3. Tag der Konservierung größer (3. Tag: 0,25 *p*H-Einheiten), um nachfolgend zum Ende der Lagerung wieder abzunehmen. In Abb. 2 ist das Verhalten des zellulären ATP, ADP und AMP im Verlaufe der Lagerung bei 25 °C in ACD- und CD-Blut dargestellt. Von einem mittleren Ausgangswert von etwa 0,220 Mol ATP/Mol fällt der ATP-Gehalt nach 6 Tagen auf etwa 20% des Ausgangswertes. Auffällig ist die deutlich bessere Erhaltung von ATP im CD-Blut bei *p*H-Konstanthaltung bis zum 5. Tag der Konservierung. Sowohl im ACD- als auch im CD-Blut treten in den ersten 12 Std. zwischenzeitlich Anstiege über den Ausgangswert hinaus auf. Die Konzentrationen von zellulärem ADP und AMP fallen von Ausgangswerten um

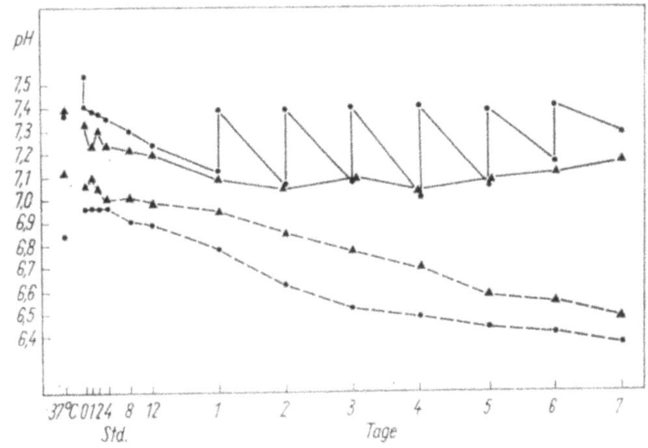

Abb. 1. Veränderungen des extrazellulären und intrazellulären *p*H-Wertes im Konservenblut bei 25 °C-Lagerung.
ACD-Blut — — —, CD-Blut ———, pH_i ●, pH_e ▲

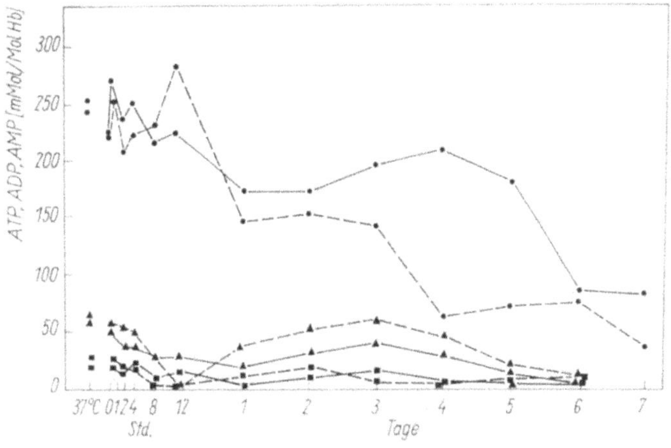

Abb. 2. Verhalten der Adeninnukleotide im Konservenblut bei 25 °C-Lagerung.
ACD-Blut — — —, CD-Blut ———, ATP ●, ADP ▲, AMP ■

etwa 0,065 Mol ADP/Mol Hb bzw. 0,025 Mol AMP/Mol Hb im Verlaufe der ersten 12 Std. stark ab, deutlicher ausgeprägt im ACD-Blut. Nachfolgend ist von diesem Zeitpunkt ab ein Anstieg der Konzentrationen von ADP und AMP zu beobachten bis zu einem Maximum am 3. Lagerungstag. Zwischen 4. und 5. Lagerungstag fallen die ADP-Konzentrationen auf Werte um 0,01 Mol ADP/Mol Hb ab. Der zelluläre 2,3-DPG-Spiegel fällt im ACD-Blut innerhalb des ersten Lagerungstages von Ausgangswerten um 1,3 Mol 2,3-DPG/Mol Hb auf 0,15 Mol 2,3-DPG/Mol Hb ab. Im CD-Blut ist der Abfall von 2,3-DPG verlangsamt, jedoch werden am zweiten Lagerungstag ebenfalls Werte um 0,15 Mol 2,3-DPG/Mol Hb gemessen. Bemerkenswert ist, daß im CD-Blut erst nach 12 Std. ein verstärkter Abbau von 2,3-DPG auftritt (Abb. 3).

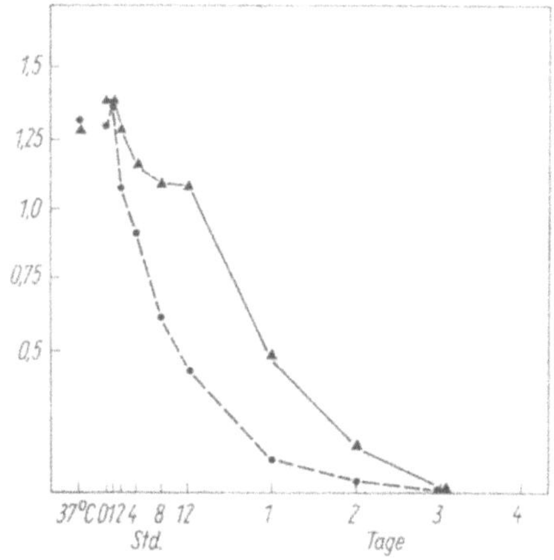

Abb. 3. Verhalten der 2,3-DPG-Konzentrationen im Verlaufe der Lagerung bei 25 °C. ACD-Blut ●– – –●, CD-Blut ▲——▲

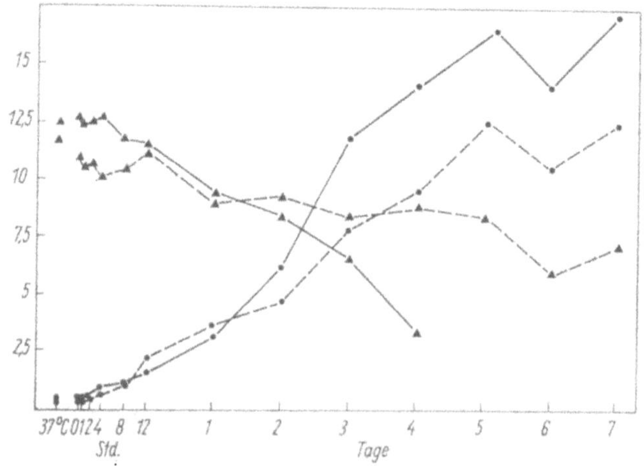

Abb. 4. Veränderungen der Laktat- und Glukosekonzentration konservierter Erythrozyten im Verlaufe der Lagerung bei 25 °C.
ACD-Blut – – –, CD-Blut ——, Glukose ▲, Laktat ●

Die Laktatkonzentrationen im CD- und ACD-Blut betragen am Anfang der Konservierung etwa 0,5 Mol Laktat/Mol Hb. Im Verlaufe der Lagerung steigen die zellulären Konzentrationen von Laktat nahezu linear im ACD-Blut auf Werte um 12,4 Mol Laktat/Mol Hb an. Eine deutlich stärkere Laktatproduktion der Zellen ist im CD-Blut ab 1. Tag zu beobachten; am 5. Lagerungstag beträgt die Laktatkonzentration 16,5 Mol Laktat/Mol Hb. Vom 5. Tag der Konservierung an ist in beiden Konservenbluten ein Sistieren der Laktatproduktion zu sehen. Im Verhalten der Glukosekonzentrationen der Konservenblute ist entsprechend vergleichbar mit der Laktatproduktion ab 1. Tag der Lagerung ein stärkerer Abfall des Glukosespiegels im CD-Blut zu beobachten (Abb. 4).

Diskussion

Durch den Abfall des *p*H-Wertes im Konservenblut ergeben sich im Verlaufe der Lagerung umfangreiche Veränderungen im Stoffwechsel der Erythrozyten, wobei besonders die Konzentrationsabnahme an zellulärem 2,3-DPG aus klinischer Sicht bedeutungsvoll ist (Aufrechterhaltung der O_2-Transportfunktion des Hämoglobins). Eine Konstanthaltung des *p*H-Wertes während der Lagerung sollte diesen Veränderungen entgegenwirken. Durch die Verwendung eines leicht alkalischen CD-Mediums (*p*H 7,4) als Stabilisatorlösung und der täglichen Einstellung des extrazellulären *p*H-Wertes auf 7,4 im Konservenblut wurde versucht, eine Aufrechterhaltung des zellulären 2,3-DPG zu erreichen. ACD-Blut diente als Kontrollwert und zum Vergleich.

Ein neutraler oder leicht alkalischer *p*H-Wert des Konservenblutes hat auf die Erhaltung mit 2,3-DPG konservierter Erythrozyten deutlichen Einfluß. Verschiedene Autoren [1, 2, 8—10] berichten über eine verbesserte Erhaltung von zellulärem 2,3-DPG in ACD- und CPD-Blut mit erhöhtem Ausgangs-*p*H-Wert und führen dies auf eine verminderte Aktivität der DPGase sowie auf die Aktivierung von Schlüsselenzymen der Glykolyse (wie Hexokinase und Phosphofruktokinase) zurück. Wie in Abb. 3 ersichtlich ist, wird im CD-Blut durch tägliche *p*H-Einstellung auf einen Ausgangswert von 7,4 bei 25 °C eine Erhaltung des 2,3-DPG der Zellen innerhalb der ersten 12 Std. der Lagerung möglich. Der nach diesem Zeitpunkt parallel zum ACD-Blut eintretende stärkere Abfall ist möglicherweise auf den *p*H-Abfall im Konservenblut zurückzuführen, da sowohl der intrazelluläre als auch der extrazelluläre *p*H-Wert nach 12 Std. Werte um 7,2 erreicht hat und der intrazelluläre *p*H-Wert trotz täglicher *p*H-Einstellungen des Konservenblutes auf 7,4 im Verlaufe der Lagerung weiter abfällt. Die *p*H-Konstanthaltung im CD-Blut hat einen deutlichen Einfluß auf die Stoffwechselaktivität der konservierten Erythrozyten im Gegensatz zum ACD-Blut. Vom ersten Tag der Lagerung an ist sowohl eine stärker ausgeprägte Laktatbildung als auch ein größerer Abfall der Glukosekonzentration zu beobachten. Vom 5. Lagerungstag an scheint ein Versiegen der Laktatproduktion und der Glukoseverwertung in beiden Konservenbluten vorzuliegen, die zumindest nicht für CD-Blut mit der fast vollständigen Hemmung der Glykolyse ab *p*H-Werten von 6,6 zu erklären sind [11]. Parallel zu diesem Verhalten ist die Abnahme der ATP-Konzentration zum 6. Lagerungstag hin anzusehen.

Veränderungen im Verhalten der Adeninnukleotide (ATP, ADP und AMP) sind besonders in der ersten Phase der Konservierung ersichtlich und als Adaptation an die in der Konserve vorliegenden Bedingungen aufzufassen. ATP-Anstiege über den Ausgangswert hinaus werden durch den schnellen Abfall des zellulären 2,3-DPG erklärlich (deutlicher ausgeprägt im ACD-Blut [12]). Die längere Erhaltung des zellulären ATP-Spiegels im CD-Blut mit *p*H-Konstanthaltung ist Vergleich mit ACD-Blut besonders zwischen dem 2. und 5. Lagerungstag ausgeprägt.

An Hand der ermittelten Daten ist ersichtlich, daß sich durch die gewählte Methode zur Aufrechterhaltung des *p*H-Wertes auf 7,4 deutliche Unterschiede im Verhalten der gemessenen biochemischen Parameter ergeben, jedoch eine Aufrechterhaltung sowohl von ATP-Gehalt als auch 2,3-DPG-Konzentration der Zellen über den Lagerungzeitraum hinweg nicht möglich wird, da die Konzentrationen von 2,3-DPG bereits nach 12 Std. stark absinken.

Literatur

[1] Dawson, R. B., M. R. Loken u. D. H. Crater: Transfusion **12**, 46 (1972)
[2] Dawson, R. B., F. R. Camp u. N. F. Conte: Milit. Med. **139**, 300 (1974)
[3] Kleine, N., u. H. Schmitt: Blut **8**, 145 (1962)
[4] Strauss, D., S. Tänzer, F. Schmutzler, L. Gülke, H. P. Rettig, H. Roigas u. H. J. Raderecht: Folia haemat., Lpz. **101**, 243 (1974)
[5] Strauss, D. u. H. J. Raderecht: Folia haemat., Lpz. **101**, 232 (1974)
[6] Bartel, U., L. Manitz, G. Matthes u. F. Weber: Dissertation. Humboldt-Universität Berlin, Med. Fakultät 1974
[7] Nygaard, S. F., u. M. Rørth: Scand. J. clin. Lab. Invest. **24**, 399 (1969)
[8] Asakura, T., Y. Sato, S. Minakami u. H. Yoshikawa: Clinica chim. Acta **14**, 840 (1966)
[9] Beutler, E., A. Meul u. L. A. Wood: Transfusion **9**, 109 (1969)
[10] Dawson, R. B., W. F. Kocholaty u. J. L. Gray: Transfusion **10**, 299 (1970)
[11] Blum, K. U. in: Modern Problems of Blood Preservation. W. Spielmann u. S. Seidl (Hrsg.). Fischer, Stuttgart 1970, S. 21
[12] de Verdier, C. H., C. Högman, L. Garby u. J. Killander: Acta physiol. scand. **60**, 141 (1964)

Anschrift der Verfasser: Dr. Gert Matthes, Humboldt-Universität Berlin, Bereich Medizin (Charité), Abteilung für Transfusions- und Transplantationswesen 104 Berlin, Schumannstr. 20/21; Dr. Dorothea Strauss, Bezirksinstitut für Blutspende- und Transfusionswesen Berlin, 113 Berlin, Atzpodienstraße 11—13

Acta biol. med. germ., Band 36, Seite 537—542 (1977)

Bezirksinstitut für Blutspende- und Transfusionswesen, 113 Berlin, DDR

Einfluß von Bikarbonat auf den 2,3-DPG-Gehalt konservierter Erythrozyten

V. STIGGE, E. STIGGE und D. SCHWANKE

Zusammenfassung

ACD-AG-Vollblutkonserven (Hämatokritwert 36%) und Erythrozytenresuspensionen (Hämatokritwert 80%) wurden unter Zusatz von Bikarbonat bei 4 °C gelagert. Bei den Erythrozytenresuspensionen erfolgte gleichzeitig ein Zusatz von Xylit und anorganischem Phosphat. Die erythrozytäre 2,3-DPG-Konzentration wurde in Abhängigkeit von der Bikarbonatkonzentration in Vollblutkonserven länger erhalten. In Erythrozytenresuspensionen ist die Erhaltung von 2,3-DPG darüber hinaus noch wesentlich verbessert.

Aus den Ergebnissen und Literaturangaben wird geschlossen, daß Bikarbonat an Metabolite des Pentosestoffwechsels fixiert wird und somit direkt am Stoffwechsel der Erythrozyten beteiligt ist.

Die Produkte der Bikarbonatfixierung können den 2,3-DPG-Haushalt der Erythrozyten beeinflussen.

Einleitung

Die Bedeutung des 2,3-Diphosphoglyzerates (2,3-DPG) für die Sauerstofftransportfunktion des Hämoglobins hat zu Versuchen geführt, das schnelle Absinken der 2,3-DPG-Konzentration in den Erythrozyten während der Blutkonservierung zu verhindern. Durch geeignete Zusätze und *p*H-Veränderungen sollte der Stoffwechsel der roten Blutzellen so beeinflußt werden, daß der Abbau des 2,3-DPG gehemmt und/oder die 2,3-DPG-Synthese gefördert werden. Insbesondere *p*H-Erhöhungen, Zusätze von Pyruvat, Fruktose, Dihydroxyazeton und anderer Stoffe wurden erprobt.

BEUTLER und WOOD [1] versuchten mit Erfolg, durch Zusätze von Bikarbonat zu im Verhältnis 1 + 1 resuspendierten Erythrozyten eine positive Stoffwechselwirkung herbeizuführen. Sie benutzten CO_2-diffusible Plastebeutel zur Lagerung des Blutes und beobachteten starke 2,3-DPG-Veränderungen. Eine Pufferwirkung des leichtflüchtigen Kohlensäure-Bikarbonat-Systems in den Plastebeuteln wurde nicht ausgeschlossen.

Material und Methodik

Es wurden zwei Untersuchungskomplexe bearbeitet:

A. Venenblut von fünf unausgewählten, klinisch gesunden Blutspendern wurde bei der Abnahme aus der Vena cubitalis in je fünf 120 ml-Blutkonservenflaschen mit den verschiedenen Konservierungslösungen im Verhältnis 4 + 1 gemischt. Der Hämatokritwert lag bei 36%.

Bei den Konservierungslösungen handelte es sich um ACD-AG-Lösungen üblicher Zusammensetzung [2] mit unterschiedlichen Zusätzen von Bikarbonat. Durch Variation des Zitronensäure/Zitrat-Verhältnisses wurde vor der Sterilisation der *p*H-Wert der Lösung so eingestellt, daß der Anfangs-*p*H-Wert in jedem Konservenblut — die Kontrolle ausgenommen — bei 7,4 lag.

Tabelle 1
Zusammensetzung der in Versuch B verwendeten Lösungen

Lösung Nr.	1	2	3	4
Saccharose	234	204	–	–
Glukose	30	30	30	–
Xylit	–	30	30	30
Sorbit	–	–	–	30
Na_3-Zitrat	9,5	9,5	–	–
$NaHCO_3$	–	–	100	100
Na_2HPO_4	–	–	25	25
Adenin	2,5	2,5	2,5	2,5
Guanosin	2,5	2,5	2,5	2,5
Anfangs-*p*H	6,99	7,07	7,32	7,40

Angaben in mMol/l Resuspensionslösung

B. Erythrozytenkonzentrate aus ÄDTA-Blutkonserven wurden sofort nach Abhebern des Plasmas zum Zwecke der Plasmafraktionierung auf vier 120 ml-Blutkonservenflaschen aufgeteilt und durch Zusatz der Resuspensionslösungen unterschiedlicher Zusammensetzung (Tab. 1) im Verhältnis 4 + 1 resuspendiert. Es resultierte ein Hämatokritwert von etwa 80%. Der Stichprobenumfang betrug $n = 3$.
Die Sterilisation aller verwendeten Lösungen erfolgte in Blutkonservenflaschen nach dem Verfahren a 1 [2].
Die Blutkonserven wurden bei 4 ± 2 °C über 6 Wochen gelagert. In wöchentlichen Abständen wurden nach gründlichem Mischen unter aseptischen Bedingungen Blutproben zur Untersuchung entnommen.
Die *p*H-Messung wurde bei 25 °C mit einer Kapillarglaselektrode durchgeführt, die 2,3-DPG-Bestimmung nach dem modifizierten Verfahren von Nygaard and Rørth [3] im gekoppelten optischen Test [4]. Der methodische Fehler in der Serie lag bei 5%.
Der 2,3-DPG-Gehalt wurde auf die Hämoglobinkonzentration bezogen. Die dazu erfoderliche Hämoglobinbestimmung erfolgte als Zyanmethämoglobin mit einem mittleren methodischen Fehler in der Serie von 2% [2].
Signifikanzberechnungen wurden für die im Abschnitt A beschriebenen Untersuchungen mit dem *t*-Test, für die im Abschnitt B mit dargestellten dem *U*-Test durchgeführt. Zugrundegelegt wurde eine Irrtumswahrscheinlichkeit von 5%.

Ergebnisse

Bei einem Anfangs-*p*H-Wert von 7,0, wie er in üblichen ACD-AG-Blutkonserven besteht, fällt die erythrozytäre 2,3-DPG-Konzentration innerhalb von 2 Wochen Kältelagerung auf weniger als 5% der Anfangskonzentration ab. Bei Erhöhung des Anfangs-*p*H-Wertes auf 7,4 bleibt die 2,3-DPG-Konzentration für eine Woche vollständig erhalten. Ein Wert von 5% der Ausgangskonzentration wird erst nach 4 Wochen erreicht (Abb. 1). In Abhängigkeit von der Bikarbonatkozentration in den Vollblutkonserven wird zusätzlich zu diesem *p*H-Effekt die 2,3-DPG-Erhaltung signifikant verlängert (Tab. 2). Die *p*H-Werte unterscheiden sich während der Lagerung in diesen Bluten nicht (Abb. 2).
Im Falle der Erythrozytenresuspensionen mit einen Hämatokritwert von 80% sinkt die 2,3-DPG-Konzentration bei vergleichbaren Anfangs-*p*H-Werten etwas langsamer als in Vollblutkonserven (Abb. 3). Unter Zusatz von Xylit wird 2,3-DPG darüber hinaus signifikant länger erhalten, was den Befunden in Vollblut

Tabelle 2

Erythrozytäre 2,3-DPG-Konzentration nach 2 Wochen 4 °C-Lagerung in ACD-AG-Vollblutkonserven in Abhängigkeit von der Bikarbonat-Endkonzentration im Blut bei einem Anfangs-*p*H-Wert von 7,4

Nr.	Bikarbonatkonzentration (mMol/l Blut)	2,3-DPG-Konzentration (% des Ausgangswertes)
1*	0	4
2	0	52
3	4	73
4	10	90
5	20	102

* = Kontrolle mit Anfangs-*p*H von 7,0

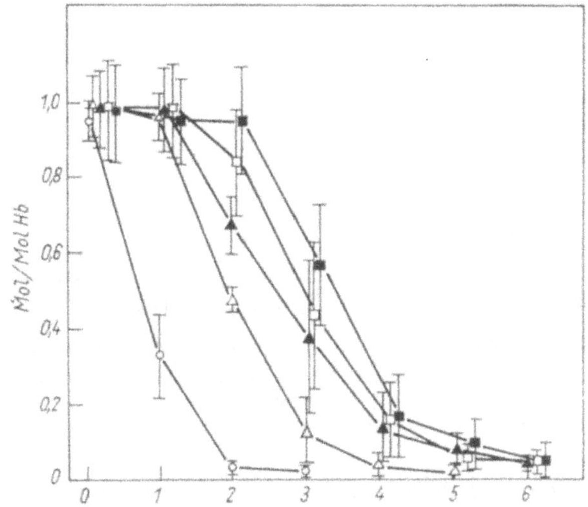

Abb. 1. Abhängigkeit der 2,3-DPG-Konzentration im ACD-AG-Konservenblut von der Bikarbonatkonzentration während einer 4 °C-Lagerung ($\bar{x} \pm s$; $n = 5$). ○ Nr. 1, △ Nr. 2, ▲ Nr. 3, □ Nr. 4, ■ Nr. 5 (vgl. Tab. 2)

entspricht [5]. Gleichzeitiger Zusatz von Bikarbonat hat zusätzlich eine vollständige Erhaltung der 2,3-DPG-Konzentration bis zu 5 Lagerungswochen zur Folge.

Diskussion

Nachdem schon 1940 von Scudder und Smith [6] und 1961 von Nashat et al. [7] bei der Aufbewahrung von Erythrozyten in CO_2-Atmosphäre eine positive Beeinflussung einiger wichtiger Eigenschaften von Erythrozyten (Überlebensrate, Spontanhämolyse, Kationenaustausch) festgestellt wurde und nachdem durch Übertragung der Erkenntnisse von Jakoby et al. [8] über die CO_2-Fixierung in Pflanzen auf den Erythrozytenstoffwechsel durch Fortier et al. [9] die Enzyme

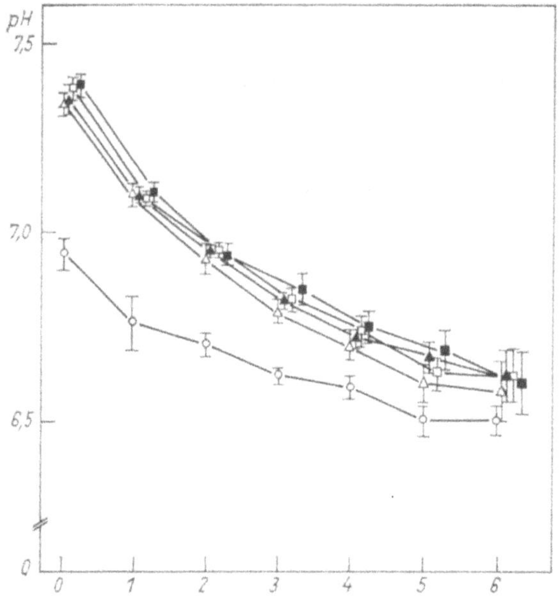

Abb. 2

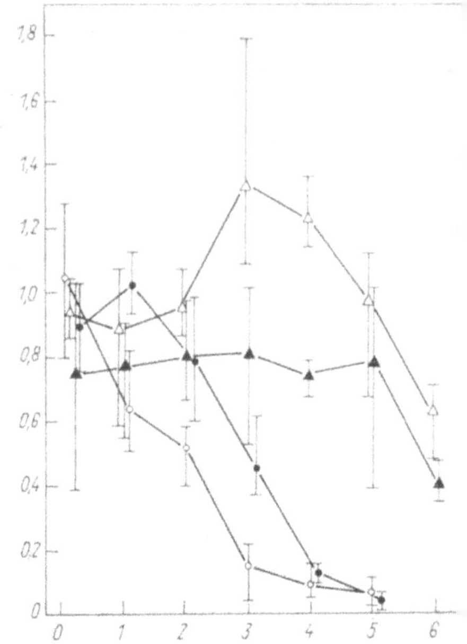

Abb. 3

Abb. 2. Verhalten des *p*H-Wertes in ACD-AG-Blutkonserven mit verschiedenen Bikarbonat-Konzentrationen ($\bar{x} \pm s$; $n = 5$). Zeichenerklärung vgl. Abb. 1

Abb. 3. Verhalten der 2,3-DPG-Konzentration in Erythrozytenresuspensionen aus ÄDTA-Blut bei unterschiedlicher Zusammensetzung der Resuspensionslösungen während der 4 °C-Lagerung.
○ Nr. 1, ● Nr. 2, △ Nr. 3, ▲ Nr. 4 (vgl. Tab. 1). $n = 3$

Phosphoribulokinase und Ribulosediphosphatkarboxylase in Erythrozyten nachgewiesen wurden, konnte die Teilnahme von CO_2 am Stoffwechsel der roten Blutzellen wahrscheinlich gemacht werden.

Nach JAKOBY et al. [8] wird CO_2/Bikarbonat an Ribulosediphosphat unter Bildung eines C6-Körpers fixiert, der schnell zu Glyzerinaldehydphosphat (GAP) und Phosphohydroxypyruvat zerfallen soll. Während GAP in der Glykolyse die Bereitstellung von 1,3-DPG für die 2,3-DPG-Bildung sichert, wird beim Umsatz von Phosphohydroxypyruvat zu 3-Phosphoglyzerat (3-PG) das in der Glykolyse entstehende NADH wieder oxydiert, so daß die Laktatbildung in den Erythrozyten unter diesen Bedingungen stark vermindert ist [9].

Der Einfluß dieser Reaktionen auf den NADH/NAD^+-Quotienten und die Effektorwirkung des 3-PG auf den Diphosphoglyzerat-Zyklus könnten eine gesteigerte 2,3-DPG-Bildung und verminderten 2,3-DPG-Abbau zur Folge haben. Die signifikant verbesserte Erhaltung des 2,3-DPG unter Zusatz von Bikarbonat könnte durch diese Zusammenhänge verursacht sein.

Xylit als Substrat für Erythrozyten kann nach Dehydrierung und Phosphorylierung zu Xylulose-5-phosphat die Bedingungen für eine CO_2-Fixierung an Ribulosediphosphat verbessern helfen. In Gegenwart von Xylit kommt es in Erythrotenresuspensionen mit Zusatz von Bikarbonat zu wesentlich längerer Erhaltung der 2,3-DPG-Konzentration als in den xylitfreien, aber bikarbonathaltigen Vollblutkonserven. Einseits ist der Stoffwechsel des Xylit nicht so empfindlich gegen das im Konservenblut während der Lagerung absinkende *p*H, andererseits wird bei der Fixierung des Bikarbonats infolge des NADH-Verbrauchs durch Phosphohydroxypyruvat das beim Xylitabbau entstehende NADH oxydiert. NADH kann damit nicht den Umsatz des GAP zu 1,3-DPG behindern.

Die wahrscheinliche Beteiligung des NADH/NAD^+-Systems an den Stoffwechselprozessen zur Erhaltung des 2,3-DPG geht auch aus der veränderten Situation bei Ersatz der Glukose durch Sorbit hervor. Sorbit kann nur nach Dehydrierung am Erythrozytenstoffwechsel teilnehmen. Dabei in geringem Maße zusätzlich entstehendes NADH muß den NADH/NAD^+-Quotienten verschieben. Da die LDH-Reaktion dann am NADH-Umsatz stärker beteiligt wird, erfolgt eine Beeinflussung der 3-PG-Konzentration, und die 1,3-DPG-Bildung wird möglicherweise durch leichten NAD^+-Mangel in der GAPD-Reaktion gestört. Daraus resultiert eine verminderte 1,3-DPG-Bereitstellung für die 2,3-DPG-Bildung und eine geringere Effektorwirkung des 3-PG auf den Diphosphoglyzerat-Zyklus.

Es ist zwar kein Anstieg des 2,3-DPG während der Lagerung wie beim Glukose/Xylit-Umsatz, zu verzeichnen eine langfristige Erhaltung des 2,3-DPG beim Ausgangswert mit seinen positiven Auswirkungen auf die Sauerstofftransportfunktion des Hämoglobins kann aber trotzdem beobachtet werden.

Den Erythrozytenresuspensionen wurde neben Bikarbonat gleichzeitig anorganisches Phosphat hinzugesetzt. Obwohl nach Untersuchungen von SCHWANKE [10] ein Zusatz von Phosphat allein in der gewählten Konzentration (25 mol/l Lösung = 5 mMol/l Suspension) eine Verschlechterung der 2,3-DPG-Erhaltung verursacht, könnte in Gegenwart von Bikarbonat eine Phosphatsubstitution wegen des intensiveren Stoffwechsels für die GAPD-Reaktion möglicherweise sogar erforderlich werden und damit für optimale Bedingungen zur Erhaltung der 2,3-DPG-Konzentration über mehrere Wochen mitverantwortlich sein.

Literatur

[1] BEUTLER, E., u. L. A. WOOD: Vox sang. **20**, 403 (1971)
[2] Deutsches Arzneibuch der DDR 7. Aufl. Akademie-Verlag, Berlin 1968
[3] NYGAARD, S. F., u. M. RØRTH: Scand: J. clin. Lab. Invest. **24**, 399 (1969)
[4] STIGGE, E.: Dissertation. Akademie für ärztlliche Fortbildung, Berlin 1976
[5] STIGGE, V., D. STRAUSS, H. ROIGAS u. H. J. RADERECHT: Dte. Gesundheits Wes. **28**, 1805 (1973)
[6] SCUDDER, J., u. SM. SMITH: Cold Spring Harb. Symp. quant. Biol. **8**, 269 (1940)
[7] NASHAT, F. S., A. K. YASSIN u. S. SUBHIYAH: Archs. int. Pharmacodyn. Thér. **130**, 86 (1961)
[8] JAKOBY, W. B., D. O. BRUMMOND u. S. OCHOA: J. biol. Chem. **218**, 811 (1956)
[9] FORTIER, N. L., L. GALLAND u. F. J. LIONETTI: Archs. Biochem. Biophys. **119**, 69 (1967)
[10] SCHWANKE, D.: Ingenieur-Arbeit. Ingenieur-Schule für Chemie, Berlin 1975

Für die Autoren. Dr. V. STIGGE, Bezirksinstitut für Blutspende- und Transfusionswesen, 113 Berlin, DDR, Atzpodienstr. 11—13

Summary

V. STIGGE, E. STIGGE, and D. SCHWANKE: Effect of bicarbonate on the content of 2,3-DPG of stored erythrocytes

Both whole blood in ACD-adenine-guanosine solution (PCV about 36%) and resuspensions of erythrocytes from EDTA-blood (PCV about 80%) were stored at 4 °C with an admixture of sodium bicarbonate. To the resuspensions simultaneously were added xylitol and inorganic phosphate.

The erythrocyte concentration of 2,3-disphosphoglycerate in whole blood will be preserved the longer the more bicarbonate is added. Compared with whole blood in resuspensions of erythrocytes the preservation of 2,3-DPG is significantly prolonged, probably because xylitol was added.

From these results and from literature data it is concluded that bicarbonate is fixed to metabolites of the pentose metabolism. The products of the fixation of bicarbonate may influence the 2,3-DPG metabolism of the erythrocytes.

Acta biol. med. germ., Band 36, Seite 543—548 (1977)

Bezirksinstitut für Blutspende- und Transfusionswesen, 113 Berlin DDR,
und Institut für Laboratoriumsdiagnostik im Städtischen Klinikum, 1115 Berlin-Buch, DDR

Verhalten von Lebensfähigkeit, ATP- und 2,3-DPG-Gehalt resuspendierter Erythrozyten während mehrwöchiger Kältelagerung[1]

D. Strauss, W. Meurer und D. Dekowski

Zusammenfassung

Die aus ACD- bzw. ACD-AG-(Adenin-Guanosin) Blut hergestellten leukozyten- und thrombozytenarmen Erythrozytenkonzentrate wurden durch Resuspendieren auf einen Hämatokrit von ca. 55% eingestellt und 6 Wochen bei 4 °C gelagert. Die Resuspensionslösung enthielt Xylit, anorganisches Phosphat, Bikarbonat, Adenin und Guanosin. In einer Variante wurde zusätzlich Glukose, Zitrat und Saccharose zugefügt, in einer weiteren zusätzlich Sorbit.

Im Sorbit-Xylit-Medium blieben der 2,3-Diphosphoglyzerat- und ATP-Gehalt länger erhalten als im Glukose-Xylit-Medium. Alle resuspendierten Eythrozytenkonzentrate wiesen höhere ATP-Werte als nichtresuspendierte auf. Zusätze von Adenin und Guanosin bereits zum Ausgangsblut (ACD-AG-Blut) waren bezüglich des ATP-Gehaltes effektiver als spätere Zusätze mit der Resuspensionslösung.

Resuspendierte Erythrozyten aus einem Glukose-Adenin-Guanosin-Medium, das gleichfalls Zitrat und Saccharose enthielt, hatten nach 3 Wochen Lagerung eine für therapeutische Zwecke ausreichend hohe Überlebensrate von 80 bis 85%.

Einleitung

Die klinischen Anforderungen an eine hochwertige Erythrozytensubstitution ohne Leukozyten- und Thrombozytenbeimengungen bei insgesamt nur geringer Kreislaufbelastung des Transfusionsempfängers durch extrazelluläre Flüssigkeitszufuhr können durch die gegenwärtig üblichen Erythrozytenkonzentrate mit einem Plasmarest und „buffy coat", teilweise mit Zusatz des gleichen Volumens einer zucker-zitrat-haltigen Resuspensionslösung nicht in ausreichendem Maße erfüllt werden. Ein wesentlicher Nachteil besteht vor allem im schnellen Abfall des 2,3-DPG-Gehaltes auf Werte unterhalb 20% des Normalgehaltes innerhalb von 1—2 Wochen Lagerung, wodurch eine Sauerstoffmangelversorgung von Gewebe durch konservierte Erythrozyten insbesondere nach Massivtransfusionen mit 2,3-DPG-verarmten Erythrozyten auftreten kann. Ein weiterer Nachteil besteht in einem hohen Anteil lebensunfähiger Zellen bis zu 30% nach 3—6 Wochen Lagerung.

Ziel der vorliegenden Arbeit war es, durch Zusätze von Xylit, anorganischem Phosphat und Bikarbonat bzw. bei Ersatz der Glukose-Saccharose-Zitrat-Resuspensionslösung durch eine Sorbit-Xylit-Lösung bei gleichzeitiger Erhöhung des Anfangs-*p*H-Wertes in der Zellsuspension den ATP- und 2,3-DPG-Gehalt während der Lagerung für einen längeren Zeitraum zu erhalten.

[1]) Ergebnisse aus dem medizinischen Forschungsprojekt „Blut-Gewebe-Konservierung und -Typisierung" des Ministeriums für Gesundheitswesen der DDR

Material und Methoden

Aus frisch entnommenem Spenderblut, versetzt mit dem Humanzitratstabilisator mit und ohne Adenin und Guanosin (2. AB-DDR) wurde „buffy coat"-freies Erythrozytenkonzentrat hergestellt, portionisiert und mit 1/5 Volumen Resuspensionslösung A bzw. B versetzt (Tab. 1). Lösung B wurde bereits von STIGGE [1] zur Resuspension von Erythrozytenkonzentrat aus ÄDTA-Blut verwendet. Die *p*H-Werte der deplasmatisierten Zellkonzentrate betrugen 7,1. Es trat keine *p*H-Veränderung nach Zusatz der Resuspensionslösung auf.
Die Lagerung der resuspendierten Erythrozytenkonzentrate erfolgte bei 4 °C. In wöchentlichen Abständen wurden ATP (Test Combination der Firma Boehringer, Mannheim), 2,3-DPG nach NYGAARD und RÖRTH [2] und der Hämoglobingehalt bestimmt sowie der *p*H-Wert bei 25 °C gemessen.
Die 24-Std. Überlebensrate von Erythrozyten wurde an gesunden Versuchspersonen mit ^{51}Cr-markierten Zellen bestimmt. Die Erythrozyten waren 3 bzw. 6 Wochen in 1/5 Volumen Resuspensionslösung (Endkonzentrationen in der Zellsuspension: Glukose 12, Adenin 0,5, Guanosin 0,5, Zitrat 2, Saccharose 47 mM) gelagert worden.

Tabelle 1
Endkonzentrationen der Zusätze im resuspendierten Erythrozytenkonzentrat

Substanzen	Medium A (mM)	Medium B (mM)	Substanzen	Medium A (mM)	Medium B (mM)
Glukose	6	—	Adenin	0,5	0,5
Xylit	6	6	Guanosin	0,5	0,5
Sorbit	—	6	Na_3-Zitrat	2	—
Na_2HPO_4	5	5	Saccharose	47	—
$NaHCO_3$	20	20			

Ergebnisse

2,3-DPG-Gehalt

Der 2,3-DPG-Gehalt in heparinisiertem Spenderblut betrug 1,03 ± 0,09 Mol/Mol Hb ($n = 8$). Beim Erythrozytenkonzentrat aus ACD-Blut deutete sich nach

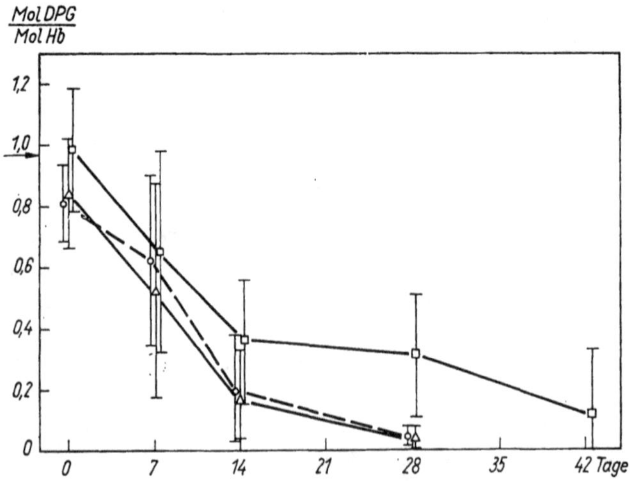

Abb. 1. Verhalten des 2,3-DPG-Gehalts resuspendierter Erythrozyten aus ACD-Blut während der Lagerung bei 4 °C in Lösung A bzw. B. ($\bar{x} \pm s$, $n = 4$).
○— — —○ Erythrozytenkonzentrat, △———△ resuspendierte Erythrozyten in Lösung A, □———□ resuspendierte Erythrozyten in Lösung B, → Normalwert $\bar{x}$

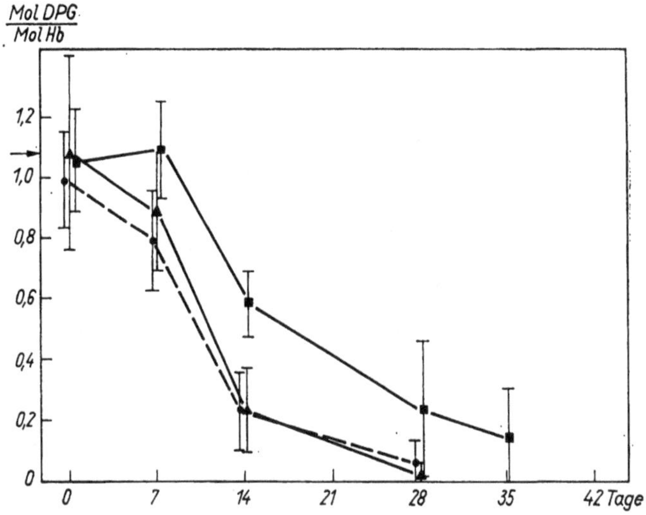

Abb. 2. Verhalten des 2,3-DPG-Gehalts resuspendierter Erythrozyten aus ACD-AG-Blut während der Lagerung bei 4 °C in Lösung A bzw. B (Erläuterungen s. Abb. 1)

Resuspension mit Lösung B eine verbesserte 2,3-DPG-Erhaltung an (Abb. 1), die jedoch nur am 28. Lagerungstag signifikante Unterschiede ($p < 0{,}05$) gegenüber nicht resuspendiertem Zellkonzentrat und Resuspensionsmedium A aufwies.

Wurden Erythrozytenkonzentrate aus ACD-AG-Blut hergestellt, trat der bei Konzentraten aus ACD-Blut beobachtete Anfangsverlust im 2,3-DPG-Gehalt bei der Herstellung der Zellkonzentrate nicht auf. Während der 1. Lagerungswoche betrug der Abfall nur 10—20%, im Konzentrat aus ACD-Blut jedoch 40 bis 50% vom Normalwert (Abb. 2). Im Resuspensionsmedium B blieb der 2,3-DPG-Gehalt während der 1. Lagerungswoche auf dem Ausgangswert erhalten, nach 2 Wochen lagen noch 60% vor. Die Unterschiede zu Medium A sind signifikant ($p < 0{,}05$).

ATP-Gehalt

Der ATP-Gehalt im heparinisierten Spenderblut betrug 0,23 ± 0,05 Mol/Mol Hb ($n = 8$). In den resuspendierten Erythrozytenkonzentrataten aus ACD-Blut wurden während des gesamten Lagerungszeitraumes signifikant höhere ATP-Konzentrationen als in nicht resuspendierten gemessen. Die Maximalwerte am 7. Lagerungstag betrugen 140% des Ausgangswertes (Abb. 3). Signifikante Unterschiede ($p \leqq 0{,}1$) im ATP-Verhalten zwischen den geprüften Varianten A und B traten bei resuspendierten Konzentraten aus ACD-AG-Blut am 14. Lagerungstag auf (Abb. 4).

Überlebensrate

Nach 3 Wochen Lagerung betrug die Überlebensrate von Erythrozyten aus einem in Glukose-Saccharose-Zitrat-AG-Medium resuspendierten Konzentrat 82 ± 2% ($n = 5$), nach 6 Wochen Lagerung 77 ± 7% ($n = 5$).

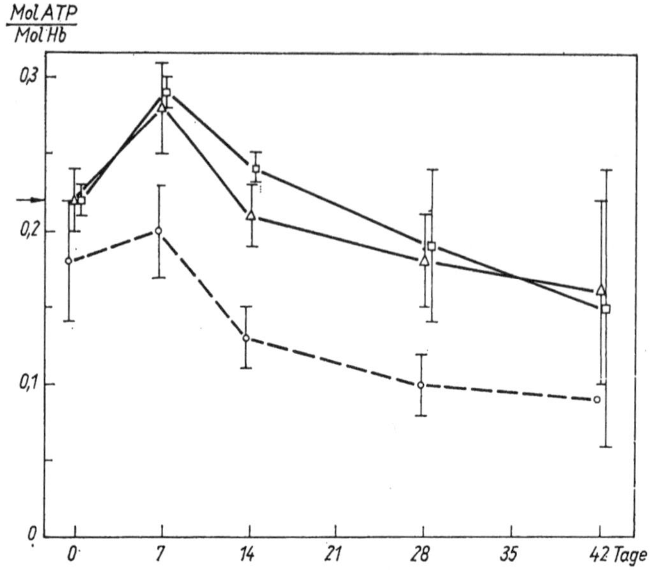

Abb. 3. Verhalten des ATP-Gehalts resuspendierter Erythrozyten aus ACD-Blut während der Lagerung bei 4 °C in Lösung A bzw. B.
(Erläuterungen s. Abb. 1)

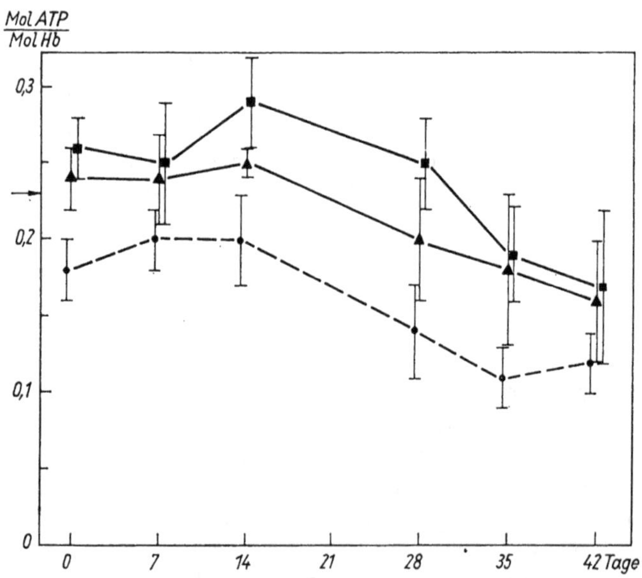

Abb. 4. Verhalten des ATP-Gehalts resuspendierter Erythrozyten aus ACD-AG-Blut während der Lagerung bei 4 °C in Lösung A bzw. B.

Diskussion

Bei der Optimierung der Konservierungslösung für Erythrozyten sind eine Reihe von Faktoren zu berücksichtigen, wie geringe Verdünnung des Zellkonzentrates, Sterilisierbarkeit der Lösung unter Hitzeeinwirkung ohne Karamelisierung, Verträglichkeit der zugesetzten Substanzen, da ein Entfernen der Resuspensionslösung vor der Transfusion unter den derzeitigen Bedingungen in der Praxis kaum möglich ist [3]. Daraus leiten sich u. a. die Bemühungen um den Ersatz von karamelisierenden Zuckern in einer neutralen Konservierungslösung durch nichtkaramelisierende Zuckeralkohole bzw. andere Substrate der Glykolyse des Erythrozyten ab. Dabei kann gleichzeitig der Nachteil einer verminderten Glukoseverwertung während der Lagerung infolge des *p*H-Abfalls umgangen werden.

Bei der Zusammenstellung der in dieser Arbeit verwendeten Resuspensionslösungen wurde von verschiedenen Überlegungen ausgegangen. Über die Verwendung von Xylit und Sorbit als Substrate für die Erythrozytenglykolyse unter Konservierungsbedingungen lagen bereits Ergebnisse vor [1,4—6]. Der 2,3-DPG-Gehalt konnte mit diesen Substraten im Vergleich zur Glukose länger erhalten werden. Gleichzeitiger Zusatz von Pyruvat zur Erhöhung des $NAD/NADH^+$-Quotienten zu Beginn der Lagerung erhöht diesen Effekt [7—9]. Zusatz von Bikarbonat ließ einen erhöhten 2,3-DPG-Gehalt auf Grund der von E. Stigge [1] und V. Stigge et al. [10] diskutierten CO_2-Fixierung an Metabolite des oxydativen Pentosephosphatweges erwarten. Die gewählte niedrige Konzentration von anorganischem Phosphat im Konservierungsmedium (5 mM) zielte ebenso wie Adenin- und Guanosinzusätze auf einen erhöhten ATP-Gehalt während der Lagerung hin. Anorganisches Phosphat erhöht einerseits das Trioseangebot durch eine Aktivierung der Phosphofruktokinase, andererseits auch die 1,3-DPG-Bildung als Reaktionspartner der GADPH-Reaktion. Ein Einfluß auf die phosphorolytische Spaltung des Guanosins ist ebenfalls gegeben.

Die Kombination von Xylit, Sorbit, Phosphat, Bikarbonat, Adenin und Guanosin (Medium B) im Konservierungsmedium ließ einen deutlichen Effekt auf den ATP-Gehalt erkennen, weniger stark ausgeprägt war er auf den 2,3-DPG-Gehalt. Da bei den vorliegenden Untersuchungen der Anfangs-*p*H im resuspendierten Erythrozytenkonzentrat nur bei 7,1 lag, blieben die zu erwartenden stärker ausgeprägten Effekte auf den 2,3-DPG-Gehalt aus. Stigge [1] fand bei einem höheren Anfangs-*p*H von 7,3 noch nach 5 Wochen Lagerung resuspendierter Erythrozyten aus ÄDTA-Blut in der von uns mit B bezeichneten Lösung 2,3-DPG-Werte von 80% des Normalwertes.

Aus den vorliegenden Ergebnissen ist eine Überlegenheit der Sorbit-Xylit-Lösung gegenüber der Glukose-Xylit-Lösung — alle anderen Zusätze waren in Art und Konzentration gleich — sowohl im ATP- wie auch im 2,3-DPG-Verhalten zu erkennen. Der positive Effekt des Sorbits könnte durch eine erhöhte Triosebildung aus Sorbit über Fruktose, Fruktose-1-phosphat und Dihydroxyazetonphosphat zusätzlich zur Bildung aus Xylit erklärt werden, insbesondere bei erniedrigtem *p*H-Wert, wenn das Hexokinase-Phosphofruktokinase-System gehemmt ist.

Der Vorzug der hier gepüften Resuspensionslösungen gegenüber der bisher in der Praxis verwendeten Lösung — Glukose-Saccharose-Zitrat-AG — bei großer Verdünnung läßt sich aus den Überlebensraten ableiten. Der gefundene Anteil von 20—30% lebensunfähiger Erythrozyten entsprach einem Verlust bis zu 50% der ATP-Konzentration. Die neue Xylit-Sorbit-Lösung mit den beschriebenen Zusätzen gewährleistete eine ATP-Erhaltung auf dem Ausgangsniveau und darüber während der ersten 3 Lagerungswochen und einen Abfall bis zur 6. Woche um nur 25%, woraus auf eine hohe Überlebensrate über 80% zu schließen ist. Aus den vorliegenden Ergebnissen ist zu schlußfolgern, daß die gewählte Zusammensetzung der Resuspensionslösung für Erythrozytenkonzentrat, die die bisherigen Erkenntnisse über den Stoffwechsel von Erythrozyten unter Konservierungsbedingungen — Adeninnukleotidverhalten, Anteil von Glukoseumsatz über den oxydativen Pentosephosphatweg, Regulation des 2,3-DPG-Zyklus — berücksichtigen, Vorteile für die Erhaltung von Lebensfähigkeit und Sauerstofftransportfunktion der Erythrozyten beinhaltet, jedoch ein optimal erscheinender Anfang-*p*H- von 7,3 bis 7,4 die Vorteile noch verstärken kann.

Literatur

[1] STIGGE, E.: Dissertation. Akademie für Ärztliche Fortbildung, Berlin 1976

[2] NYGAARD, S. F., u. M. RÖRTH: Scand. J. clin. Lab. Invest. **24**, 299 (1969)

[3] STRAUSS, D.: Probl. Hämat. Transfus. Transplant. **2**, 8 (1974/75)

[4] ASAKURA, T., H. NINOMIYA, S. MINAKAMI u. H. YOSHIKAWA in: International Symposium on Metabolism, Physiolgy, and Clinical Use of Pentoses and Pentitols, Hakone, Japan, 1967. B. L. HORECKER, K. LANG u. Y. TAKAGI (Hrsg.) Springer, Berlin, Heidelberg, New York, 1969, S. 158

[5] STIGGE, V., D. STRAUSS, H. ROIGAS u. H. J. RADERECHT: Dte. GesundheitsWes. **28**, 1850 (1973)

[6] LUN, A., V. STIGGE, S. ZIEMER, D. STRAUSS u. H. J. RADERECHT: Folia haemat., Lpz. **102**, 335 (1975)

[7] PANIKER, N. V., u. E. BEUTLER: J. Lab. clin. Med. **78**, 472 (1971)

[8] DEUTICKE, B., J. DUHM u. R. DIERKESMANN: Pflügers Arch. ges. Physiol. **326**, 15 (1971)

[9] STIGGE, V., A. LUN, S. ZIEMER, D. STRAUSS u. H. J. RADERECHT: Folia haemot., Lpz. **101**, 256 (1974)

[10] STIGGE, V., E. STIGGE u. D. SCHWANKE: Acta biol. med. germ. **36**, 537 (1977)

Summary

D. STRAUSS, W. MEURER, and D. DEKOWSKI: Survival, ATP, and 2,3-DPG content of resuspended erythrocytes during storage at 4 °C.

Leukocyte- and thrombocyte-poor packed red cells obtained from ACD or. ACD-AG blood were resuspended to a hematocrit of about 55% and stored at 4 °C. The resuspension solution consisted of xylitol, inorganic phosphate, bicarbonate, adenine (A) and guanosine (G) solved in water. In one case glucose, citrate and sucrose were also added, in another one, sorbitol.

The 2,3-DPG and the ATP level remained for a longer period in the sorbitol-xylitol-medium than in the glucose-xylitol-medium. The ATP content in the red cell suspension was higher than in packed cells. Higher ATP values were obtained in red blood cells from whole blood with adenine and guanosine.

The survival rate of resuspended red blood cells in glucose-AG-citrate-sucrose medium was about 80—85% after 3 weeks of storage and 77% after 6 weeks with a higher range.

Acta biol. med. germ., Band. 36, Seite 549—553 (1977)

Bezirksinstitut für Blutspende- und Transfusionswesen, 113 Berlin, DDR

Verhalten von Adeninnukleotiden im Konservenblut mit Adenin- und Guanosinzusätzen

V. Stigge und B. Seidel

Zusammenfassung

Unter üblichen Lagerungsbedingungen wurde ACD-Konservenblut mit Adenin, Guanosin und beiden Purinderivaten gleichzeitig substituiert. In wöchentlichen Abständen wurden die Konzentrationen von ATP, ADP und AMP bestimmt.
Aus den gewonnenen Werten ergibt sich die Schlußfolgerung daß die Substitution des Konservenblutes mit Adenin und Guanosin wahrscheinlich zu einer Synthese aller drei Adeninnukleotide führt.
Die scheinbare additive Wirkung von Guanosin und Adenin auf die ATP-Konzentration muß wahrscheinlich wegen der gleichzeitigen Erfassung von GTP auf die Unspezifität der ATP-Bestimmungsmethode zurückgeführt werden. GTP hätte dann Bedeutung für die Überlebensrate von Erythrozyten, die in ACD-AG-Konservierungslösung gelagert wurden.

Einleitung

Die ACD-AG-Konservierungslösung für Blut wurde in der gesamten DDR eingeführt, nachdem in umfangreichen Untersuchungen die Wirksamkeit der Purinzusätze auf den Stoffwechsel der Erythrozyten überprüft worden war. Durch den Zusatz von Adenin und Guanosin zur ACD-Konservierungslösung konnte die Lagerdauer von Konservenblut bei 4 °C von bisher 3 Wochen auf 6 Wochen verlängert werden, was den Blutkonservendepots bedeutende Vorteile bietet. Darüber hinaus konnte durch die Zusätze eine wesentliche qualitative Verbesserung des Konservenblutes insbesondere in den ersten Tagen und Wochen der Lagerung erzielt werden.
Die Zusätze Adenin und Guanosin bewirken eine signifikant verbesserte Erhaltung der ATP-Konzentration der konservierten Zellen [1]. Die Konzentration des zellulären ATP korreliert unter bestimmten Bedingungen über einen längeren Zeitraum eng mit der Überlebensrate der Erythrozyten im Empfängerorganismus [2].
Diese 24-Std.-Überlebensrate wird durch den Adeninzusatz signifikant verbessert. Gleichzeitiger Zusatz von Guanosin führt zu weiterer signifikanter Verbesserung der Überlebensrate [3].
Ziel der vorliegenden Untersuchungen soll es sein, über die Wirkungsweise der Purinzusätze (Synthesesteigerung oder Abbauhemmung der Adeninnukleotide), über deren Einfluß auf die Konzentration der einzelnen Adeninnukleotide und über den Anteil des Guanosins an diesen Wirkungen weitere Kenntnisse zu erlangen.

Material und Methodik

Das Spenderblut wurde unmittelbar bei der Abnahme aus der V. cubitalis klinisch gesunder Spender in vier 120 ml-Blutkonservenflaschen eingeleitet. In den Blutkonservenflaschen wurden die Konservierungslösungen ACD, ACD-A, ACD-G und ACD-AG vorgelegt. Die Sterilisation der Lösungen erfolgte nach dem Verfahren a1 des DAB 7-DDR.
Die Mischung des Blutes mit den Konservierungslösungen wurde im Verhältnis 4 + 1 vorgenommen, so daß Adenin- und Guanosinendkonzentrationen von je 0,5 mMol/l Blut resultierten. Den Erythrozyten standen damit umgerechnet etwa 300 mMol jedes Purinderivates pro Mol Hämoglobin zur Verfügung.
Die Lagerung des Konservenblutes erfolgte bei 4 ± 2 °C über 7 Wochen. Der Stichprobenumfang betrug jeweils $n = 10$.
In wöchentlichen Abständen wurden aus den Konservenflaschen nach gründlichem Mischen unter aseptischen Bedingungen Blutproben zur Untersuchung entnommen. Die Bestimmung der ATP-Konzentration erfolgte mit der Test-Combination der Firma Boehringer, Mannheim, im gekoppelten optischen Test.

$$\text{3-PG} + \text{ATP} \xleftrightarrow{\text{PGK}} \text{1,3-DPG} + \text{ADP}$$

$$\text{1,3-DPG} + \text{NADH} + \text{H}^+ \xleftrightarrow{\text{GAPD}} \text{GAP} + \text{NAD}^+ + \text{P}_i$$

$$\text{GAP} \xleftrightarrow{\text{TIM}} \text{DHAP}$$

$$\text{DHAP} + \text{NADH} + \text{H}^+ \xleftrightarrow{\text{GDH}} \text{Gly-1-P} + \text{NAD}^+$$

Die methodische Variabilität lag in unserem Labor bei 7% in der Serie.
Die ADP- und AMP- Konzentrationen wurden ebenfalls mit einer Test-Combination der Firma Boehringer, Mannheim, bestimmt. Für beide Parameter lag die methodische Variabilität in der Serie bei 5%.
Als Bezugsgröße diente die Hämoglobinkonzentration, die nach der Zyanmethämoglobin-Methode (DAB 7-DDR) gemessen wurde. Die methodische Variabilität in der Serie lag bei 2%.
Für Signifikanzberechnungen wurde der t-Test mit einer Irrtumswahrscheinlichkeit von 5% angewendet.

Ergebnisse

Die Zusätze Adenin und Guanosin wirkten additiv auf die gemessene ATP-Konzentration während einer Lagerungsperiode von 7 Wochen. Durch Kombination von Adenin und Guanosin konnte die ATP-Konzentration 2 Wochen vollständig erhalten bleiben, um dann unter den vorliegenden Bedingungen stetig auf etwa ein Viertel nach 7 Wochen abzufallen (Abb. 1).
Zusatz von Adenin hatte eine signifikante Erhöhung der ADP-Konzentration im Blut zur Folge mit einem Maximum in der 4. Lagerungswoche. Zusatz von Guanosin hatte keinen Einfluß auf die ADP-Konzentration (Abb. 1). Im Vergleich zum adeninfreien Blut erfolgte bei Zusatz von Adenin zu ACD-Konservenblut ein signifikant stärkerer Anstieg der AMP-Konzentration mit einem Maximum um die 5. Lagerungswoche (Abb. 1).
Während der gesamten Lagerungsperiode konnten keine Unterschiede zwischen den pH-Werten in den verschiedenen Konservenbluten festgestellt werden.
Die Summe der Adeninnukleotide (ATP + ADP + AMP) in ACD-AG-Blut zeigte bis zur 3. Lagerungswoche einen signifikanten Anstieg gegenüber den adeninfreien Konservenbluten, unterschied sich aber nicht von der Summe in ACD-A-Blut. Bei Zusatz von Guanosin scheint eine Tendenz zu besserer Erhaltung der Summe der Adeninnukleotide zu bestehen (Abb. 2).

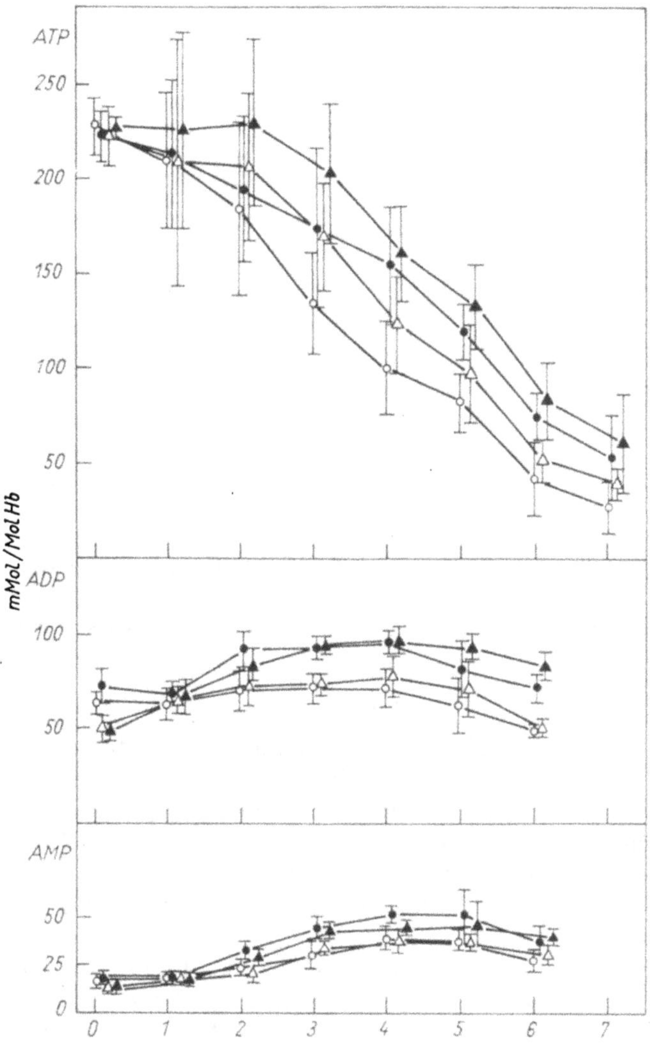

Abb. 1. Verhalten der ATP-, ADP- und AMP-Konzentrationen im Konservenblut mit Adenin- und Guanosinzusätzen während der Lagerung bei 4 °C.
○ Kontrolle (ACD-Blut);
● Zusatz von Adenin (ACD-A-Blut);
△ Zusatz von Guanosin (ACD-G-Blut);
▲ Zusätze von Adenin und Guanosin (ACD-AG-Blut);

Diskussion

Der Mechanismus einer Wirkung des Guanosins auf die Konzentration der Adeninnukleotide läßt sich aus den bisherigen Kenntnissen über den Erythrozytenstoffwechsel nicht erklären. Eine schnelle Aufspaltung des Guanosins mit Freisetzung von Guanin (Bogs; Stigge, unveröffentlichte Befunde) sowie die Bildung von GTP in den konservierten Erythrozyten wurden im Konservenblut chromatographisch nachgewiesen [4].

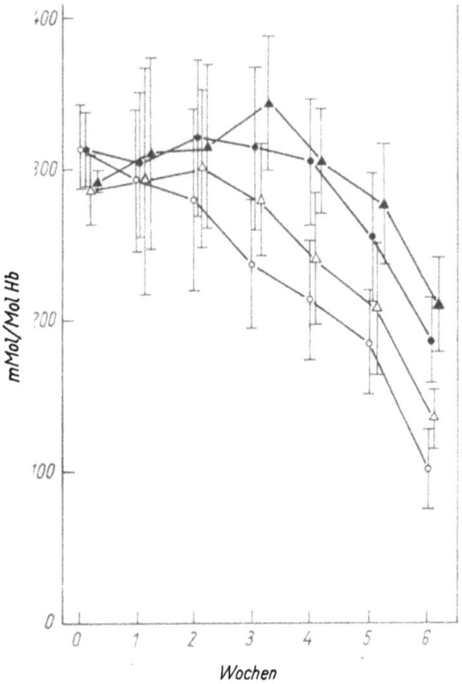

Abb. 2. Verhalten der Summe der Konzentrationen von ATP, ADP und AMP im Konservenblut (vgl. Legende zu Abb. 1)

Die verwendete enzymatische ATP-Bestimmungsmethode ist nicht spezifisch für ATP. Auch GTP wird von Phosphoglyzeratkinase umgesetzt [5]. Im guanosinsubstituierten Konservenblut kommt es zu einer GTP-Bildung, die nennenswerten Umfang erreichen und damit die ATP-Bestimmung mit PGK im Sinne falsch hoher Ergebnisse beeinflussen könnte. So ist möglicherweise das additive Verhalten von Adenin und Guanosin auf die von uns bestimmte ATP-Konzentration zu erklären.

Eine meßbare GDP-Konzentration, die wegen der Unspeziftät der ADP-Nachweisreaktion in der ADP-Konzentrationskurve erscheinen müßte, kann in den guanosinsubstituierten Konservenbluten nicht beobachtet werden.

Wenn tatsächlich die verbesserte ATP-Erhaltung in ACD-AG-gegenüber ACD-A-Blut nur scheinbar verbessert ist, und in Wirklichkeit die Summe ATP + GTP bestimmt wurde, dann muß die signifikant verbesserte Überlebensrate in ACD-AG-Blut im Vergleich zu ACD-A-Blut auf gebildetes GTP zurückgeführt werden. Die mögliche Bedeutung des gebildeten GTP könnte für die Erythrozyten darin bestehen, daß es bei Guanosinzusatz ebenso wie ATP durch Phosphorylierung aus GDP in der Glykolyse entsteht und als Energieüberträger im Zwischenstoffwechsel wirksam wird.

Ein Vergleich des bei 4 °C-Lagerung in Vollblutkonserven über mehrere Wochen zu beobachtenden Verhaltens der Adeninukleotide mit Daten, die aus Untersuchungen mit gewaschenen Erythrozyten bei 37 °C gewonnen wurden, ist nicht unmittelbar möglich.

Die Synopsis der drei Adeninnukleotide macht es wahrscheinlich, daß die längere Erhaltung der ATP-Konzentration in adenin- und guanosinsubstituiertem Konservenblut eher durch verstärkte Nukleotidsynthese als durch gehemmten Nukleotidabbau zustande kommt.

Hohe ATP-Konzentrationen zu Beginn der Lagerung bei ausreichender Adenylatkinaseaktivität können einen schnellen Umsatz des gebildeten AMP zu ADP zur Folge haben, so daß der Anstieg der gemessenen AMP-Konzentrationen erst dann erfolgt, wenn die ATP-Konzentration deutlich vermindert und möglicherweise die Aktivität der Adenylatkinase lagerungsbedingt abgefallen ist.

Der frühzeitige Anstieg der ADP-Konzentrationen während der Lagerung weist darauf hin, daß die glykolytische Aktivität der Erythrozyten schon früher abfällt, mit der Folge einer verminderten Phosphorylierung des ADP zu ATP. Andererseits kann an dem früheren Anstieg des ADP auch die Bildung aus AMP beteiligt sein.

Eine verminderte Abbaurate des zellulären ATP aufgrund von Hemmwirkungen durch Zusätze der Purinderivate läßt sich nicht völlig ausschließen. So differenzierte Reaktionen der verschiedenen Adeninnukleotide auf die Adenin- und Guanosinzusätze wären dann aber kaum zu erwarten.

Literatur

[1] Bishop, C.: Proc. Soc. exp. Biol. Med. **122**, 424 (1966)
[2] Dern, R. J., G. J. Brewer u. J. J. Wiorkowski: J. Lab. clin. Med. **69**, 968 (1967)
[3] Roigas, H., D. Strauss u. H. J. Raderecht: Z. ges. inn. Med. **24**, 850 (1969)
[4] Strauss, D.: Dissertation. Humboldt-Universität, Berlin, Biowissenschaftliche Fakultät 1970
[5] Jaworek, D., W. Gruber u. H. U. Bergmeyer in: Methoden der enzymatischen Analyse, 2. Aufl .H. U. Bergmeyer (Hrsg.). Akademie-Verlag, Berlin 1970, S. 2024

Summary

V. Stigge and B. Seidel: The behaviour of adenine nucleotides in stored blood with admixtures of adenine and guanosine

ACD stored blood was substituted with adenine, guanosine or both during usual storage conditions. In weekly intervals concentrations of ATP, ADP and AMP were estimated. From the obtained values it was suggested that the substitutions of adenine and guanosine probably lead to synthesis of the three measured adenine nucleotides.

The apparent additive effect of adenine and guanosine upon ATP concentration probably is caused by the simultaneous measurement of GTP and ATP in the used nonspecific method. Under these conditions GTP would have significance for the survival of erythrocytes stored in ACD-AG-medium.

Acta biol. med. germ., Band 36, Seite 555—560 (1977)

Physiologisch-Chemisches Institut der Medizinischen Akademie Magdeburg, 301 Magdeburg, DDR

Studies on NAD^+ permeability in intact mitochondria from rabbit reticulocytes

I. STEINBRECHT and W. AUGUSTIN

Summary

Functionally intact mitochondria from rabbit reticulocytes are characterized by a low NAD^+ level after the preparation (0.29 nmoles NAD^+ + NADH/mg protein).
They are apparently impermeable for NADH and exhibit a slow net uptake of NAD^+. From the increase of O_2-uptake in state 3 and the increase of NADH concentration in state 4 of respiration after the addition of NAD^+ we concluded that 3—10 min are necessary for the saturation with NAD^+ at 23 °C. 2 mM NAD^+ extramitochondrially are not sufficient to saturate the mitochondria with NADH and probably NAD^+, too.
Because of the net uptake of NAD^+ we assume that reticulocyte mitochondria lose NAD^+ during their preparation. If they are incubated with the physiological concentration of 300 μM NAD^+, which was found in reticulocytes, a value of 1.9 nmoles NAD^+ + NADH mg protein was calculated.
At an extramitochondrial NAD^+ concentration of 300 μM, reticulocyte mitochondria exhibit an almost maximal O_2-uptake in the presence of oxaloacetate or α-ketoglutarate.
It is concluded that the mitochondria in intact reticulocytes contain the "normal" complement of NAD^+ + NADH.

ntroduction

Functionally intact mitochondria from rabbit reticulocytes exhibit an active respiration with NAD^+-dependent substrates only after addition of NAD^+ to the incubation medium. The reason for this property of reticulocyte mitochondria may be seen in their low NAD^+ level [1]. The question arose whether this low NAD^+ content of the reticulocyte mitochondria is either a damage brought about by the fractionation procedure or is an intrinsic property of the reticulocyte mitochondria. The latter possibility does not seem to be very likely in view of the active respiration of intact reticulocytes with different NAD^+-dependent substrates and the well preserved functional properties of the isolated mitochondria [1].
We may assume that reticulocyte mitochondria lose NAD^+ during the preparation. In order to check this we made experiments concerning the permeability of reticulocyte mitochondria for NAD^+.

Materials and methods

NAD^+, NADH, ATP, α-ketoglutaric acid and oxaloacetic acid were obtained from Boehringer, Mannheim, alcohol dehydrogenase from VEB Arzneimittelwerk, Dresden, and nicotinamide from VEB Leunawerke.
Reticulocyte mitochondria were prepared by the slightly modified method of KUNZENDORF and AUGUSTIN [2]. NAD^+ and NADH were assayed by the method of STEINBRECHT and KUNZ [3].

The medium used for the incubation of the mitochondria had the following composition: 135 mM mannitol, 12 mM $MgCl_2$, 30 mM KH_2PO_4, 0.5 mM EDTA, 60 mM KCl triethanolamine buffer, *p*H 7.4.
Protein was determined by a modification of the method of BODE et al. [4]. The biuret reagent used contained 2 per cent Triton X-100, and the measurements were made at 557 nm. The pyridine nucleotide content was related to nonhemoglobin protein.

Results

As the direct measurement of NAD^+ uptake by reticulocyte mitochondria is rather difficult, because only a very small part of the added NAD^+ is taken up by the mitochondria, we used two indirect methods.
The first one was a polarographic measurement of the rate of oxygen consumption in state 3 with α-ketoglurate after the addition of NAD^+ (Fig. 1). The time

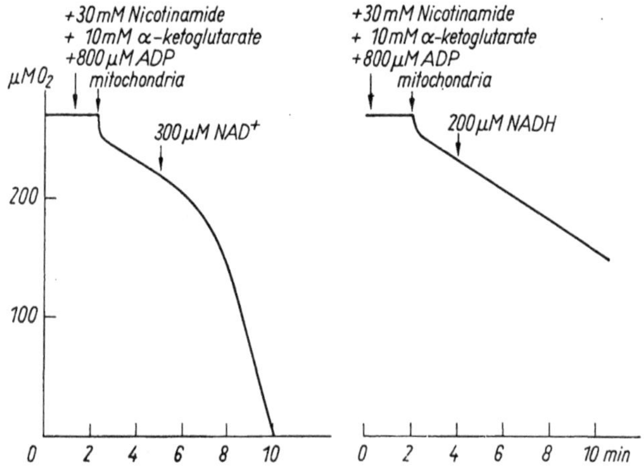

Fig. 1. Polarographic measurement of the O_2-uptake by reticulocyte mitochondria after addition of NAD^+ and NADH.
A collodium covered platinum electrode in a quasi closed system was used according to LIESE and JUNG [5]. Nicotinamide was added for the inhibition of NAD^+ glycohydrolase

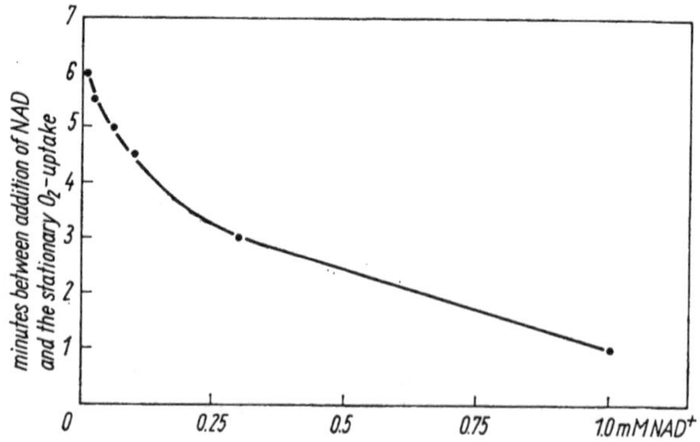

Fig. 2. Time between addition of NAD^+ and the stationary O_2-uptake in the presence of 10 mM α-ketoglutarate and different concentrations of NAD.
The suspension contained 800 μM ADP and 30 mM nicotinamide

between the addition of NAD^+ and the maximal O_2-uptake depended on the concentration of NAD^+ (Fig. 2). 6 min were found after the addition of 20 μM NAD^+ and 3 min after the addition of 300 μM NAD^+.

The second method used for the assay of NAD^+ uptake was the determination of NADH increase after the addition of NAD^+. This method seems to be justified, since reticulocyte mitochondria are impermeable for NADH as well as other mitochondria. Addition NADH did not increase the endogenous respiration and the respiration in the presence of α-ketoglutarate (Fig. 1). After the addition NADH to reticulocyte mitochondria the NADH concentration (100 μM) remained constant during 10 min. NADH formed after the addition of NAD^+ was exclusively found inside the mitochondria.

A maximal formation of NADH after the addition of 300 μM NAD^+ in state 4 with α-ketoglutarate was measured after 8—10 min at 23 °C (Fig. 3). If the formation of NADH is plotted versus the extramitochondrial NAD^+ concentration,

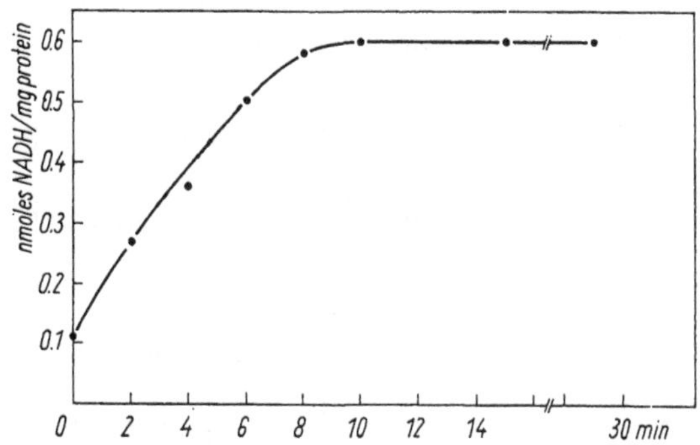

Fig. 3. Kinetics of NADH formation in mitochondria from rabbit reticulocytes after addition of 300 μM NAD^+.
The suspension contained 10 mM α-ketoglutarate, 1 mMATP, 30 mM nicotinamide

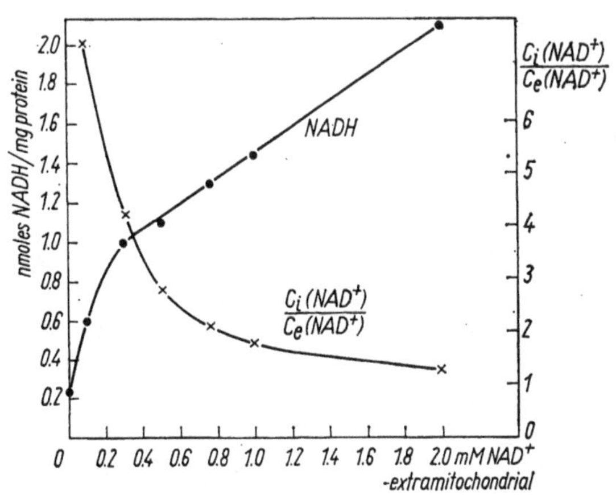

Fig. 4. NADH formation in reticulocyte mitochondria after addition of different concentrations of NAD^+.
Experimental conditions: 10 mM α-ketoglutarate, 1 mM ATP, 30 mM nicotinamide, incubation time 12 min

a biphasic curve results (Fig. 4). The increase of the NADH formation becomes smaller at higher NAD^+ concentrations than 300 μM. Even at an extramitochondrial NAD^+ concentration up to 2 mM the mitochondria do not seem to be saturated with regard to the NADH level.

The mitochondrial NAD^+ content may be calculated from their NADH content and the degree of reduction of NAD^+. In reticulocyte mitochondria the degree of reduction for the endogenous NAD^+ in the presence of α-ketoglutarate in state 4 was determined to be 0.45 (0.37—0.49) (Table 1). The same value was reported for rat liver under such conditions by Klingenberg and Slenczka [7]. The results calculated on this basis show that at low extramitochondrial NAD^+ concentrations the concentration inside the mitochondria is apparently higher than outside. At high extramitochondrial NAD^+ concentrations the intramitochondrial concentration approaches the extramitochondrial one (Fig. 4).

Table 1

Pyridine nucleotides in mitochondria from rabbit reticulocytes and rat liver and in reticulocytes

	NAD^+ + NADH	$NADP^+$ + NADPH	$\frac{NADH}{NAD^+ + NADH}$ state 4, substrate: α-ketoglutarate
	nmoles/mg protein	nmoles/mg protein	
Reticulocyte mitochondria immediately after the preparation	0.29 (0.14—0.59) $n = 14$	0.44 (0.20—0.63) [1] $n = 8$	0.45 (0.37—0.49) $n = 5$
after incubation with 300 μM NAD^+	1.0 (1.6—2.2) $n = 3$		
Liver mitochondria	3.8 (3.3—4.7) [6]	5.2 (3.6—7.5) [6]	0.43 (0.32—0.55) [7]
	nmoles/ml cells	nmoles/ml cells	
Reticulocytes	207	64	
Reticulocytes after the preparation of mitochondria (recovery in all fractions)	42	37	

After the incubation with 300 μM NAD^+ a value of 1.9 (1.6—2.2) nmoles NAD^+ + + NADH/mg protein for reticulocyte mitochondria could be calculated (Tab. 1). Since the NAD^+ concentration of the intact reticulocytes was determined with about 300 μM, this value can be supposed to respresent the in vivo NAD^+ content of the reticulocyte mitochondria.

The concentration of 300 μM NAD^+ is sufficient to yield nearly the maximal respiration of reticulocyte mitochondria. In the presence of oxaloacetate a half maximal oxgen consumption is obtained with 70 μM NAD^+ (Fig. 5), in the presence of α-ketoglutarate with 20 μM NAD^+ (Fig. 6). In consequence at the physiological concentration of 300 μM NAD^+, 81% of the maximal respiration are found if oxaloacetate is the substrate, and 94% if α-ketoglutarate is oxidized.

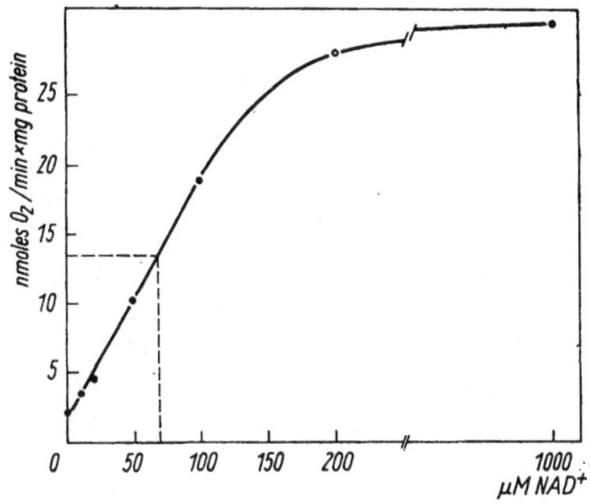

Fig. 5. O_2-Uptake by reticulocyte mitochondria in the presence of 10 mM oxaloacetate and different concentrations of NAD+.
The suspensions contained 30 mM nicotinamide and 800 µM ADP

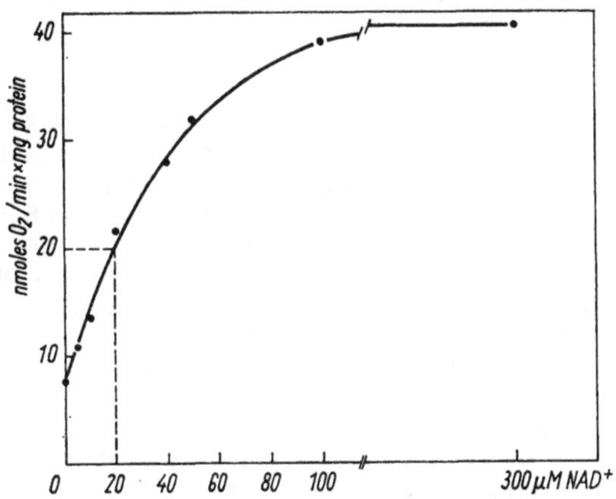

Fig. 6. O_2-Uptake by reticulocyte mitochondria in the presence of 10 mM α-ketoglutarate and different concentrations of NAD+.
The suspensions contained 30 mM nicotinamide and 800 µM ADP

Discussion

Purvis and Lowenstein [8] incubated isolated rat liver mitochondria with labelled NAD^+ in state 3 at 25 °C for 60 min without obtaining maximal labelling. This time requirement can be compared to the time between the addition of NAD^+ and the stationary O_2-uptake of reticulocyte mitochondria in state 3 with α-ketoglutarate. We observed 6 min at the lowest extramitochondrial concentration (20 µM) and 1 min at the highest extramitochondrial concentration (1 mM) At the concentration of 1 mM the time for the saturation with NAD^+ must be longer than the time till the maximal O_2-uptake, as the mitochondria exhibit an almost maximal respiration before they are saturated with NAD^+. At the concentration of 300 µM NAD^+ the time between the addition of NAD^+ and the maximal respiration amounts to 3 min. The time for the maximal formation of NADH after the addition of 300 μM NAD^+ in state 4 is 8—10 min. This difference might be explained by different rates of NAD^+ uptake in the active and controlled state of respiration.

Though PURVIS and LOWENSTEIN [8] found a slow exchange of uptake of NAD^+ by rat liver mitochondria, NAD^+ can leak out rapidly under special conditions, too. BREMER et al. [9] report that rat liver mitochondria lose 80—90% of their $NAD^+ + NADP^+$ within 6 min, if they are incubated at 30 °C in a medium containing high concentrations of inorganic phosphate.

The observed uptake of NAD^+ by reticulocyte mitochondria may explain that they lose NAD^+ during the preparation, which is favoured by a splitting of the NAD^+ in the homogenate by the very active NAD^+ glycohydrolase of the outer cell membrane [10, 11], (Table 1).

At the physiological NAD^+ concentration of 300 μM we calculated a value of 1.9 nmoles $NAD^+ + NADH$/mg protein. This value is in the range of the $NAD^+ + NADH$ content of mitochondria from other species [12]. Thus we may conclude that reticulocyte mitochondria in the intact cell contain the "normal" complement of pyridine nucleotides necessary for their function.

References

[1] AUGUSTIN, H. W., and H.-J. KUNZENDORF: VI. Internationales Symposium über Struktur und Funktion der Erythrozyten, Berlin 1970. (= Abh. Akad. Wiss. DDR). S. M. RAPOPORT and F. JUNG (Eds.). Akademie-Verlag, Berlin 1972, p. 459

[2] KUNZENDORF, H.-J., and H. W. AUGUSTIN: VI. Internationales Symposium über Struktur und Funktion der Erythrozyten, Berlin 1970. (= Abh. Akad. Wiss. DDR). S. M. RAPOPORT and F. JUNG (Eds.) Akademie-Verlag, Berlin 1972, p. 455

[3] STEINBRECHT, I., and W. KUNZ: Acta biol. med. germ. **29**, 495 (1972)

[4] BODE, Ch., H. GOEBELL, and E. STÄHLER: Z. klin. Chem. klin. Biochem. **6**, 418 (1968)

[5] LIESE, W., and K. JUNG: Z. med. Labortechn. **12**, 9 (1971)

[6] STEINBRECHT, I., and W. KUNZ: Acta biol. med. germ. **25**, 731 (1970)

[7] KLINGENBERG, M., and W. SLENCZKA: Biochem. Z. **331**, 486 (1959)

[8] PURVIS, J. L., and J. M. LOWENSTEIN: J. biol. Chem. **236**, 2794 (1961)

[9] BREMER, J., A. WOJTCZAK, and S. SKREDE: Europ. J. Biochem. **25**, 190 (1972)

[10] ALIVISATOS, S. G. A., S. KASHKET, and O. F. DENSTEDT: Can. J. Biochem. Physiol. **34**, 46 (1956)

[11] HOFMANN, E. C. G.: Biochem. Z. **327**, 273 (1955)

[12] KLINGENBERG, M. in: Funktionelle und morphologische Organisation der Zelle. Wissenschaftliche Konferenz der Gesellschaft deutscher Naturforscher und Ärzte in Rottach-Egern 1962. P. KARLSON (Ed.). Springer, Berlin, Göttingen, Heidelberg 1963 p. 72.

Address of the authors: I. STEINBRECHT and W. AUGUSTIN, Physiologisch-Chemisches Institut der Medizinischen Akademie Magdeburg, DDR 301 Magdeburg, Leipziger Str.44

Acta biol. med. germ., Band, 36, Seite 561—563 (1977)

Institut für Physiologische und Biologische Chemie der Humboldt-Universität, 104 Berlin, DDR

Untersuchungen zur Antimyzin A-resistenten Atmung erythroider Zellen

M. Müller, R. Dumdey, B. Saffert, und V. Löffler

Summary

In bone marrow of rabbits and in erythroid cells of other species there exists an Antimycin A-resistant respiration. Its mechanism corresponds to that postulated for the Antimycin A-resistant respiration of rabbit reticulocytes.
There exist only quantitative differences compared with rabbit reticulocytes.

Für verschiedene Zellarten wurden sauerstoffverbrauchende Prozesse nachgewiesen, die nicht durch die Atmungskette katalysiert werden [1—4].
Unsere Ergebnisse zur Antimyzin A-resistenten Atmung von Kaninchenretikulozyten lassen sich nicht vereinbaren mit bekannten Systemen eines atmungskettenunabhängigen O_2-Verbrauchs [5].
Nur 80% des O_2-Verbrauchs der Retikulozyten erfolgt über die Atmungskette. 20% des Gesamt-O_2-Verbrauchs sind Antimyzin A-resistent. Die strenge Korrelation zwischen Antimyzin A-resistenter Atmung und mitochondrialer Atmung läßt auf eine Beteiligung mitochondrialer Funktionen am Antimyzin A-resistenten O_2-Verbrauch schließen.

Tabelle 1
Antimyzin A-resistente Atmung in verschiedenen Zellarten

Zellart	Substrat	$\frac{^{14}CO_2 \cdot 100}{^{14}CO_2\text{-Kontrolle}}$	$\frac{\Delta O_2 \cdot 100}{\Delta O_2\text{-Kontrolle}}$
Knochenmark (Kaninchen)	1-[^{14}C]-Glukose (20 mM)	95	
	1,4-[^{14}C]-Sukzinat (10 mM)	16	7
	1-[^{14}C]-Oktanoat (tracer)	<1	
Retikulozyten 62% (Ratte)	1-[^{14}C]-Glukose (20 mM)	115	
	1,4-[^{14}C]-Sukzinat (10 mM)	10	16
	1-[^{14}C]-Oktanoat (tracer)	2	
Erythrozyten (Ente)	1-[^{14}C]-Glukose (20 mM)	103	35
	1-[^{14}C]-Oktanoat (tracer)	<1	

Tabelle 2
Wirkung von Malonat (10 mM) auf die Antimyzin A-resistente Atmung (1,4-[^{14}C]-Sukzinat, 10 mM)

Zellart	$\frac{^{14}CO_2 \cdot 100}{^{14}CO_2\text{-Kontrolle}}$		$\frac{\Delta O_2 \cdot 100}{\Delta O_2\text{-Kontrolle}}$	
	Antimyzin A	Antimyzin A + Malonat	Antimyzin A	Antimyzin A + Malonat
Retikulozyten 62% (Ratte)	10,0	2,0	16,0	15,5
Retikulozyten > 50% (Ente)	10,6	6,0	11,2	11,5

Zu der gleichen Schlußfolgerung gelangten wir durch Studien, die zur Charakterisierung der Antimyzin A-resistenten CO_2-Bildung aus zytosolischen und mitochondrialen Substraten durchgeführt wurden.
Alle Ergebnisse, die wir an intakten Zellen erhalten haben, sprechen für eine NADP-Abhängigkeit des Antimyzin A-resistenten O_2-Verbrauchs und für einen Transport reduzierender Äquivalente aus den Mitochondrien in das Zytosol. An diesem Transport reduzierender Äquivalente sind wahrscheinlich Transhydrogenasen und der "Malat-shuttle" beteiligt.
Die Antimyzin A-resistente Atmung ist keine Besonderheit der Kaninchenretikulozyten. In wesentlichen charakteristischen Merkmalen, wie der Antimyzin A-resistenten CO_2-Bildung aus Glukose und Sukzinat, der nahezu völligen Hemmung der CO_2-Bildung aus Oktanoat (Tab. 1) sowie der Verminderung der Antimyzin A-resistenten CO_2-Bildung aus Sukzinat durch Malonat bei unverändertem Antimyzin-A-resistenten O_2-Verbrauch (Tab. 2) besteht Übereinstimmung zwi-

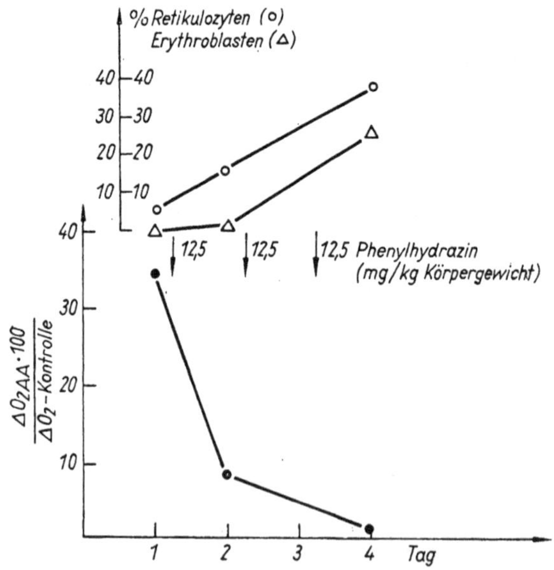

Abb. 1. Antimyzin A(AA)-resistenter O_2-Verbrauch in erythroiden Zellen der Ente

schen allen untersuchten Zellarten. Es existieren lediglich quantitative Unterschiede (Tab. 1).
Im Vergleich zu Kaninchenretikulozyten ist der prozentuale Anteil des Antimyzin A-resistenten O_2-Verbrauchs am Gesamt-O_2-Verbrauch im Kaninchenknochenmark wesentlich kleiner und etwa gleich groß in Rattenretikulozyten. Bei erythroiden Zellen der Ente steigt der Antimyzin A-resistente O_2-Verbrauch von Erythroblasten über Retikulozyten zu den Erythrozyten auf einen Anteil von 35% am Gesamt-O_2-Verbrauch an (Abb. 1). Offentsichtlich liegt innerhalb der erythroiden Reihe eine Reifungsabhängigkeit vor, die noch weiter untersucht werden muß.

Literatur

[1] ULLRICH, V.: Angew. Chemie **84**, 701 (1972)
[2] BOYD, G., u. R. SMELLIE (Hrsg.): Biological Hydroxylation Mechanisms. Biochemical Society Symposium No. 34, Edinburgh 1971. Academic Press, London u. New York 1972
[3] BOVERIS, A., N. OSHINO u. B. CHANCE: Biochem. J. **128**, 617 (1972)
[4] LOSCHEN, G., L. FLOHÉ u. B. CHANCE: FEBS Lett. **18**, 261 (1971)
[5] RAPOPORT, S., M. MÜLLER, A. HARTWIG u. R. DUMDEY: Acta biol. med. germ. **34**, 1301 (1975)

Acta biol. med. germ., Band 36, Seite 565—566 (1977)

Institut für Physiologische und Biologische Chemie der Humboldt-Universität, 104 Berlin, DDR

Charakterisierung von Retikulozyten des Menschen[1]

S. K. N. Richter-Rapoport, R. Dumdey, I. Uerlings und S. Rapoport

Die Retikulozyten des Menschen wurden bisher biochemisch noch wenig untersucht. Im allgemeinen wird angenommen, daß die Eigenschaften der „Stress"-Retikulozyten der Kaninchen (hohe gekoppelte Atmung; Pasteur-Effekt von 20%; 80% Antimycin-A-empfindliche Atmung) mit denen der Retikulozyten des Menschen übereinstimmen. Dagegen sprechen Untersuchungen an Retikulozyten des Menschen, bei denen ein minimaler Sauerstoffverbrauch festgestellt wurde. Außerdem wurde aus Studien über die Kapazität der Atmungskettenteilkomplexe an menschlichen Retikulozyten vermutet, daß die Kapazität der Zytochrom c-Oxydase kaum noch eine Steigerung der Atmung bei diesen Retikulozyten zuläßt. Um den Reifegrad der Retikulozyten des Menschen festzustellen, wurde der Sauerstoff-Verbrauch, die Größe des Pasteur-Effektes, das Ausmaß der intra- und extramitochondrialen Atmung, die Kopplung des Elektronentransports mit der oxydativen Phosphorylierung und elektronenmikroskopisch die Zahl und Struktur der Mitochondrien bestimmt.

Für diese Untersuchungen wurden 5 Blute von Neugeborenen, deren Retikulozyten durch Dichtegradientenzentrifugation angereichert wurden, und 24 Blute von verschiedensten Retikulozytosen des Menschen verwandt. Um die Frage der Speziesbedingtheit in den Unterschieden zwischen den Retikulozyten des Menschen und Kaninchen zu beantworten, wurde an Retikulozyten von normalen Kaninchen, die angereichert worden waren, der Pasteur-Effekt und die Größe der Atmung bestimmt.

Der Pasteur-Effekt, der als Prozentsatz der aeroben an der anaeroben Glykolyserate unter gleichen Bedingungen definiert wird, war nur klein oder aufgehoben. Eine Ausnahme bildete das Blut eines Frühgeborenen mit extremer rh-Inkompatibilität, das in der 30. Schwangerschaftswoche geboren wurde und kurz darauf verstarb. Der schon früher festgestellte geringe Sauerstoffverbrauch der Retikulozyten des Menschen konnte von uns bestätigt werden. Bezüglich der Aktivitäten der Atmungskettenteilkomplexe besteht der größte Unterschied zwischen Mensch und Kaninchen in der Aktivität der Zytochrom c-Oxydase.

Der Prozentsatz der Antimyzin-A-empfindlichen Atmung an der Gesamtatmung ist bei den untersuchten Bluten sehr unterschiedlich. Durch Zugabe von Oligomyzin, das die oxydative Phosphorylierung hemmt, wurde der Kopplungsgrad der Atmung an 3 Bluten von klinischen Retikulozyten bestimmt. Aus den elektronenmikroskopischen Untersuchungen ging hervor, daß die meisten Retikulozyten keine Mitochondrien mehr enthalten, wie schon die Zytochrom c-Oxydase-Aktivität vermuten ließ. Die noch vorhandenen Mitochondrien befinden sich zum großen

[1] Vgl. Acta biol. med. germ. **36**, 53—64 (1977)

Teil im Stadium der Auflösung. Es wird gezeigt, daß auch bei den normalen Retikulozyten der Kaninchen der Pasteur-Effekt und die Atmung gering sind.
Wie aus unseren Untersuchungen hervorgeht, gelangen die Retikulozyten des Menschen bei fast allen klinischen Zuständen, die mit einer erhöhten Erythropoese einhergehen, schon sehr reif ins strömende Blut. Diese Retikulozyten sind durch eine geringe Sauerstoffzehrung, die bei allen untersuchten Bluten teilweise noch intramitochondrialer Herkunft ist, charakterisiert. Der Pasteur-Effekt, der als ein Reifekriterium betrachtet werden kann, ist nur schwach ausgeprägt. Der intramitochondriale Sauerstoff-Verbrauch dieser Retikulozyten ist schon zum großen Teil entkoppelt. Sie besitzen zwar noch wenige Mitochondrien, die zum großen Teil auch qualitativ verändert sind.
Jedoch konnten wir auch bei einem der untersuchten Patientenblute unreife Retikulozyten nachweisen. Sie waren durch einen den Stressretikulozyten vergleichbaren Pasteur-Effekt gekennzeichnet. Das bedeutet, daß auch beim Menschen unter außergewöhnlichen Stressbedingungen dem Kaninchen ähnliche unreife Retikulozyten ausgeschüttet werden.
Auch die normalen Retikulozyten des Kaninchens gelangen schon ziemlich reif ins strömende Blut. Ein genereller Artunterschied zwischen den Retikulozyten des Menschen und des Kaninchens besteht also nicht. Die Frage, ob gleich starke Reize zur Bildung gleicher Retikulozytenpopulationen bei Mensch und Kaninchen führen, oder ob das Kaninchen schon auf schwächere Reize mit der Bildung von unreifen Retikulozyten reagiert, bleibt offen.

Acta biol. med. germ., Band 36, Seite 567—570 (1977).

Physiologisch-Chemisches Institut der Medizinischen Akademie Magdeburg, 301 Magdeburg, DDR
Instytut Biologii Doświadczalnej im. M. Nenckiego, P.A.N. Warszawa, Poland

Phospholipid composition and some reactions of phospholipid synthesis and degradation in mitochondria and other subcellular fractions from rabbit reticulocytes

I. Wiswedel, G. Lutze, J. Barańska, J. Zborowski, and W. Augustin

1. Phospholipid composition

Mitochondria of rabbit reticulocytes contain a higher proportion of phospholipids per mg protein than rat liver mitochondria (0,34 and 0,17 μmoles phospholipid phosphorus per mg protein, respectively) [1]. The phospholipid pattern however

Table 1
Fatty acid pattern of phospholipids from rabbit reticulocyte mitochondria (% of total fatty acid)

Fatty acid	Sphingo-myelin	Phosphatidyl-choline	Phosphatidyl-ethanolamine	Phosphatidyl-inositol	Cardio-lipin
16:0	24.32	37.11	16.68	7.35	2.23
16:1	1.91	2.22	1.64	0.65	3.28
17:0	1.37	0.91	1.17	1.22	—
16:2/DNA	0.44	0.81	0.70	0.86	0.35
18:0	15.25	4.44	11.93	28.17	1.25
18:1	9.18	14.88	18.46	10.06	25.11
10:0	0.59	0.21	—	0.56	—
18:2	3.75	29.21	26.85	18.27	59.96
20:0	1.02	0.22	0.11	0.40	0.13
20:1/18:3	1.10	2.27	2.51	2.53	1.83
21:0	0.20	+	0.07	+	0.10
20:3/20:2	0.30	1.34	2.02	5.58	2.48
20:3	0.15	+	0.28	+	0.09
22:0	3.04	0.25	0.51	2.00	1.96
22:1/20:4	2.09	3.76	7.87	12.71	0.13
23:0/20:4/22:2	1.98	0.02	+	0.17	+
20:5/22:3	0.70	0.70	0.86	0.81	0.08
24:0	1.36	0.06	0.01	0.07	0.33
24:1/22:4	27.66	0.21	0.95	1.72	0.06
22:5/24:2	2.93	0.01	0.42	0.38	0.41
22:5	0.34	0.90	4.12	4.07	0.16
26:0/22:6	0.28	0.47	3.04	2.52	0.06
26:1/24:4	0.04	+	+	+	+
C_1/ FA	28.18	48.53	57.04	56.50	86.32
saturated FA (%) / unsaturated FA (%)	0.90	0.77	0.44	0.67	0.06

is similar to liver mitochondria with the exception of a high proportion of sphingomyelin. This sphingomyelin cannot be accounted for by membrane impurities in the mitochondrial fraction. The percentage of cardiolipin is lower than in liver mitochondria, but in absolute values there is no difference.

The fatty acid composition of the individual phospholipids of reticulocyte mitochondria (Table 1) is comparable with the fatty acid composition of phospholipids from other mitochondrial species. Compared to the fatty acid pattern of reticulocyte outer cellular membranes there are differences in the fatty acid composition of the sphingomyelin. In connection with the stability of this sphingomyelin the existence of a specific mitochondrial sphingomyelin in reticulocytes may be assumed.

2. *Phospholipid synthesis*

Reticulocytes exhibit an active phospholipid turnover [2—4]. We have studied the incorporation of ^{14}C-labelled fatty acids, [^{14}C]-sn-glycerol-3-phosphate and [^{14}C]-CDP-choline in the lipids of different subcellular fractions from reticulocytes [5].

Fatty acids are incorporated into mitochondrial phospholipids by acylation of endogenously generated N-containing lysophospholipids (mainly lysophosphatidylcholine and lysophosphatidylethanolamine) and by acylation of sn-glycerol-3-phosphate yielding phosphatidic acid.

Measuring the incorporation of ^{14}C-labelled sn-glycerol-3-phosphate into mitochondrial phospholipids, the activity was exclusively recovered in the phosphatidic acid fraction. The time course of this reaction indicates that lysophosphatidic acid may be the first product of acylation, which is further acylated to phosphatidic acid as the main product of this pathway in reticulocyte mitochondria (Fig. 7). There is a high initial rate of lysophosphatidic acid formation followed by a rapid decrease of its formation and practically no further changes after ten minutes of in-

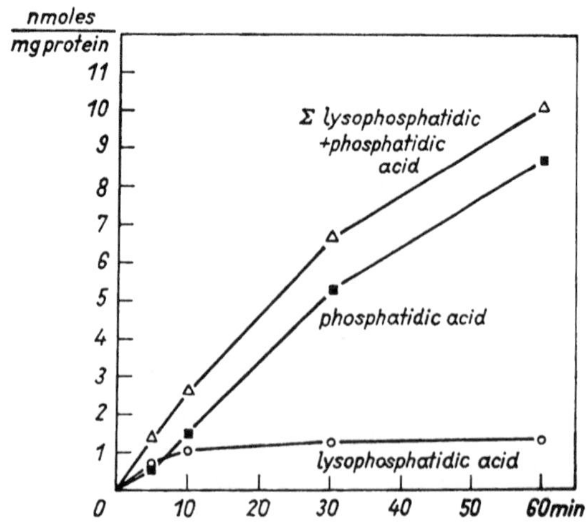

Fig. 1. Time course of lysophosphatidic, phosphatidic and lysophosphatidic + phosphatidic acid formation after standard incubation with [^{14}C]-sn-glycerol 3-phosphate

cubation. The activity of phosphatidic acid formation increases continuously up to 60 minutes.

Looking to the considerable capacity for phosphatidic acid synthesis in the reticulocytes the question of de novo synthesis of phospholipids according to the Kennedy-pathway arises. We could confirm the activity of CDP-choline:1,2-diglyceride phosphocholine transferase in the mitochondria, outer cellular membranes and postmitochondrial fractions as well. If the activities of CDP-choline:1,2-diglyceride phosphocholine transferase are plotted against the activities of NADPH-cytochrome c-reductase of identical fractions a significant correlation can be demonstrated (Fig. 2). Thus the activity of lecithin synthesis goes parallel with the activity of the marker enzyme for the microsomal fraction.

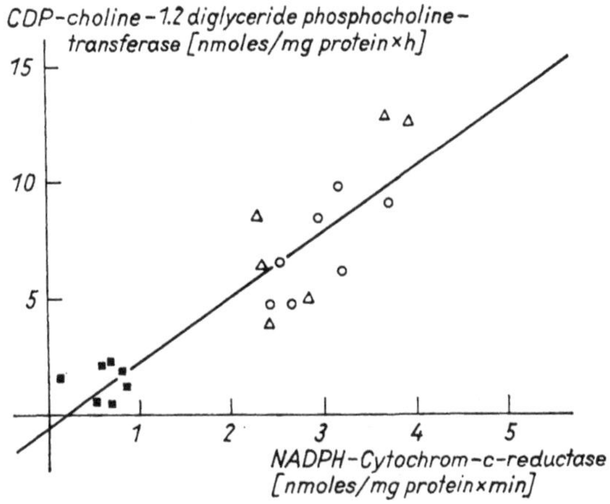

Fig. 2. Correlation of NADPH-cytochrome-c-reductase activity and CDP-choline: 1,2-diglyceride phosphocholinetransferase activity.
The two enzyme activities of membranes, mitochondria and postmitochondrial fractions (■, ○, △) are plotted against each other

3. *Preliminary data about degradation of phospholipids*

Phospholipase A activity could be detected in mitochondria and cytosol from rabbit reticulocytes as well [6, 7]. This was confirmed by the liberation of fatty acids and the increase of lysophospholipids during incubation of reticulocyte mitochondria with and without cytosol. Preincubation of mitochondria with nupercaine inhibited the acylation of endogenous N-containing lysophospholipids.

The decrease of total and individual phospholipids was higher than could be accounted for by lysophosphatic formation. Obviously a further splitting of monoacylphosphatides takes place. During incubation of mitochondria with cytosol and purified phosphatides acid soluble phosphate was liberated and phosphatidic acid and diglycerides exhibited a small increase. The possibility of a phospholipase activity resembling phospholipase D in reticulocytes is currently under investigation.

References

[1] LUTZE, G., D. KUNZE, G. REICHMANN, I. WISWEDEL, and W. AUGUSTIN (manuscript to be submitted for publication)
[2] O'DONNELL, V. J., P. OTTOLENGHI, A. MALKIN, O. F. DENSTEDT, and R. D. HEARD: Can. J. Biochem. Physiol. **36**, 1125 (1958)
[3] MARKS, P. A., A. GELLHORN, and C. KIDSON: J. biol. Chem. **235**, 2579 (1960)
[4] SLOVITER, H. A., and R. K. BOSE: Biochim. biophys. Acta **116**, 156 (1966)
[5] ZBOROWSKI, J., J. BARAŃSKA, I. WISWEDEL, W. AUGUSTIN, and L. WOJTCZAK (manuscript to be submitted for publication)
[6] PARRA, M. G., T. SCHEWE, and S. M. RAPOPORT: Acta biol. med. germ. **34**, 1075 (1975)
[7] WISWEDEL, I., G. LUTZE, and W. AUGUSTIN: 5th International Colloquium on Bioenergetics and Mitochondria, Varna, Bulgaria 1975

Acta biol. med. germ., Band 36, Seite 571—577 (1977)

Physiologisch-Chemisches Institut der Medizinischen Akademie Magdeburg,
301 Magdeburg, DDR

Studies on the functional significance of mitochondrial bound hexokinase in rabbit reticulocytes

F. N. Gellerich and H. W. Augustin

Summary

Mitochondria from rabbit reticulocytes contain about 50% of the total reticulocyte hexokinases. The proportion of mitochondrial hexokinases may be changed under different metabolic conditions. Mitochondrial bound and soluble hexokinases exhibit different kinetic properties ($K_{M\,ATP}$ and glucose-6-phosphate inhibition). The respiratory rate of isolated reticulocyte mitochondria in the presence of glucose depends on the glucose-6-phosphate concentration, as the ADP generation by the endogenous hexokinases is strongly inhibited by glucose-6-phosphate. In the experimental system all intermediary states of mitochondrial respiration can be adjusted between the state of maximal activity (state 3 or active state) and the controlled or resting state (state 4) by different glucose-6-phosphate levels. The stationary levels of the extramitochondrial adenine nucleotides in this experimental system have been measured. The rate of mitochondrial respiration and ATP formation depends on the extramitochondrial ATP/ADP ratio. At ratios of about 10 and lower the mitochondria are in their maximum phosphorylation state, at higher ratios the mitochondrial ATP formation is controlled by the extramitochondrial ATP/ADP ratio. It is postulated that the close interconnection between the mitochondrial hexokinase and the mitochondrial ATP forming system in reticulocytes is of functional significance for mitochondrial-cytosolic interactions in rabbit reticulocytes and probably in other types of cells with mitochondrial hexokinases, too.

Introduction

Hexokinases in rabbit reticulocytes exhibit a dual localisation in vitro and in vivo as well: one compartment of mitochondrial bound hexokinases and one compartment of soluble, cytosolic hexokinases [1]. The question about the functional significance of such a distribution seems not only interesting but also important in view of the widespread occurrence of mitochondrial hexokinases [2]. Our results, in part reported previously, point to a contribution of the hexokinase distribution between mitochondria and cytosol to the regulation of glucose phosphorylation and glycolytic flux in rabbit reticulocytes under different metabolic conditions [3]. In order to evaluate the possible implications of the mitochondrial hexokinases for the regulation of mitochondrial ATP formation, experiments were performed lending some support to the assumption of close connections between the mitochondrial ATP generating system and the ATP consumption by the glucose phosphorylation mediated by the mitochondrial hexokinases in reticulocytes. The results obtained in this context furthermore provide data about the relationship between the extramitochondrial adenine nucleotide pattern and the mitochondrial ATP-synthesis.

Material and methods

Chemicals and enzymes used (yeast hexokinase (EC 2.7.1.1), glucose-6-phosphate dehydrogenase (EC 1.1.1.49), lactate dehydrogenase (EC 1.1.1.27), adenylate kinase (EC 2.7.4.3), py-

ruvate kinase (EC 2.7.1.40), phosphoenolpyruvate, ATP, ADP, α-ketoglutarate, NAD, NADH, NADP) were purchased from Boehringer, Mannheim, bovine serum albumin and 2-deoxy-D-glucose (2-de glucose) from Serva, Heidelberg.
The reticulocyte mitochondria were prepared by the slightly modified method of KUNZENDORF and AUGUSTIN [4].
Only preparations with respiratory control ratios better than 5 with α-ketoglutarate were used for the experiments. The medium used for the incubation of the mitochondria had the following composition: 135 mM mannitol, 20 mM inorganic phosphate, 0.5 mM EDTA, 1 mM triethanolamine, 12 mM $MgCl_2$, 2 mM NAD and 1 mg/ml bovine serum albumine. The *p*H of the standard medium was adjusted to *p*H 7,4. The additions were 10 mM α-ketoglutarate, 2 mM ATP and 3.3 mM 2-deoxy-D-glucose.
The oxygen consumption was measured polarographically in a quasi-closed system using either a collodium covered gold-electrode or a Clark-type electrode, respectively. Glucose-6-phosphate (G-6-P), ATP, ADP and AMP were assayed by standard enzymic procedures [5]. G-6-P and ATP were measured with hexokinase and G-6-P dehydrogenase, ADP and AMP with pyruvate kinase, lactate dehydrogenase and adenylate kinase.

Results

If reticulocyte mitochondria were incubated with proper respiratory substrates as for example α-ketoglutarate typical state 3/state 4 transitions can be observed after the addition of small concentrations of ADP. This coupled respiration is obtainable only in glucose free media. Because of the considerable hexokinase content of reticulocyte mitochondria the addition of its substrate glucose leads to a strong increase of the oxygen consumption by the rapid ADP formation during the endogenous hexokinase reaction, if ATP is present in the incubation medium. The respiratory rate shortly after the start with glucose approaches the rate of the active state of mitochondrial respiration (state 3). After the initial increase the respiratory rate slows down, reaching the controlled rate of the mitochondrial respiration, e.g. the state 4 according to CHANCE and WILLIAMS [6] after longer times of incubation. The reason for this decrease of the rate of oxygen consumption with time is the accumulation of G-6-P, which is inhibitory to animal hexokinases thus decreasing the rate of ADP formation by the hexokinase reaction.
Neither with glucose and added yeast hexokinase which is not inhibited by G-6-P, nor with 2-de glucose in the case of endogenous hexokinases (its product 2-de G-6-P is practically not inhibitory to animal hexokinases) a decrease of the active respiration rate induced by the substrate can be detected. These observations allowed us to set up an experimental system, the principle of which is depicted in Fig. 1. The mitochondrial hexokinase was used as an endogenous ADP-regenerating system with 2-de glucose as a substrate. The reaction velocity of the hexokinase can be adjusted by G-6-P. Consequently, additions of increasing concentrations of G-6-P to the incubation mixture leads to an increasing inhibition of the 2-de glucose phosphorylation and consequently ADP formation, resulting in a lowering of the respiratory rate. Fig. 2 shows an example of this type of experiments. With this system it was possible to adjust apparently steady states of respiration and ATP formation by oxidative phosphorylation for some minutes. All intermediary states of mitochondrial oxygen consumption between the full active state (state 3) and the controlled state (state 4) can be adjusted in this system with most preparations of reticulocyte mitochondria. Sometimes, however, further additions of yeast hexokinase were necessary to obtain the fully active state, as

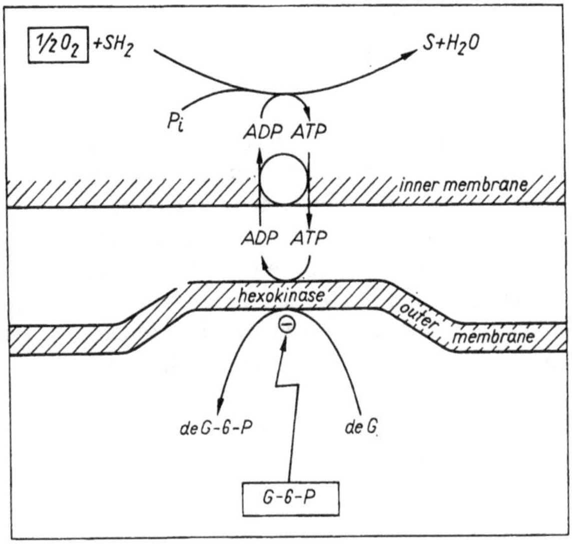

Fig. 1. Scheme of the regulation of mitochondrial respiration by G-6-P inhibition of mitochondrial hexokinases

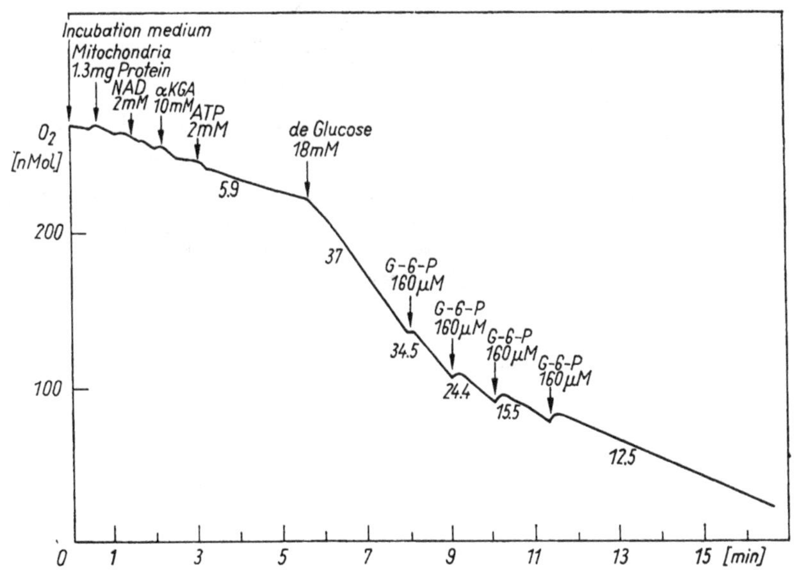

Fig. 2. Inhibition of the mitochondrial hexokinase induced respiration of reticulocyte mitochondria by G-6-P. The incubation was performed in a volume of 4 ml in a closed cell at 25 °C. The numbers on the curve denote the respiratory rate in nmoles O_2/min per mg protein

the capacity of mitochondrial hexokinases was not sufficient for attaining a steady state level of the adenine nucleotide pattern allowing the maximal stimulation of respiration. Within the steady state samples were withdrawn from the incubation mixture into perchloric acid and after immediate neutralization kept in liquid nitrogen until the analysis of the metabolites.

Fig. 3 shows the steady state levels of analytically determined G-6-P in experiments from different mitochondrial preparations. It becomes evident that the mitochondrial preparation is increasingly inhibited by growing concentrations of G-6-P. 50% inhibition was observed between G-6-P concentrations of about 150—400 μM. These values may be compared with the overall inhibition constants for the mitochondrial hexokinases. Taking into account the high concentration of inorganic phosphate they are comparable to values obtained by measuring mitochondrial hexokinases themselves in a comparable medium by *p*H-stat technique. Thus the steady state levels of the adenine nucleotides are governed by the actual activity of endogenous hexokinases in the experimental system, mainly regulated by the G-6-P level in the incubation mixture. The concentrations of the adenine nucleotides in the full active, in the controlled and in different intermediate states,

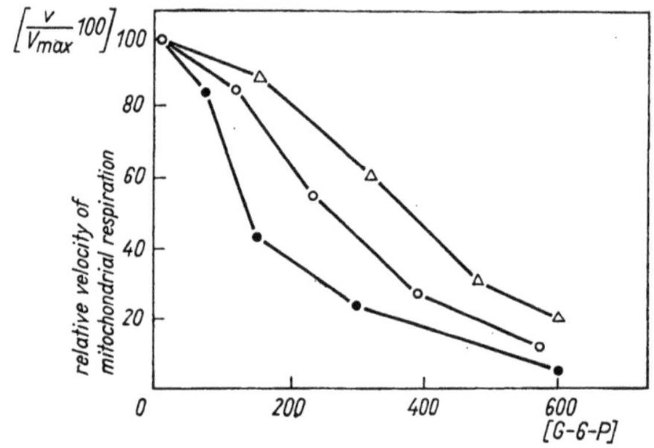

Fig. 3. Dependence of the respiration of reticulocyte mitochondria on the G-6-P concentration. Conditions as in Fig. 2. The samples for analysis of G-6-P were taken after the steady states were established. The 3 curves denote 3 experiments with different mitochondrial preparations

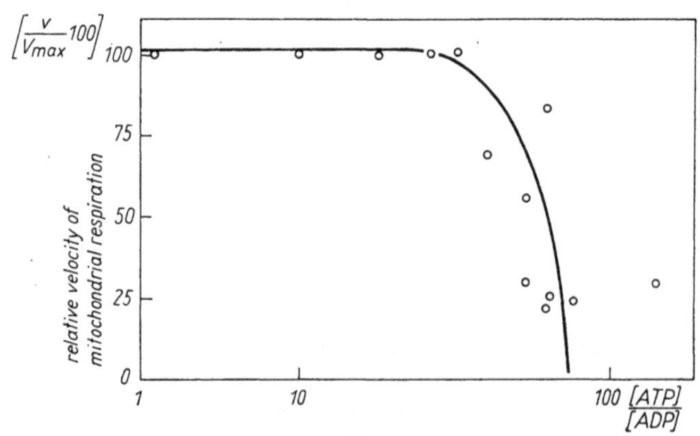

Fig. 4. Dependence of the respiratory rate of rabbit reticulocyte mitochondria on the extramitochondrial ATP/ADP ratio.

Conditions as in Fig. 2. The points represent the analysis of one set of experiments from one mitochondrial preparation as well, were determined in the above system. In view of the low mitochondrial concentration during the incubation the adenine nucleotide concentrations practically represent the extramitochondrial ones. Plotting the respiration values from different experiments in different ways it became obvious that the relationship between the respiratory rate and different concentrations of adenine nucleotides can be represented by one curve, if the relative respiratory rate is plotted against the ratio of log ATP/ADP (Fig. 4). The active state of respiration that means maximal mitochondrial phosphorylation is attained up to ATP/ADP ratios of about 10 and lower. Within higher ATP/ADP ratios up to about 100 the rate of respiration and phosphorylation becomes dependent on the ATP/ADP ratio. It should be emphasized that the numbers given for this relationship at present are to be taken as a qualitative description under the experimental conditions used and probably will be the subject of further refinement by the work currently under way.

Discussion

We proposed to name enzymes existing and acting within intact cells both in bound and soluble forms "semimembrane enzymes". There are other candidates besides the hexokinases which may belong to this group [7, 8].
The second group of enzymes in this context may be named "eumembrane enzymes". These enzymes exist and act within the intact cell exclusively in the bound form representing the typical marker enzymes for cellular membranes [9, 10]. Different kinetic and allosteric properties of enzymes from the latter group in bound and solubilized forms [9] are assumed to be without significance for the regulation of cell metabolism, whereas semimembrane enzymes exhibiting different kinetic properties in bound and released forms may play specific roles in metabolic regulation.
In rabbit reticulocytes the mitochondrial bound forms of hexokinases are the more active forms of these enzymes [3]. Whereas the maximal velocity of hexokinase remained unchanged upon detachment of the mitochondrial enzyme, the overall-$K_{M\,ATP}$ increased from 0.34 to 0.72 mM. Released hexokinases are about four times more sensitive against G-6-P-inhibition than the bound enzymes (overall-K_i = 7 μM and 31 μM G-6-P, respectively). This difference of both forms decreases, however, with increasing concentrations of inorganic phosphate.
Using different methods for the rapid fractionation of undiluted reticulocyte homogenate we obtained preliminary data, which indicate that the attachment of the hexokinase to the mitochondria is influenced by the metabolic state of the cells. It was observed that the portion of mitochondrial hexokinases increased in rabbit reticulocytes after the transition from aerobic to anaerobic conditions of incubation. If such a redistribution with a higher proportion of mitochondrial hexokinase under anaerobic conditions takes place, one has to consider the contribution of this mechanism for the higher rate of glucose phosphorylation under anaerobic conditions, e.g. its contribution to the Pasteur-effect of the reticulocyte [11, 12]. With some simplifying assumptions it was calculated from one set of experiments that the rate of glucose phosphorylation should increase from 0.285 (aerobically) to 0.538 (anaerobically) of the maximal rate caused by the hexokinase redistribution mechanism. At present it seems at least very likely that the semi-

membrane enzyme nature of hexokinases may be of functional significance for the regulation of reticulocyte glycolysis under different metabolic conditions.

If the partition of reticulocyte hexokinases between mitochondria and cytosol plays a role in the regulation of reticulocyte glycolysis, the question about a possible functional role of the mitochondrial hexokinase for the regulation of mitochondrial functions becomes obvious. In particular, the possible implication of mitochondrial hexokinase in the regulation of mitochondrial respiration and ATP-formation was investigated. We have shown that the hexokinase of the reticulocyte mitochondria can serve as an endogenous ATP-trap in phosphorylating mitochondria. An experimental system was evolved by which it was possible to adjust the respiratory rate and consequently the rate of ATP-formation of well coupled mitochondria to all intermediary rates between the maximal rate of respiration and the rate of respiration in the resting state of the mitochondria.

The respiratory activity of the mitochondria in the in vitro system seems to depend on the G-6-P level directly. It should be emphasized that this relationship must not have a meaning for the function of the mitochondria within the intact reticulocyte, since the ATP-consumption by the hexokinase reaction forms only a small part of total ATP-consuming reactions in the intact cell. Consequently the dependence of the mitochondrial ATP-synthesis on the extramitochondrial adenine nucleotide pattern was measured. It was found that the dependence of the mitochondrial respiratory rate from the extramitochondrial adenine nucleotides could be described by the extramitochondrial ATP/ADP-ratio rather than by the ADP-concentration or the phosphorylation potential. The shape of the curve (Fig. 4) and the control range of the extramitochondrial ATP/ADP-ratio found in the reticulocyte mitochondria is in agreement with the results obtained by rat liver mitochondria [13] and rat heart mitochondria (SCHALLER, unpublished results). Thus this regulation characteristic may be a general one for the regulation of mitochondrial ATP synthesis by the adenine nucleotide pattern of the extramitochondrial compartment under conditions of no substrate limitation. The explanation for this type of control characteristics is given by the properties of the adenine nucleotide translocator of the mitochondrial inner membrane [13].

The question if this regulatory principle is working within the intact cell as in the in vitro system cannot be answered at present; especially more information about the cytosolic ATP/ADP ratios is needed. Preliminary data for the bulk phase of cytosol from rabbit reticulocytes would suggest that reticulocyte mitochondria are working nearly maximally under different conditions, which at least is not very likely.

It is postulated that the close spatial interconnection of hexokinases with the mitochondrial ATP forming system is an important functional element of the mitochondrial-cytosolic interaction in all cells containing mitochondrial hexokinase.

Implications of this functional relationship were already put forward for the explanation of the Crabtree-effect [14]. Furthermore it was shown that the mitochondrially formed ATP could be used more efficiently with apparent higher affinity than cytosolic ATP and the mitochondria in turn respond more rapidly to ADP formed by the mitochondrial hexokinases than by the soluble enzyme [15]. The possibility of a functional adenine nucleotide compartment in a membrane near space of the mitochondria should be seriously considered in the next future. For such type of compartmentation the mitochondrial hexokinase may play an important role, especially by its ability to act as an efficient coupling site between the glycolytic reaction sequence and the mitochondrial ATP synthesis under different energetic demands of the cell.

Current experiments concerning the time dependence of ATP and G-6-P synthesis in tracer kinetic studies on isolated mitochondria and intact reticulocytes seem to support this tentative hypothesis.

References

[1] GELLERICH, F. N., and H. W. AUGUSTIN: VII. Internationales Symposium über Struktur und Funktion der Erythrozyten, Berlin 1973. (= Abh. Akad. Wiss. DDR). S. M. RAPOPORT and F. JUNG (Eds). Akademie-Verlag, Berlin 1975, p. 625

[2] CRANE, R. K., and A. SOLS: J. biol. Chem. **210**, 597 (1954)

[3] GELLERICH, F. N., and H. W. AUGUSTIN: 9th FEBS-Meeting, Budapest 1974, Abstracts p. 224

[4] KUNZENDORF, H.-J., and H. W. AUGUSTIN: VI. Internationales Symposium über Struktur und Funktion der Erythrozyten, Berlin 1970. (= Abh. Akad. Wiss. DDR). S. M. RAPOPORT and F. JUNG (Eds). Akademie-Verlag, Berlin 1972, p. 455

[5] BERGMEYER, H. U. (Ed.): Methoden der enzymatischen Analyse. 2. Aufl., Akademie-Verlag, Berlin 1970

[6] CHANCE, B., and G. R. WILLIAMS: J. biol. Chem. **217**, 409 (1955)

[7] CLARKE, F. M., and C. J. MASTERS: Archs Biochem. Biophys. **153**, 258 (1972)

[8] MAYER, R. J.: J. Neurochem. **19**, 2127 (1972)

[9] RACKER, E.: Fedn Proc. Fedn Am. Socs exp. Biol. **26**, 1335 (1967)

[10] STETTEN, M. R., and F. F. BURNETT: Biochim. biophys. Acta **128**, 344 (1966)

[11] JACOBASCH, G., J. MATUSCH, G. SCHÖNIAN, CH. GERTH, and S. M. RAPOPORT: VII. Internationales Symposium über Struktur und Funktion der Erythrozyten, Berlin 1973. (= Abh. Akad. Wiss. DDR). S. M. RAPOPORT and F. JUNG (Eds). Akademie-Verlag, Berlin 1975, p. 617

[12] GHOSH, A. K., and H. A. SLOVITER: J. biol. Chem. **248**, 3035 (1973)

[13] KÜSTER, U., R. BOHNENSACK, and W. KUNZ: Biochim. biophys. Acta **440**, 391 (1976)

[14] CHANCE, B., and B. HESS: Ann. N.Y. Acad. Sci. **63**, 1008 (1956)

[15] GOTS, R. E., and S. P. BESSMAN: Archs Biochem. Biophys. **163**, 7 (1974)

Address of the authors: F. N. GELLERICH and H. W. AUGUSTIN, Physiologisch-Chemisches Institut der Medizinischen Akademie Magdeburg, DDR 301 Magdeburg, Leipziger Str. 44

Acta biol. med. germ., Band 36, Seite 579 (1977)

Physiologisch-Chemisches Institut, Bereich Medizin der Karl-Marx-Universität Leipzig, 701 Leipzig, DDR

Association behaviour of human erythrocyte phosphofructokinase: Dependence of molecular weight on enzyme concentration at *p*H 8.0

G. Zimmermann and W. Schellenberger

The *self-association* of human erythrocyte phosphofructokinase has been studied at *p*H 8.0 and at 4 °C in the presence of 2 mM fructose-6-phosphate by means of frontal gel chromatography. Enzyme concentrations from 0.7 μg/ml to 0.7 mg/ml were applied.

Experimental data about this phosphofructokinase, weight average partition coefficient and apparent molecular weight, were estimated in correlation to reference proteins, assuming a logarithmic function and a random distribution function. The basic associating unit is the tetramer of about 330000 composed of subunits with a molecular weight of about 83000.
These data have been compared with various *association models* by making the following assumption: a) phosphofructokinase undergoes a reversible and open association process, b) the rate of equilibrium between the different associated states is fast in comparison to the penetration of the protein through the gel.
The dependence of the experimental partition coefficients on the enzyme concentration is characterized by *two groups of parameters*: a) the association constants determining the distribution coefficients of the various enzyme forms, b) the Stokes radius or the molecular weight, respectively, of the basic associating unit and the frictional coefficient, characterizing the structural properties of the different enzyme forms.

Model variants differing in the number of association constants involved in the association process and in respect to the association geometry have been tested. The best fit was obtained by assuming two association constants, one refering to the tetramer-octamer transition step and the other to the propagation step forming associated beyond the octameric state. The ratio between these two association constants points to a weak negative cooperativity.
These results demonstrate qualitative differences in the association behaviour between *p*H 7.1 and *p*H 8.0. A correlation to the kinetics and association characteristics of other mammalian phosphofructokinases is given.

Acta biol. med. germ., Band 36, Seite 581—585 (1977)

[1] Institut für Physiologische und Biologische Chemie der Humboldt-Universität zu Berlin, 104 Berlin, DDR
[2] Inštitut za biofiziko in Inštitut „J. Stefan", 61105 Lipičeva 2, SFRJ

Properties of the hexokinase-phosphofructokinase system on the basis of an extended PFK-model

M. Otto[1], G. Jacobasch[1] and S. Svetina[2]

Summary

The dependence of the glycolytic flux through the hexokinase(HK)-phosphofruktokinase (PFK) system on the ATP concentration has been constructed by means of kinetic equations of the HK and the PFK which include the most important effectors of both enzymes. The dependence of the flux on the ATP concentration displays a maximum. Under "in-vivo" conditions the system operates at an ATP concentration which already inhibits the flux. Furthermore, the influence of the sum of adenine nucleotides on the dependence of the flux on ATP has been investigated. At total adenine nucleotide concentrations equal or larger than 2 mM only a slight increase in the maximal flux through the HK-PFK system is observed while a decrease in the concentration of total adenine nucleotides leads to a drastic decrease of the flux.

Introduction

The HK-PFK system is a control unit of the glycolysis with respect to the glycolytic flux [1, 2]. The mathematical description of its behaviour with respect to several effectors is based on the description of both enzymes. Though the molecular mechanism of the phosphofructokinase is not known in detail, a rough description of the HK-PFK system on the basis of simplified models which are valid for well-defined conditions should be possible. This paper presents such calculations for the HK-PFK system of red blood cells taking the most important effectors of both enzymes into account.

Material and methods

1. *Hexokinase*

The data for the HK were taken from a paper by Gerber et al. [3]. The main points are briefly summarized here: The enzyme was 680 fold purified to a specific activity of 0.3 µmole/min · mg from human erythrocytes. The kinetics studies were performed at 37 °C and *p*H 7.2. The equation for the derived kinetic model is:

$$v = \frac{1}{\mathrm{N}} \cdot \left(V_1 \cdot \frac{\mathrm{MgATP}}{K_{\mathrm{MgATP}}} + V_2 \cdot \frac{\mathrm{MgATP} \cdot \mathrm{Mg}}{K_{\mathrm{MgATP}} \cdot K_{\mathrm{MgATP,Mg}}} \right)$$

$$N = 1 + \frac{\mathrm{MgATP}}{K_{\mathrm{MgATP}}} + \frac{\mathrm{Mg}}{K_{\mathrm{Mg}}} + \frac{\mathrm{MgATP} \cdot \mathrm{Mg}}{K_{\mathrm{MgATP}} \cdot K_{\mathrm{MgATP,Mg}}} + \left(\frac{\text{Gluc-6-P}}{K_{\mathrm{Gluc,Gluc6P}}} + \frac{\mathrm{Gluc} \cdot 1{,}6\,\mathrm{P_2}}{K_{\mathrm{Gluc,GlucP_2}}} \right) \times$$
$$\times \left(1 + \frac{\mathrm{Mg}}{K_{\mathrm{Mg}}}\right) + \frac{\mathrm{P_2G}}{K_{\mathrm{P_2G}}} + \frac{\mathrm{Mg}}{K_{\mathrm{Mg}}} \cdot \frac{\mathrm{P_2G}}{K_{\mathrm{Mg,P_2G}}}$$

where V_1 and V_2 are the maximal velocities in the absence and in the presence of saturating magnesium concentrations. K_A is the dissociation constant of a given enzyme-effector (A) complex. For the numerical values of the constants see [3, 4].

2. *Phosphofructokinase*

PFK from rat erythrocytes was purified 400fold according to KÜHN et al. [5]. The specific activity amounted to 35 U/mg protein. The conditions for the kinetical assay are described in detail elsewhere [6, 7]. For the calculation a simplified version of a model [8] based both on an allosteric transition and association-dissociation features of phosphofructokinase was used. It could be shown [8] that for low enough enzyme concentrations the following simplified equation can be used for the description of the PFK kinetics:

$$V = \frac{V_M \cdot \text{Fru-6-P}}{K_{F6P} + \text{Fru-6-P}} \cdot \frac{\text{MgATP}}{K_{MgATP} + \text{MgATP}} \cdot \frac{1}{1 + L'}$$

$$L' = \frac{L}{2\,K_{Ass}^{PFK} \cdot E_t}$$

$$L = L_0 \cdot \frac{\left(1 + \frac{I}{K_I}\right)^4}{\left(1 + \frac{A}{K_A}\right)^4 \cdot \left(1 + \frac{\text{Fru-6-P}}{K_{\text{Fru-6-P}}}\right)^4}$$

where K_{Ass}^{PFK} represents an association constant of phosphofructokinase and E_t is the total enzyme concentration. L_0 is the allosteric constant describing the equilibrium between the active and the inactive enzyme conformation in the absence of positive and negative effectors A and I, respectively. In the calculation of the HK-PFK system, L' was adjusted in such a way that at "in-vivo" concentrations of the effectors the velocity of the PFK equals the experimentally determined glycolytic flux. A value of $L' = 0.02$ was obtained. For V_M a value of 250 μmoles/ml cells · h was assumed. Concerning the numerical values of the other parameters compare [4, 6, 7]. The flux through the HK-PFK system of human red blood cells was calculated under the assumption that the kinetic parameters of the phosphofructokinase from human and rat erythrocytes are approximately equal.
The calculations are based on the following assumptions:

1. Glucose-6-phosphate and fructose-6-phosphate are nearly at equilibrium K = glu-6-P/fru-6-P = 0.4 [9].
2. The sum of adenine nucleotides is constant (2 mM) [10].
3. The adenylate kinase equilibrium holds
$K = \frac{ADP^2}{ATP \cdot AMP} = 0.5$. The value of $K = 0.5$ is valid for a variety of experimental conditions investigated [11–13].
4. The total magnesium concentration is 2 mM [14].
5. Only the complex formation between magnesium and ATP is taken into account. $K_{ass} = 12\,300\ M^{-1}$ [15].

Results

As shown in Fig. 1 the dependence of the glycolytic flux through the HK-PFK system on the ATP concentration can be constructed by determining the intersection points of the curves v_{HK} versus ATP and v_{PFK} versus ATP at several constant glucose-6-phosphate levels. As can be seen from the figure the dependence of the flux on the ATP concentration displays a maximum at ATP ≈ 1.4 mM. In Table 1 a comparison between the theoretical and the experimental values [2, 15] for the flux and some metabolite concentrations under standard conditions (*p*H 7.2, 37 °C) is given. With the exception of AMP in all cases a good agreement between the theoretical and the experimental values is obtained. The disagreement in the AMP values might be caused by the neglection of the magnesium dependence of the adenylate kinase [12, 13].

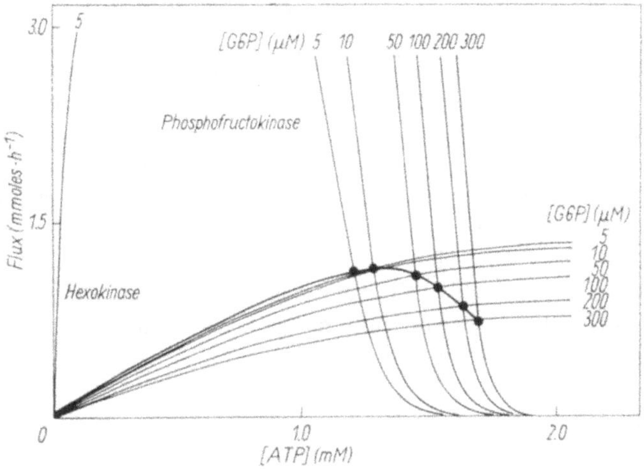

Fig. 1. Construction of the dependence of the glycolytic flux through the HK-PFK system on the ATP concentration. For details see text

Table 1

Comparison between the theoretical and the experimental values for the flux and some metabolite concentrations

	Theor. value (mmoles/h)	Exp. value (mmoles/h)
Flux	1.05	1.1
MgATP	1.35	1.44
ATP_{free}	0.17	0.18
Mg_{free}	0.64	0.67
Gluc-6-P	0.10	0.06
5′-AMP	0.14	0.07

The properties of the HK-PFK system presented here confirm the hypothesis that ATP is the main regulatory factor in the control of the red blood cell glycolysis. Although in the streaming blood the concentration of the adenine nucleotides in the erythrocytes is maintained at a relatively constant level, changes in its concentration are known, e.g. in the case of several metabolic diseases, under the influence of drugs or under several experimental conditions in vitro. A decrease in the adenine nucleotide concentration was observed where the formation of ATP was less than its consumption due to an inhibition of the glycolysis, e.g. during an increase of the energy-dependent reactions in connection with the cation transport.

An investigation of the properties of the HK-PFK system at different adenine nucleotide concentrations permits not only an analysis of the effect on the energy metabolism of human red blood cells but is also suited for studies on the glycolytic metabolism of red blood cells from other sources considerably differing in their adenine nucleotide concentration. At present there are no indications that there exist differences in the properties of HK and PFK of red blood cells from different origin.

Fig. 2 shows the dependence of the flux through the HK-PFK system on the ATP concentration at 4 different concentrations of the adenine nucleotides. In all calculations the magnesium concentration was assumed to be 2 mM. As can be seen from Fig. 2 in the range of 2—4 mM of adenine nucleotides only slight changes occur in the maximal glycolytic flux. The magnitude of the maximal glycolytic flux is mainly determined by the concentration of MgATP. Below concentrations of adenine nucleotides of 2 mM a drastic decrease in the maximal glycolytic flux is observed, though the course of the curves is maintained. As a consequence of an increase in the concentration of adenine nucleotides higher concentrations of ATP are required for inhibition of the flux.

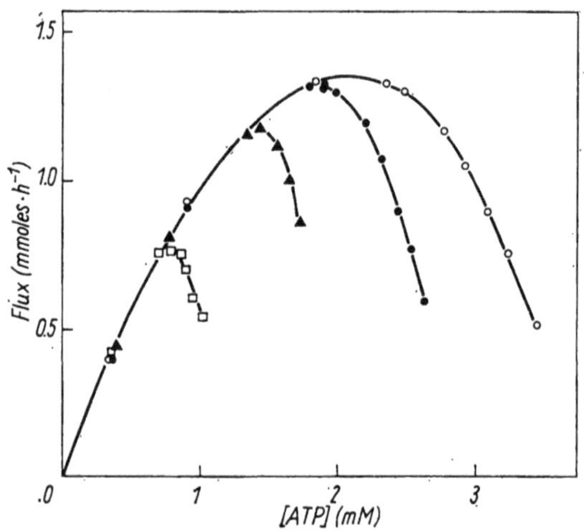

Fig. 2. Dependence of the flux on the ATP concentration at four different concentrations of adenine nucleotides: □ 1 mM, ▲ 2 mM, ● 3 mM, ○ 4 mM. The magnesium concentration is equal to 2 mM

The correlation between the magnitude of the maximal glycolytic flux and the hexokinase capacity is maintained as could be shown on the basis of calculations where the hexokinase capacity was reduced by 50%.

Correspondingly the maximal glycolytic flux was decreased by about 50% while the inhibition of the glycolytic rate by ATP remained nearly constant. Experimental observations in the cases of renal rachitis of humans and during a long-time magnesium deficiency of dogs where a drastic reduction of the hexokinase was not accompanied by a decrease of the glycolytic flux have to be explained by means of additional factors which at present have not been included in the model of the HK-PFK system [16].

Discussion

The results presented here were obtained on the basis of rather simplified models for both enzymes. The application of these models is justified by the fact that essential features of the HK-PFK system could be well described making the simplifying assumptions stated above. Nevertheless the following properties of both enzymes should be included in further calculations: it has been reported that the hexokinase is activated by inorganic phosphate [17]. In the case of phosphofructokinase our model is based both on an allosteric transition and association-dissociation phenomena but it cannot be decided at present to which extent association-dissociation effects of the enzyme take part in the regulation of the PFK activity under approximated intracellular conditions. To come to a conclusion on this point, it would be necessary to know the corresponding association constant(s). In our calculations we assumed that the simplified version of the "hybrid model" is valid which holds for $K_{ass}^{PFK} \cdot E_t \ll 1$ [8].

References

[1] RAPOPORT, T. A., R. HEINRICH, and S. RAPOPORT: Biochem. J. **154**, 449 (1976)
[2] JACOBASCH, G., S. MINAKAMI, and S. RAPOPORT in: Cellular and Molecular Biology of Erythrocytes. H. S. YOSHIKAWA and S. M. RAPOPORT (eds.). University of Tokyo Press, Tokyo 1974, p. 55
[3] GERBER, G., H. PREISSLER, R. HEINRICH, and S. M. RAPOPORT: Eur. J. Biochem. **45**, 39 (1974)
[4] RAPOPORT, T. A., M. OTTO, and R. HEINRICH: Acta biol. med. germ. **36**, 461 (1977)
[5] KÜHN, B., G. JACOBASCH, and S. RAPOPORT: Acta biol. med. germ. **23**, 1 (1969)
[6] OTTO, M., R. HEINRICH, B. KÜHN, and G. JACOBASCH: Eur. J. Biochem. **49**, 169 (1974)
[7] OTTO, M., R. HEINRICH, G. JACOBASCH, and S. RAPOPORT: In preparation
[8] OTTO, M., and S. SVETINA: Studia biophys. (in press)
[9] MINAKAMI, S., and H. YOSHIKAWA: J. Biochem., Tokyo **59**, 139 (1966)
[10] JACOBASCH, G.: Ph. D. Thesis, Humboldt-Universität zu Berlin, Math.-naturwiss. Fak. 1970
[11] EGGLESTON, L. V., and R. HEMS: Biochem. J. **52**, 156 (1952)
[12] HORN, A., R. HUSUNG, M. SCHROEDER, and H. BOERNIG: Acta biol. med. germ. **27**, 821 (1971)
[13] ROSE, I. A.: Proc. natn. Acad. Sci. U.S.A. **61**, 1079 (1968)
[14] BERNSTEIN, R. E.: J. clin. Invest. **38**, 1572 (1959)
[15] GERBER, G., H. BERGER, G.-R. JÄNIG, and S. RAPOPORT: Eur. J. Biochem. **38**, 563 (1973)
[16] JACOBASCH, G., I. SYLLM-RAPOPORT, K. WOLF, and J. RYMON: Folia haemat., Lpz., **83**, 310 (1965)
[17] ROSE, I. A.: Expl Eye Res. **11**, 264 (1971)

Acta biol. med. germ., Band 36, Seite 587—596 (1977)

Institut für Physiologische und Biologische Chemie der Humboldt-Universität Berlin,
104 Berlin, DDR

Kontrolle der Glykolyse bei Magnesiummangel: Untersuchungen an intakten roten Blutzellen und Hämolysaten

G. Jacobasch, Ch. Gerth und P.-G. Fabricius

Zusammenfassung

Das Verhalten der Glykolyserate und glykolytischen Metabolitkonzentrationen wurde unter Bedingungen des Magnesiummangels untersucht. Der Mg-Mangel wurde im Tierexperiment diätetisch und im Hämolysat durch Zusatz eines Komplexbildners erzeugt. Die Ergebnisse belegen, daß die Verminderung des freien Mg-Spiegels durch den Abbau des ATP und 2,3-Diphosphoglyzerats (2,3-DPG) abgeschwächt wird. Die Analyse der Kontrollstärken läßt die Schlußfolgerung zu, daß die Drosselung der Glykolyserate auf eine Hemmung des Hexokinase-Phosphofruktokinase-Kontrollsystems zurückzuführen ist. Die Abnahme des MgATP- und freien Mg^{++}-Spiegels erklärt eine verminderte Glukosephosphorylierung durch die Hexokinase. Die Hemmung der Phosphofruktokinase durch ATP ist durch einen leichten Anstieg des freien ATP-Spiegels bei gleichzeitiger Senkung des Fru-6-P-Spiegels verstärkt. Die Erhöhung des PEP-Spiegels ist nicht Ausdruck einer größeren Kontrollstärke der Pyruvatkinase, sondern auf eine Abnahme der freien Mg^{++}- und MgADP-Konzentration zurückzuführen.

Einleitung

Der Energiestoffwechsel der Erythrozyten wird durch die Höhe des glykolytischen Stoffwechsels bestimmt. Eine Vielzahl von Faktoren, die die Glykolyse aktivieren oder hemmen können, beeinflussen ihre Kontrolle [1]. Zu diesen Faktoren zählt das Magnesium. Wie aus tierexperimentellen Untersuchungen bekannt ist, wird durch Senkung der zellulären Magnesiumkonzentration eine Verminderung der Glykolyserate und des ATP-Spiegels hervorgerufen [2, 3]. Diese Hemmung des Energiestoffwechsels unter Magnesiummangelbedingungen ist die Ursache für die im Tierexperiment beobachtete hämolytische Anämie. Durch eine verstärkte Ausschüttung von Retikulozyten kann sie weitgehend kompensiert werden [4, 5].

In der vorliegenden Arbeit werden Ergebnisse vergleichender Untersuchungen an roten Blutzellen von Ratten unter Magnesiummangel und Hämolysaten, in denen durch Zusatz eines Komplexbildners in vitro eine Senkung des Magnesiumspiegels erreicht wurde, gegenübergestellt.

Durch diese Untersuchungen wurde geprüft, ob die Senkung des Adeninnukleotidspiegels Ausdruck eines verstärkten Adeninnukleotidabbaus ist oder auf eine Hemmung der Adeninnukleotidsynthese in den unter Magnesiummangelbedingungen gebildeten Zellpopulationen zurückzuführen ist.

Ein weiteres Anliegen der Arbeiten bestand darin, auf der Grundlage von Metabolitbestimmungen Aussagen über den zeitlichen Verlauf von Konzentrationsveränderungen machen zu können. Darüber hinaus sollte eine Analyse der Reaktionspartner der drei Mg-abhängigen Kontrollenzyme Hexokinase, Phosphofruktokinase und Pyruvatkinase Schlußfolgerungen darüber erlauben, ob der Magnesiummangel die Kontrollstärken dieser Enzyme beeinflußt.

Material und Methoden

Zur Ergänzung des tierexperimentellen Mg-Mangels wurden 230 50 g schwere Albinoratten ca. 14 Tage mit einer Mg-Mangeldiät nach den Angaben von KRUSE et al. ernährt [6]. Für jeweils 25 Ratten wurde unter Zusatz von Aqua bidest. ein Nahrungsbrei zubereitet, der folgende Zusammensetzung aufwies: 80 g Kasein; 224 g Stärke; 32 g Butter, 23,9 g Salzmischung. — Die Salzmischung enthielt 6 g $CaCO_3$, 4 g KCl, 2,8 g $NaHCO_3$, 6,8 g KH_2PO_4, 2 g NaCl, 2,3 g $Fe_2(SO_4)_3(NH_4)_2SO_4 \cdot 24\,H_2O$. Zusätzlich wurde jeden 2. Tag ein halbes Dragee des Polyvitaminpräparates „Summavit forte" des VEB Jenapharm pro Tier gegeben. Die Tiere hatten freien Zugang zum Futter und zu Aqua bidest.
Der Mg-Gehalt der Diät lag mit 0,2—0,3 mg/100 g Diät weit unter dem mit 56 mg/100 g Futter angegebenem Bedarf [7]. Die Mg-Mangeltiere wurden unter annähernd Mg-freien Kautelen in Einzelkäfigen gehalten.
Die Kontrolltiere wurden z. T. mit dem gleichen Nahrungsbrei wie oben angegeben unter Zusatz optimaler $MgCl_2$-Mengen ernährt; einige Kontrolltiere erhielten im Handel angebotenes Pellet-Standardfutter. Die Tiere beider Kontrollgruppen wiesen in den überprüften Parametern keine signifikanten Unterschiede auf.
Der Zeitpunkt der Tötung wurde an Hand der klinischen Befunde des Magnesiummangelsyndroms festgelegt. Die Blutgewinnung erfolgte in Äthernarkose aus dem Herzen. 2 mal in 0,9%iger NaCl-Lösung gewaschene rote Blutzellen wurden vor der 3. Waschung in isotonem Tris-NaCl-Puffer, *p*H 7,2 oder *p*H 8,5, suspendiert, anschließend abzentrifugiert. Die Zellen wurden dann mit dem entsprechenden Suspensionsmedium zu etwa gleichen Teilen versetzt. Das Medium bestand aus einem isotonem Tris-NaCl-Puffergemisch, *p*H 7,2 bzw. 8,5, das weiterhin Phosphat in einer Endkonzentration von 1 mM und 8 mM Glukose enthielt. Der *p*H-Wert der Suspension von 7,2 bzw. 8,2 wurde mit einer Glaselektrode überprüft.
Die Laktatbestimmung erfolgte nach der von BARKER und SUMMERSON [8] angegebenen Methode, nachdem durch Trichloressigsäure die Eiweiße gefällt waren. Für alle übrigen Metabolitbestimmungen wurde eine Enteiweißung mit 0,6 N $HClO_4$ durchgeführt.
Die Metabolite Gluc-6-P, Fru-1,6-P, DOAP, GAP, 3PG, 2PG, PEP, Pyr, ATP, ADP, AMP wurden gering modifiziert nach den in BERGMEYER angegebenen enzymatischen Methoden durchgeführt [9].
Für die Hämolysatversuche wurde Menschenblut verwendet. 1 ml gewaschene Zellen wurden mit 2 ml Krebs-Ringer-Bikarbonat-Phosphatpuffer, *p*H 7,4, entsprechend den Angaben von OSKI und SMITH [10], versetzt und die Hämolyse durch dreimaliges Frieren und Tauen erzeugt. Anschließend wurde das Stroma abzentrifugiert; NAD und Glukose wurden in einer Endkonzentration von 0,1 bzw. 15 mM dem stromafreien Hämolysat zugesetzt. Die Senkung des Mg-Spiegels im Hämolysat wurde durch eine definierte Zugabe von Chelating Resin A_1 (Dowex, Fluka AG) erreicht.
Zur Kalzium- und Magnesiumbestimmung diente die komplexrimetrische Titration [11].

Ergebnisse

Die Überprüfung einiger Vollblutparameter im Verlaufe der Experimente ergab, daß mit zunehmender Hypomagnesiämie der roten Blutzellen die Mortalitätsrate der Versuchstiere anstieg. Die Ratten wurden deshalb für Stoffwechseluntersuchungen entblutet, wenn ca. 40% der Tiere unter den typischen Zeichen der Magnesiummangeltetanie ad exitum gekommen waren.
Zu diesem Zeitpunkt ließen sich im Vollblut der Ratten charakteristische Metabolitveränderungen als Ausdruck der Magnesiummangelbedingungen nachweisen (Tab. 1).
Die durchschnittliche Abnahme des zellulären Mg-Spiegels betrug 45%. Parallel dazu trat eine Verminderung des Adeninnukleotidspiegels um ca. 30% auf, die auf eine Abnahme der ATP-Konzentration zurückzuführen ist. Die Überprüfung der Hämatokrit- und Retikulozytenwerte zeigte, daß zu diesem Zeitpunkt bereits eine mehr oder weniger stark ausgeprägte hämolytische Anämie bei den

Tabelle 1
Verhalten einiger Glykolysemetabolite im Vollblut von Ratten unter Magnesiummangelbedingungen

Metabolit	Mg-Mangel ($n = 16$)		Kontrolle ($n = 9$)	
	µMol/l Zellen			
	$\bar{x}$	$\pm s$	$\bar{x}$	$\pm s$
Mg^{++}	1,79	0,56	3,26	0,73
ATP	1032	325,3	1517	305,9
ADP	189	33,4	178	178,1
AMP	97	34,0	112	28,9
$\sum$ Adeninnukleotide	1318		1807	
Gluc-6-P	132	2,0	150	6,0
Fru-1,6-P	34,3	4,0	72	2,0
DOAP + GAP	42,3	3,5	62	3,1
PEP	49,0	1,0	21	3,0
Pyr	347,0	7,0	372	12,0

Versuchstieren bestand. Die höchsten Retikulozytenwerte lagen bei 35%. Wie aus Abb. 1 ersichtlich ist, besteht eine deutliche Abhängigkeit zwischen den Hämatokritwerten der Vollblute und den zellulären ATP-Spiegeln. Je ausgeprägter die Anämie war, desto höher war auf Grund des Retikulozytengehaltes dieser Blute der ATP-Spiegel. Obwohl Retikulozyten im Vergleich zu Erythrozyten sowohl durch höhere Adeninnukleotidkonzentrationen als auch Hexosephosphat- (Gluc-6-P, Fru-6-P, Fru-1-6-P) und Triosephosphatspiegel (DOAP, GAP) charakterisiert sind, waren typische Konzentrationsveränderungen von Glykolysemetaboliten als Folge des Mg-Mangels deutlich nachweisbar. Die signifikante Veränderung des Adeninnukleotidspiegels ist durch die Abnahme der ATP-Konzentration bestimmt. Da das ATP ein entscheidender Kontrollparameter des HK-PFK-Systems ist, ist der ATP-Abfall mit einer Verminderung des Gluc-6-P, Fru-1,6-P und Triosephosphatspiegels verbunden. Charakteristisch für den Mg-Mangel ist weiterhin eine deutliche Erhöhung des PEP-Spiegels und eine Abnahme des 2,3-DPG und 1,3-DPG. Die gleichen Konzentrationsveränderungen glykolytischer Metabolite wie im Vollblut sind auch bei Untersuchungen von Zellsuspensionen in vitro nachweisbar. Der ATP-Abfall zeigt dabei eine lineare Beziehung zu Konzentrationsabnahmen des zellulären Mg-Spiegels. Dieser ist um so größer, je höher der *p*H-Wert ist. Nach einstündiger Inkubation bei 37 °C wurde bei *p*H 7,2 für Kontrolltiere ein ATP-Spiegel von 1350 µM ($n = 5$) und für Mg-Mangelratten von 1030 µM ($n = 6$) bestimmt; bei *p*H 8,2 betrugen die entsprechenden Werte 471 µM ($n = 5$) bzw. 227 µM ($n = 5$) ATP.

Die Glykolyserate ist sowohl bei *p*H 7,2 als auch bei *p*H 8,2 um 40—50% vermindert. Die *p*H-Abhängigkeit der Glykolyserate bleibt dabei voll erhalten ebenso wie die Aktivierbarkeit der Glykolyse durch Phosphat.

Abb. 2. zeigt eine zusammenfassende Darstellung der prozentualen Konzentrationsveränderungen der Glykolysemetabolite und der Glykolyserate bei *p*H 7,2 und *p*H 8,2. Die Ergebnisse der Metabolitbestimmungen sind in Übereinstimmung mit der Arbeitshypothese, daß die Glykolyserate im Mg-Mangel durch

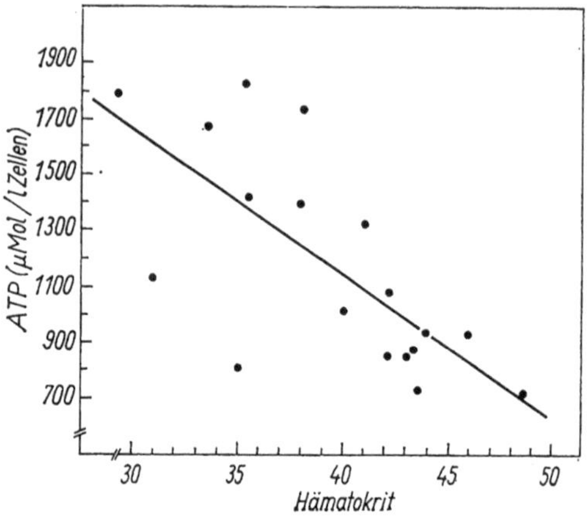

Abb. 1

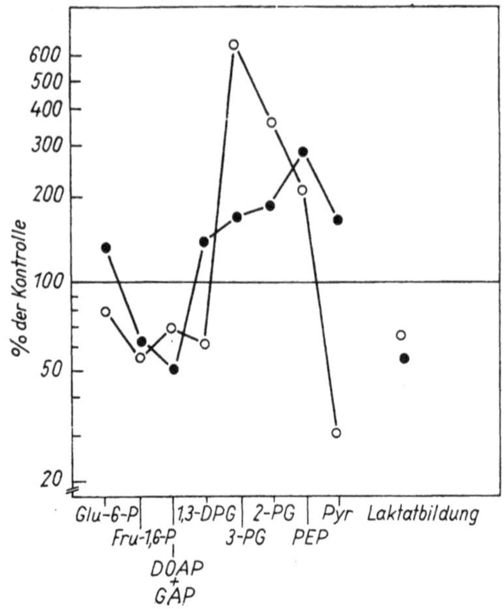

Abb. 2

Abb. 1. Wechselbeziehungen zwischen Hämatokritwert und ATP-Spiegel im Vollblut von Ratten im Magnesiummangel

Abb. 2. Prozentuale Änderungen der Glykolyserate und der Metabolitkonzentrationen von roten Blutzellen der Ratte im Magnesiummangel.
Die Bestimmungen wurden nach 60minütiger Inkubation der Zellsuspensionen bei *p*H 7,2 (○) und *p*H 8,2 (●) bei 37 °C durchgeführt. Die Ergebnisse wurden mit denen entsprechender Kontrollversuche verglichen

eine Hemmung des HK-PFK-Systems gedrosselt ist. Das beweist die Abnahme des Hexose- und Triosephosphatspiegels. In Abhängigkeit vom *p*H-Wert treten für Pyruvat entgegengesetzte Konzentrationsverhältnisse auf. Bei *p*H 8,2 wurde in Übereinstimmung zur Abnahme des Triosephosphatspiegels und als Ausdruck eines verminderten NADH/NAD-Quotienten ein Pyruvatanstieg festgestellt. Der Abfall des Pyruvatspiegels bei *p*H 7,2 ist dagegen ebenso wie im Vollblut nur deutbar, wenn NADH durch außerhalb der Glykolyse ablaufender Prozesse oxydiert wird.

Die 2,3-DPG-Synthese zeigt eine lineare Abhängigkeit zum 1,3-DPG, dessen Höhe weitgehend durch den ATP/ADP-Quotienten bestimmt wird. Unter der Annahme eines thermodynamischen Gleichgewichtes der Phosphoglyzeratkinasereaktion leitet sich unter Mg-Mangelbedingungen ein Anstieg des 3-PG/1,3-DPG-Quotienten ab. Diese Schlußfolgerung ist in Übereinstimmung mit den analysierten 3-PG- und 1,3-DPG-Konzentrationen.

Eine Reihe von Beobachtungen sprechen dafür, daß die freie Mg-Konzentration der Zelle unter normalen Bedingungen weitgehend konstant gehalten wird. In diesem Zusammenhang kommt der Adenylatkinase eine Art Pufferwirkung zu. Das Enzym, das eine Gleichgewichtsreaktion katalysiert, zeigt eine hohe Enzymkapazität und große Unterschiede in den Dissoziationskonstanten der Mg-Kom-

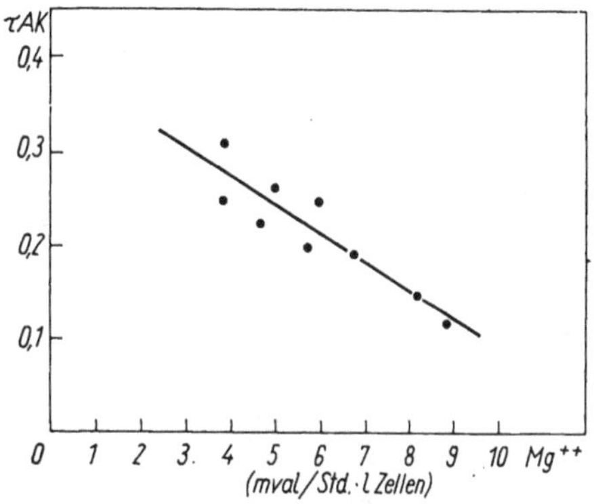

Abb. 3. Abhängigkeit des Massenwirkungsquotienten der Adenylatkinase (AK) vom Mg-Spiegel roter Blutzellen der Ratte im Vollblut

plexe [12, 13]. Die in Abb. 3 dargestellten Befunde bestätigen erwartungsgemäß den linearen Abfall des Massenwirkungsquotienten der Adenylatkinase mit steigender Mg-Konzentration.

Der Einfluß des Mg-Mangels auf den freien Mg-Spiegel sowie einige freie und Mg-komplexierte Metabolitkonzentrationen wurde anhand einer Berechnung geprüft. Berücksichtigt wurden für das Rechenprogramm 2,3-DPG, ATP, ADP, AMP, PEP und Mg. Hb wurde als Ligand vernachlässigt. Für die entsprechenden Dissoziationskonstanten wurden Literaturangaben verwendet [12—15]. Tab. 2 zeigt die Ergebnisse der durchgeführten Berechnungen für Zellsuspensionen bei 3 *p*H-Werten. Die Ergebnisse lassen folgende Schlußfolgerungen zu. Die Abnahme der freien Mg-Konzentrationen roter Blutzellen ist im Mg-Mangel auf Grund der Abnahme des ATP- und 2,3-DPG-Spiegels geringer als die des Gesamt-Mg. Die Konzentration der MgATP- und MgADP-Komplexe ist dagegen unter Mangelbedingungen stark vermindert. Der freie ATP-Spiegel ist bei *p*H 7,2 und 8,2 im Vergleich zu Kontrollwerten leicht erhöht.

Tabelle 2
Analyse Mg-komplexierter und freier Metabolitkonzentration roter Blutzellen der Ratte

Bedingungen	Mg	Mg	ATP	MgATP	ATP	ADP	MgADP	DPG	MgDPG
	μMol/l Zellwasser								
*p*H 7,2 Kontr.	5420	468	2160	1840	320	360	132	13600	2981
Mg ↓	2692	255	1758	1316	443	218	52	8120	1073
*p*H 8,2 Kontr.	6980	684	930	841	89	653	319	16000	5129
Mg ↓	3504	435	873	737	136	359	145	9600	2176
*p*H 8,4 Kontr.	6115	627	494	442	52	497	229	16000	4810
Mg ↓	2705	391	236	199	37	198	70	9600	2038

Für die Berechnungen wurden folgende Metabolite berücksichtigt: ATP, ADP, AMP, 2,3-DPG, PEP

Die durchgeführten Modelluntersuchungen an Hämolysaten sind als eine Ergänzung zu den tierexperimentellen Arbeiten aufzufassen. Da sie an weitgehend einheitlichen Zellpopulationen durchgeführt wurden, sind reifungsbedingte Stoffwechseleinflüsse auszuschließen. Ein weiterer Vorteil dieses Systems bestand darin, daß zeitliche Verlaufsstudien bei definierten Mg-Konzentrationen möglich waren. Die in Tab. 3 zusammengefaßten Ergebnisse der Hämolysatversuche bei Variation des Mg-Spiegels bestätigen die im Tierexperiment erhobenen Befunde. Eine Verringerung des Mg-Spiegels unter 19 μMol/g Hb bewirkt eine deutliche Hemmung der Glykolyse. Das Verhalten der Adeninnukleotidkonzentrationen läßt folgende Schlußfolgerungen zu. Die Senkung des Mg-Spiegels bewirkt eine Abnahme des ATP- und Zunahme des ADP- und AMP-Spiegels. Der Anstieg der ADP- und AMP-Konzentrationen wird von dem einsetzenden Adeninnukleotidabbau überlagert. Wird die Glykolyserate durch Senkung des Mg-Spiegels auf mehr als 1/4 gedrosselt, kommt es zu einem steilen Abfall der ATP-Spiegel und zu einer Erhöhung der Abbaurate der Adeninnukleotide. Die tiefen Adeninnukleotidkonzentrationen von Mg-Mangelzellen sind somit als das Ergebnis einer gesteigerten Abbaurate und nicht als Folge einer gehemmten Adeninnukleotidsynthese aufzufassen.

Tabelle 3

Verhalten des glykolytischen Stoffwechsels bei Senkung der Magnesiumkonzentration im Hämolysat

Mg^{++} (μMol/g Hb)	Laktat (mMol/g Hb · Std.)	$\sum$ Adeninnukleotide	ATP	ADP	AMP	Glu-6-P	PEP
		(μMol/g Hb)					
44	7,2	6438	5660	288	490	463	316
19	7,0	6296	5240	490	566	—	349
9	6,0	5873	4775	612	486	244	369
Chelating Resin A_1							
55 μl/g Hb	5,2	5815	4460	808	547	146	432
110	5,0	4015	2682	750	583	187	548
200	2,0	4357	2350	1185	822	167	555
250	0,6	2736	488	773	1475	45	1261
500	0,2	1095	500	317	278	—	830

Die Inkubationszeit betrug 90 min bei *p*H 7,4 und 37 °C.

Die Abnahme des Gluc-6-P- und der kontinuierliche Anstieg des PEP-Spiegels mit fortschreitender Senkung des Mg-Spiegels entsprechen, auf Grund der gehemmten Glukosephosphorylierung durch die Hexokinase und der drastischen Senkung der MgADP-Konzentration als Reaktionspartner der Pyruvatkinasereaktion, den Erwartungen. Andererseits konnte im Hämolysat im Gegensatz zu den tierexperimentellen Befunden an intakten Zellen keine Abnahme des 2,3-DPG-Spiegels im Mg-Mangel festgestellt werden. Offensichtlich ist auch ohne Verringerung des Mg-Spiegels, allein durch Senkung des *p*H-Wertes der 2,3-DPG-Abbau im Hämolysat nicht wie in intakten Zellen aktivierbar.

Das Hämolysatsystem zeigte für Kontrollansätze über mehr als 4 Std. eine nahezu gleichbleibende Glykolyserate und konstante Metabolitkonzentrationen, d. h. ATP-bildende und verbrauchende Reaktionen befanden sich in einem annähernden Fließgleichgewicht. Die Überprüfung zeitabhängiger Prozesse unter den Bedingungen des Mg-Mangels war deshalb in diesem System möglich. Abb. 4 zeigt

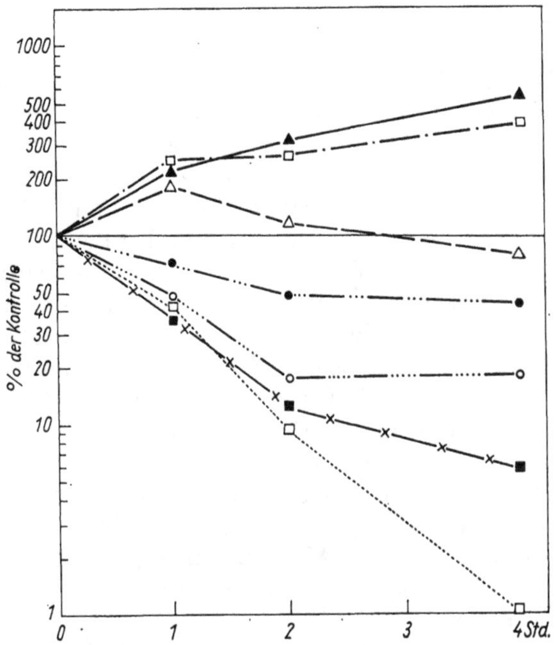

Abb. 4. Verhalten der Laktatbildung und der Metabolitkonzentrationen bei Senkung des Mg-Spiegels in Abhängigkeit von der Zeit.
Die Verminderung des Mg-Spiegels wurde durch Zusatz von 220 µl „chelating resin A_1"/g Hb bewirkt. Die Inkubation des Hämolysates erfolgte bei 37 °C, *p*H 7,4.
Laktatbildung □······□; Adeninnukleotide ●—···—●; ATP ■—×—■; ADP △———△; Gluc-6-P ○—·····—○; AMP □—·—□; PEP ▲——▲

das Verhalten der Laktatbildung und einiger Metabolitkonzentrationen bei Senkung des Mg-Spiegels durch Zusatz von 220 µl „chelating resin A_1" im Verlaufe von 4 Std. Die Ergebnisse verdeutlichen, daß nach Zugabe des Komplexbildners sowohl die Senkung des ATP-Spiegels als auch die Hemmung der Glykolyse mit der Inkubationszeit zunehmen. Das bedeutet, daß die Imbalance zwischen ATP verbrauchenden und ATP bildenden Reaktionen verstärkt wird, die schließlich zum Zusammenbruch des Energiestoffwechsels führt. Der aus dem ATP-Defizit resultierende ADP- und AMP-Anstieg erreicht dagegen schon nach 1 Std. ein Optimum, denn durch den Adenninnukleotidabbau wird das AMP auf einen höhereren, nahezu konstanten Wert eingestellt, während der ADP-Spiegel wieder auf Werte im Normbereich abfällt. Der kontinuierliche Anstieg des PEP-Spiegels ist ein Hinweis für eine zunehmende Hemmung der PK-Reaktion.

Diskussion

Die Bedeutung des Mg als essentieller Faktor der Glykolyse ist durch zahlreiche experimentelle Daten belegt [1—4]. Dabei treten 3 Reaktionsmöglichkeiten auf: 1. eine Enzymaktivierung durch Mg^{++}, wie sie für die Hexokinase und Pyruvatkinase bekannt ist [16, 17], 2. Mg^{++} als Ligand von Substraten, Produkten und Effektoren und 3. ein Hemmeffekt, wie er in höheren Mg-Konzentrationen für eine Reihe Mg-abhängiger Enzyme beschrieben wurde.

Für die Interpretation der durch den Mg-Mangel bedingten Störungen des Energiestoffwechsels kommt nur den ersten beiden Wirkungsmechanismen eine größere Bedeutung zu. Aus der Mehrpunktkontrolle des glykolytischen Stoffwechsels leiten sich 3 Mg-abhängige Reaktionen, die sich thermodynamisch nicht im Gleichim Gleichgewicht befinden, -die Hexokinase, Phosphofruktokinase und Pyruvatkinase-, als Angriffsorte für die im Mg-Mangel auftretenden Glykolysehemmung ab [1]. Da unter normalen Bedingungen nur die HK- und PFK-Reaktion für die Glykolyserate bestimmend sind [18], wurden zunächst die Kontrollstärken dieser beiden Enzyme berechnet. Die Werte stimmen für Zellsuspensionen bei *p*H 7,2 und 8,2 mit denen der Kontrolltiere überein, das bedeutet, daß im Mg-Mangel das HK-PFK-Kontrollsystem der Glykolyse als Ganzes beeinflußt wird und keine isolierte Hemmung an einem der beiden Kontrollenzyme auftritt. Aus der Analyse der Mg-Komplexe und der freien Metabolitkonzentrationen (Tab. 2) leitet sich folgende Interpretation für die Hemmung des HK-PFK-Systems ab. Der tiefe Mg-Spiegel bewirkt eine partielle Hemmung der HK, die durch die Verminderung des MgATP-Spiegels noch verstärkt wird. Die PFK wird dagegen durch die auftretenden Konzentrationsveränderungen des freien Mg^{++} und MgATP nicht direkt beeinflußt, da der MgATP-Spiegel noch im Sättigungsbereich des Enzyms liegt und eine Aktivierung der PKF durch freies Mg^{++} nicht nachweisbar ist [19]. Die Hemmung der PFK resultiert aus der Verringerung des Fru-6-P-Spiegels als Ergebnis der gedrosselten HK-Aktivität, denn die starke ATP-Hemmung der PFK roter Blutzellen ist kompetitiv zu Fru-6-P [1]. Da der freie ATP-Spiegel unter Mg-Mangelbedingungen sogar noch ansteigt, wird auf diese Weise die Hemmung der PFK durch ATP erhöht. Die Ursache für den bei *p*H 8,2 und 8,4 im Vergleich zu *p*H 7,2 größeren ATP-Abfall (Tab. 2) ist durch den im alkalischen Bereich gesteigerten Energiebedarf der roten Blutzelle zur Aufrechterhaltung von membranabhängigen Funktionen bedingt.

Der festgestellte Konzentrationsanstieg des PEP im Mg-Mangel (Tab. 1 und 3, Abb. 2) wirft die Frage auf, ob die PK als Mg-abhängiges Enzym unter diesen Bedingungen eine Bedeutung für die Kontrolle der Glykolyserate hat. Eine Berechnung zeigt, daß die Zunahme des PEP-Spiegels nicht der Ausdruck einer größeren Kontrollstärke dieses Enzyms ist, sondern sich aus dem tiefen MgADP- und Mg-Spiegel ableitet. Als einzige Energiereserve bei Flux-Verminderungen der Glykolyse steht dem Erythrozyten das 2,3-DPG zur Verfügung. Eine Abnahme des 2,3-DPG-Spiegels, die prozentual immer geringer als die des ATP ist, tritt erst dann auf, wenn bereits ein deutlicher Abbau der Adeninnukleotide nachweisbar ist [4]. Mit der Verringerung des 2,3-DPG-Spiegels ist nicht nur ein begrenzter Energiegewinn verbunden, sondern auch eine Abschwächung der 2,3-DPG-Hemmung der Kinasen, insbesondere der der HK und PFK [1].

Da ATP und 2,3-DPG die wichtigsten Mg-Puffer des Erythrozyten sind — sie binden unter physiologischen Bedingungen etwa 75% des zellulären Mg [20] — wird durch die Abnahme dieser beiden Metabolite eine teilweise Kompensation des freien Mg^{++}-Spiegels erreicht. Unterstützt wird die Kompensation des freien Mg^{++}-Spiegels durch das Adenylatsystem, da der Massenwirkungsquotient dieser Reaktion Mg-abhängig ist [1] (Abb. 3). Die Basis dafür ist wiederum der Verlust an ATP im Mg-Mangel, wodurch aus dem verminderten Anteil des

MgATP-Komplexes freies Mg^{++} verfügbar wird; denn nur ein geringer Anteil dieses Mg^{++} reagiert mit dem sehr viel schwächeren Komplexbildner ADP. Die Kompensationsmöglichkeiten reichen insgesamt jedoch nicht aus, um den freien Mg^{++}-Spiegel konstant zu halten, d. h., ihn zu normalisieren. Diese Aussage wird durch die berechneten freien und komplexierten Glykolysemetabolite belegt (Tab. 2). Die Veränderungen des freien Mg^{++}-Spiegels werden durch den Abbau von ATP und 2,3-DPG abgeschwächt. Der freie Mg^{++}-Spiegel von Menschenerythrozyten, der mit Mg-empfindlichen Elektroden und unter Zugrundelegung der Komplexbildungskonstanten für Mg^{++}, K^+, Na^+ mit den Glykolysemetaboliten bestimmt bzw. berechnet wurde, liegt bei etwa 0,5 mM [1]. Diese Angaben sind in guter Übereinstimmung mit denen für rote Blutzellen der Ratte (Tab. 2). Der Pufferungseffekt auf den freien Mg-Spiegel ist im alkalischen Bereich stärker als bei *p*H 7,2 ausgeprägt. So beträgt z. B. bei *p*H 8,4 die durchschnittliche Abnahme des zellulären Gesamt-Mg-Spiegels 56%, der Abfall der freien Mg^{++}-Konzentration jedoch nur 38%.

Das enge Zusammenspiel von Mg^{++}- und ATP-Spiegel unterstützt die Schlüsselposition, die das ATP für den Stoffwechsel, die Struktur und Ionenzusammensetzung der roten Blutzelle hat. Die Erkennung der Zusammenhänge in der Regulation des Energiestoffwechsels einschließlich der Wirkung des Mg am Beispiel des Erythrozyten ist deshalb ein Beitrag zur Aufklärung der Rolle des Mg im Zwischenstoffwechsel von Zellen mit komplizierteren Strukturen.

Literatur

[1] Jacobasch, G., S. Minakami u. S. M. Rapoport in: Cellular and Molecular Biology of Erythrocytes. H. Yoshikawa u. S. M. Rapoport (Hrsg.). University of Tokyo Press, Tokyo 1974, S. 55

[2] Syllm-Rapoport, I., I. Strassburger u. S. Rapoport: Acta biol. med. germ. **5**, 596 (1960)

[3] Jacobasch, G.: Folia haemat., Lpz. **89**, 401 (1968)

[4] Oken, M. M., M. A. Lichtman, D. R. Miller u. P. Leblond: Blood **38**, 468 (1971)

[5] Piomelli, S. V. J., u. J. Dannis: Blood **41**, 3 (1973)

[6] Kruse, D. H., E. R. Orent u. E. V. McCollum: J. biol. Chem. **96**, 519 (1932)

[7] Syllm-Rapoport, I.: Habilitationsschrift. Humboldt-Universität Berlin, Med. Fakultät 1959

[8] Barker, S. B., u. W. H. Summerson: J. biol. Chem. **138**, 6535 (1941)

[9] Bergmeyer, H. U. (Hrsg.): Methoden der enzymatischen Analyse. Verlag Chemie, Weinheim 1962

[10] Oski, F. A., u. C. Smith: Pediatrics **41**, 473 (1968)

[11] Syllm-Rapoport, I.: Dte GesundhWes. **12**, 747 (1957)

[12] Philips, R. C., S. J. P. George u. R. J. Rutman: J. Am. chem. Soc. **88**, 2631 (1966)

[13] Belaich, J. P., u. J. C. Sari: Proc. natn. Acad. Sci. U.S.A. **64**, 763 (1969)

[14] Collier, H. B., u. A. Carr: Biochim. biophys. Acta **222**, 299 (1970)

[15] Norby, J.: Acta chem. scand. **24**, 3267 (1970)

[16] Gerber, G., H. Preissler, R. Heinrich u. S. Rapoport: Eur. J. Biochem. **46**, 39 (1974)

[17] De Asua, L. J., E. Rozengurt u. H. Carminatti: J. biol. Chem. **245**, 3901 (1970)

[18] Rapoport, T. A., R. Heinrich, G. Jacobasch u. S. Rapoport: Eur. J. Biochem. **42**, 107 (1974)

[19] Otto, M., R. Heinrich, B. Kühn u. G. Jacobasch: Eur. J. Biochem. **49**, 169 (1974)

[20] Achilles, W., G. A. Cumme u. H. Hoppe: Acta biol. med. germ. **31**, 763 (1973)

Summary

G. JACOBASCH, CH. GERTH and P.-G. FABRICIUS: Control of glycolysis in Mg-deficiency studies with intact red cells and hemolysate

The behaviour of glycolytic flux and glycolytic metabolic concentrations was studied under conditions of magnesium deficiency. The Mg-deficiency was produced in whole animals (rats) by feeding a diet almost completely free of Mg and in hemolysates of men by the addition of a chelating agent. The results show that the decrease of the free Mg-level is diminished by partial destruction of ATP and 2,3-DPG. The analysis of the control strength of the overall flux leads to the conclusion that the decrease of the glycolytic rate is caused by an inhibition of the hexokinase-phosphofructokinase-control system. The decrease of the MgATP-complex and free Mg^{++}-level explains the diminished phosphorylation of glucose by the hexokinase. The ATP-inhibition of the phosphofructokinase is amplified by a small increase of free ATP-concentration and a simultaneous decrease of the Fru-6P-level. The increase of the PEP-level is caused by the diminished free Mg^{++} and MgATP-complex and does not demonstrate a larger control strenght of the pyruvate kinase.

Acta biol. med. germ., Band 36, Seite 597—610 (1977)

Physiologisch-Chemisches Institut der Friedrich-Schiller Universität, 69 Jena, DDR

Relations between ion shifting, ATP depletion and lactic acid formation in human red cells during moderate calcium loading using the ionophore A 23187

U. Till, H. Petermann, I. Wenz, and H. Frunder

Summary

Keeping constant cellular magnesium an A 23187 mediated moderate calcium loading of human red cells causes isoosmotic cell shrinkage, potassium efflux, slight decrease of cellular *p*H, ATP depletion connected with an increase of AMP, ADP and P_i and enhanced lactic acid formation. The calcium loading and accompanying effects can be abolished by EGTA or by extracellular magnesium, the latter kept more than two orders of magnitude above that of calcium which was 30 µM.
Inhibition of the (Mg^{2+} + Ca^{2+})-dependent ATPase by ruthenium red or lanthanum decreases the calcium stimulated lactic acid formation after a lag phase. However, the ATP depletion proceeds faster and is much more pronounced under these conditions. (Mg^{+2} + + Na^+ + K^+)-dependent ATPase, hexokinase, phosphofructokinase and cell shrinkage are ruled out, too, as mediators of the ATP depletion. This suggests that an unknown ATP consuming reaction, apparently not being related to the calcium pump, causes the calcium induced ATP depletion.

Introduction

Owing to the small Ca^{2+} content of red cells [1—4] and the low membrane permeability [5] the bivalent cation ionophore A 23 187 [6] progressively has been used to mediate a rapid Ca^{2+} loading of intact red cells. The main accompanying effects found are: K^+ efflux [7—15], shrinkage connected with spheroechinocytic transformation and membrane alteration [9—12, 16—20] and ATP depletion [9, 10, 13, 15, 18, 21].
The present paper mainly deals with the responsibility of different ATP consuming processes for the ATP depletion observed. Beside ATP, the corresponding alterations of ADP and AMP levels were measured. A second point investigated is the response of the lactic acid formation on Ca^{2+} loading and accompanying alterations in the adenine nucleotide pattern.
To avoid possible interferences by Mg^{2+} on the effects of enhanced cellular Ca^{2+}, experimental conditions were worked out which do not allow ionophore induced net movements of Mg^{2+} during the time of incubation. In order to enhance the cellular Ca^{2+} content to a relatively small extent, the cells were incubated in saline media without addition of Ca^{2+}. The Ca^{2+} concentration of these media results from impurities and was in a range of 20—50 µM, increasing the normal cellular Ca^{2+} content by about 50% with the help of the ionophore.

Abbreviation: DOPA: dihydroxyacetone phosphate; 2.3-DPG: 2.3-diphosphoglyceric acid; EGTA: ethyleneglycol-bis-(β-aminoethylether)-N',N'-tetraacetic acid; 1,6-FDP: fructose-1.6-diphosphate; G-6-P: glucose-6-phosphate; PEP: phosphoenolpyruvate; 3-PG: 3-phosphoglyceric acid; RC: red cells.

Materials and methods

Chemicals

Substances used were provided by: Ionophore A 23187 Lilli Laboratories, USA; [^{45}Ca]-$CaCl_2$ Zentralinstitut für Kernforschung, Rossendorf, GDR; Ouabain Merck, Darmstadt, GFR; Procaine Jenapharm, Jena, GDR; Vinblastin Gedeon Richter, Budapest, Hungary; Cytochalasin B, Ruthenium red, Colchicine and FGTA Ferak, Berlin-West. All reagents used were reagent-grade or enzyme-grade.

Preparation of red cells and incubation conditions

Human fresh blood was obtained by venipuncture and heparinized. After centrifugation (5 min, 2000 *g*), the plasma and the upper third of cell-sediment containing most of the leucocytes and reticulocytes were removed. The red cells were washed at room temperature 3 times with wash medium and resuspended with incubation medium to a hematocrit of 40–50%.
Composition of the media:
Wash medium: 5 mM KCl, 1 mM Na_2HPO_4, 5.5 mM glucose, 146 mM NaCl, 0.16 mM $MgCl_2$.
Incubation medium: 5 mM KCl, 1 mM Na_2HPO_4, 5.5 mM glucose, 122 mM NaCl, 24 mM $NaHCO_3$, 0.16 mM $MgCl_2$.
In cases of further additions to the media, the isoosmolarity was maintained by varying the NaCl concentration.
The cells were incubated in a shaking water bath at 37 °C. For purpose of comparison two or more sets from the same donor were incubated simultaneously. The preincubation time before additions amounted to 30 min.
The *p*H value was kept constant at 7.4 with a flow of a flexible and moist mixture of CO_2 and air over the surface of the sample. With triethanolamine buffered media the same results were obtained.
Ionophore A 23187 was dissolved in a mixture of dimethylsulfoxide/alcohol 1 + 1 (2.6 mg ionophore/ml mixture) and added to the incubate at a final concentration of 5 μM. The solvent was proved to be without influence on the parameters investigated. [^{45}Ca]-$CaCl_2$ was added to the incubate at the beginning of the incubation in trace amounts. Ruthenium red was dissolved in incubation medium and added to the incubate to a final concentration of 20 μM. EGTA was added in the Mg^{2+}-complexed form.

Denaturation of the samples

Denaturation was effected by 6% $HClO_4$. For the determination of lactic acid and glucose the $HClO_4$ was added to aliquots of the incubate (3:1, v/v). For the separate determination of cellular metabolites (adenine nucleotides, 2,3-DPG, P_i) and extracellular ions (K^+, Mg^{2+}, P_i) respectively, aliquots of the incubate were shortly centrifuged. The supernatants were diluted to varying extents with 0.1 N HCl. Packed cells were denaturated under vigorous shaking (1:5 *v/v*) centrifuged and the supernatant neutralized by 5 mM K_2CO_3.

Assays

Lactic acid, glucose, ATP, ADP and AMP were determined according to Bergmeyer [22] and 2,3-DPG according to Rose and Liebowitz [23].
Concentrations of Na^+, K^+, Mg^{2+} and Ca^{2+} were determined by atomic absorption (AAS 1 Carl Zeiss, Jena) and P_i according to Martin and Doty, modified by Till et al. [24]. Radioactivity measurements [^{45} Ca] were carried out by a Geiger-Mueller counter.
To measure the cellular *p*H value aliquots of the incubate were centrifuged under air exclusion. After sucking off the supernatant for the purpose of hemolysis the packed cells were frozen down to −80 °C under air exclusion for 1/2 hrs. Immediately after thawing the *p*H value was measured by a glass electrode at 37 °C.

Results

In the presence of A 23197 the total cellular Mg^{2+} concentration can be varied from 3 to 15 μmoles/ml RC water during one hour of incubation and depends on the extracellular Mg^{2+} concentration. Thus, experiments were carried out to determine that extra cellular Mg^{2+} concentration which is in equilibrium with cellular Mg^{2+}, i.e. which is not altered by Mg^{2+} movements into or out of the red cells. The results depicted in Fig. 1 clearly show that this equilibrium concentration is between 0.1 and 0.2 mM Mg^{2+}. Therefore, all experiments dealing with the effects of an A 23 187 mediated moderate Ca^{2+} loading of the cells were carried out with 0.16 mM Mg^{2+} in the medium which avoids changes of total cellular Mg^{2+}. All media were prepared without addition of Ca^{2+} except traces of [^{45}Ca]:$CaCl_2$ in order to establish only a small Ca^{2+} loading of the cells. Therefore, the extracellular Ca^{2+} content results from impurities of the salts used and was in the range of 20—50 μM. Under these conditions the ionophore causes a strong decrease of the extracellular Ca^{2+} concentration and of the extracellular total radioactivity of the added [^{45}Ca] (Fig. 2A, B solid lines). A corresponding increase of cellular total radioactivity can be found. The main Ca^{2+} influx takes place during the first 30 min. After 60 min equilibrium seems to be established (proved by long time incubation, not given in Fig. 2).

From the decrease of the extracellular Ca^{2+} concentration one can calculate that the corresponding increase of the cellular Ca^{2+} content amounts to 10—25 nmoles per ml RC water.

The ionophore causes a slight acidification within 15 min of incubation from 7.17 ± 0.03 ($n = 10$) to 7.12 ± 0.04 ($n = 10$). A decrease of the cellular *p*H after loading with small amounts of Ca^{2+} is in accordance with data of Reed [15]. In contrast loading with high Ca^{2+} concentrations leads to opposite H^+ fluxes [7, 15]. It shows that there is a more complicated relationship between Ca^{2+} and proton fluxes. Reed [15] suggests that the A 23187 induced H^+ influx at low Ca^{2+} concentrations is counterbalanced by a Mg^{2+} efflux leading to an electroneutral $Mg^{2+}/2\,H^+$ exchange. However, the results given in Table 1 show that during an A 23187 mediated moderate Mg^{2+} loading of the cells which is not able to

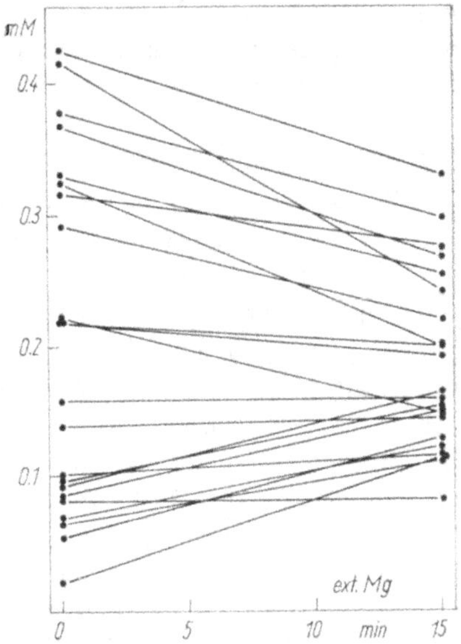

Fig. 1. Determination of the extracellular Mg^{2+} concentration not being altered by A 23187. Washed red cells were incubated with a hematocrit of about 50% at 37 °C and *p*H 7.4. The composition of the incubation medium is given in methods with the exception of different Mg^{2+} concentrations between 0.02–0.42 mM at the beginning of the incubation (ordinate). 15 min after the addition of A 23187 (5 μM), the extracellular Mg^{2+} concentration was measured again.

Table 1

ΔMg^{2+} extracellular (mM)	ΔpH cellular	ΔMg^{2+} extracellular (mM)	ΔpH cellular
−.023	−.09	−.075	−.014
−.069	−.09	−.098	−.05
−.075	−.01	−.175	−.085

6 aliquots of washed red cells were incubated at 37 °C in saline with different concentrations of Mg^{2+} between 0.219 mM and 0.417 mM. The extracellular Mg^{2+} concentration and the cellular pH were measured shortly before and 15 min after addition of A 23187 and the differences are given as Δ. The extracellular pH was kept constant at 7.4 ± 0.03 during the incubation by a flexible mixture of CO_2/air.

abolish the accompanying Ca^{2+} influx the celluar pH value decreases to about the same extent as under conditions were no Mg^{2+} movement takes place.

The ionophore induced Ca^{2+} influx seems to be connected with a rapid turnover of the (Mg^{2+} + Ca^{2+})-dependent ATPase. This can be derived from the decreasing specific radioactivity of the extracellular Ca^{2+} during the action of the ionophore (Fig. 2C, solid line). This isotope dilution is only possible when a certain amount of primarily unlabelled cellular Ca^{2+} becomes accessible and leaves the cells by the Ca^{2+} pump against the net Ca^{2+} influx. The conclusion is confirmed by further experiments with ruthenium red or $LaCl_3$, inhibitors of the+ (Mg^{2+} + Ca^{2+})-dependent ATPase: The addition of ruthenium red without A 23187 was proved to have no influence on the parameters investigated. If added together with A 23187 the Ca^{2+} influx takes place to about the same extent but markedly faster than without ruthenium red (Fig. 2A, B broken lines). The same holds if ruthenium red is added after the application of A 23187 but before Ca^{2+} influx has been ceased. Then, the extracellular Ca^{2+} concentration decreases rapidly, too. $LaCl_3$ was proved to give the same results as ruthenium red. The results are in accordance with a marked inhibition of the Ca^{2+} pumping system which otherwise would tend to retard the ionophore induced Ca^{2+} influx.

Effects of an A 23187 mediated moderate excess cellular calcium

All effects described in the following section and shown in Fig. 2 clearly can be attributed to the ionophore mediated Ca^{2+} influx into the cells: If cells are incubated with EGTA equimolar to the Ca^{2+} concentration in the medium a subsequent addition of A 23187 avoids Ca^{2+} influx and abolishes all accompanying effects. A subsequent addition of Ca^{2+} in excess over EGTA again brings about the same typical alterations. Fig. 3 shows this with respect to lactic acid formation, hematocrit and extracellular K^+ concentration. Incubation with EGTA up to a final concentration of 1 mM without A 23187 addition is proved to be without influence on the parameters investigated.

Cell shrinkage: With the hematocrit method a shrinkage of about 25% of the initial cell volume can be observed (Fig. 2D). The shrinkage corresponds accurately to values given in the literature after loading with high [10, 12] as well as with very small Ca^{2+} amounts [7]. Therefore, the phenomenon seems to

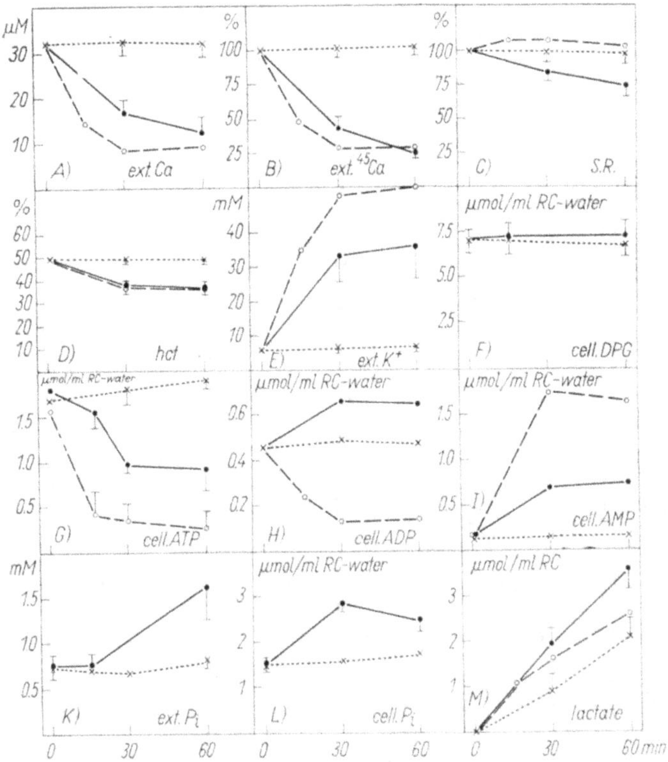

Fig. 2. Effects of A 23187 and A 23187 plus ruthenium red on ion distribution, cell volume, concentrations of adenine nucleotides and 2,3-DPG and lactic acid formation in human red cells. Washed red cells were incubated at 37 °C and *p*H 7.4 in saline with 0.16 mM $MgCl_2$ to avoid net movement of Mg^{2+} and about 30 µM $CaCl_2$ to allow Ca^{2+} influx in the presence of A 23187. For purpose of labelling traces of [^{45}Ca]-$CaCl_2$ were added.

×·········× red cells without any addition; ●———● red cells plus 5 µM A 23 187; ○— — —○ red cells plus 5 µM A 23 187 plus 20 µM ruthenium red.

For purpose of comparison parallel sets with cells from one donor were investigated simultaneously. The points with standard deviations are mean values of 5—10 determinations. Points whithout standard deviation are mean values of two estimations. With the exception of *G* and *I* all curves have been corrected to a common point at zero time. For values given in *F*—*I*, the ionophore induced shrinkage of cell volume is taken into account.

A) ext. Ca = extracellular Ca^{2+} concentration; B) ext. ^{45}Ca = total radioactivity of extracellular Ca^{2+} in % of the initial value; C) S. R. = specific radioactivity of extracellular Ca^{2+} in % of the initial value = ext. ^{45}Ca/ext. Ca; D) hct = hematocrit; E) ext. K^+ = extracellular K^+ concentration; F) cell. DPG = cellular 2.3-DPG concentration; G) cell. ATP = cellular ATP concentration; H) cell. ADP = cellular ADP concentration; I) cell. AMP = cellular AMP concentration; K) ext. P_i = extracellular P_i concentration; L) cell. P_i = cellular P_i concentration; M) lactate = virtual lactic acid concentration per ml RC, recalculated from values determined in the incubate

be completely marked by a very small Ca^{2+} influx. In accordance with this maximum shrinkage is complete within the first 15 min after A 23187 addition whereas Ca^{2+} influx proceeds.

The shrinkage is taken into account for the calculation of the metabolite concentrations per ml RC water.

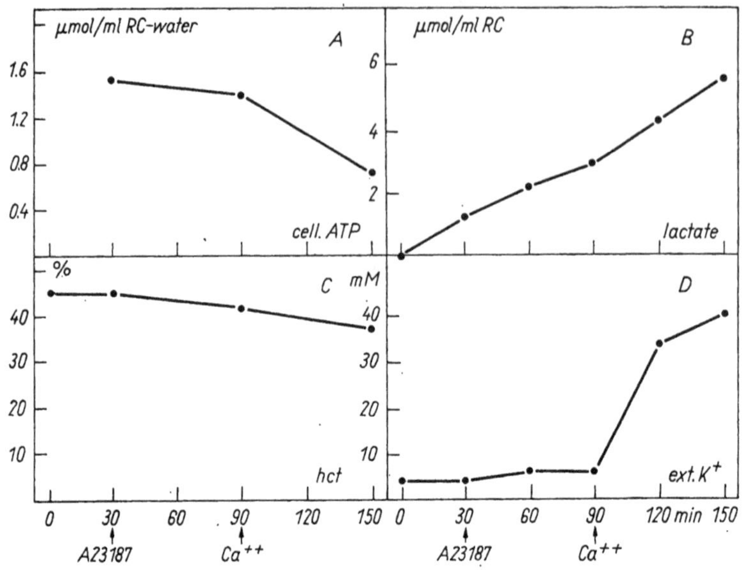

Fig. 3. Influence of EGTA on the action of A 23187.
Washed red cells were incubated at 37 °C, *p*H 7.4, in saline containing 80 μM Mg-EGTA. After 30 min A 23187 (5 μM) and after 90 min $CaCl_2$ (3 mM) were added. Mean values of two experiments. A) cell. ATP = cellular ATP concentration; B) lactate = virtual lactic acid concentration per ml RC, recalculated from values determined in the incubate; C) hct = hematocrit; D) ext. K^+ = extracellular K^+ concentration

During shrinkage the extracellular Cl^- concentration remains constant and the extracellular osmolarity is slightly increasing. Both facts point to the extrusion of cellular water plus ions in an isotonic composition during shrinkage. This is in accordance with other results on coupling of water efflux together with K^+ and Cl^- [11, 15], the preponderance of K^+ efflux over Na^+ influx [25] and a diminution of the sum of cellular K^+ und Na^+ concentration during shrinkage [15]. A close coupling of the shrinkage with the K^+ efflux seems to be very probable because it can be abolished by incubating the cells in KCl media [11]. Under our conditions a slight swelling of the cells is found in a K^+ rich medium although Ca^{2+} loading, ATP depletion and increase in lactic acid formation proceed (Fig. 4). From these findings it seems to be improbable that a Ca^{2+} induced association of red cell spectrin with certain membrane particles [26] could cause cell shrinkage. This is supported by our experiments with procaine, cytochalasin B, colchicine and vinblastin which failed to prevent the Ca^{2+} induced shrinkage.
To control how far the ATP depletion and/or enhanced lactic acid formation during ionophore induced Ca^{2+} loading (see Fig. 2G, M) are attributed to the accompanying loss of cell water a shrinkage was forced by incubation in a hypertonic saline medium (500 mosmoles/l) to the same extent as during Ca^{2+} loading. In this case ATP concentration as well as lactic acid formation are not altered. This shows that shrinkage itself is not responsible for the Ca^{2+} induced effects.
P o t a s s i u m e f f l u x: Extracellular K^+ rises from 5 mM up to about 35 mM (Fig. 2E). Probably, the steep increase during the first 30 min represents the sum of accelerated K^+ efflux and the K^+ extrusion caused by shrinkage.

The K^+ efflux is a well established effect of A 23187 mediated Ca^{2+} loading [7—12]. Our data are in accordance with an observation of REED and LARDY [7] who found the highest K^+ loss when incubating red cells in saline with very low Ca^{2+} concentrations.

In the presence of A 23187 ruthenium red accelerates and increases the K^+ loss and establishes equilibrium between intra- and extracellular space (Fig. 2E).

Lactic acid formation (Fig. 2M): Untreated cells show a normal linear lactic acid formation of 2.08 ± 0.44 ($n = 7$) µmoles/ml RC · h. With A 23 187 the rate is with $3.58 + 0.45$ ($n = 13$) µmole/ml RC · h nearly doubled despite the slight decrease of cellular *p*H value. In some cases the quotient of lactic acid formation/glucose consumption is measured and amounts to 1.78 ± 0.39 ($n = 6$) in untreated cells and to 3.06 ± 0.61 ($n = 8$) after A 23187 addition. The enhanced value after A 23187 addition points to a possible decrease of one or more glycolytic intermediates. This could be confirmed by a few measurements of the concentration of G-6-P, 1,6-FDP, DOAP, 3-PG and PEP all of which decreased considerably. 2,3-DPG can be ruled out as a source for lactic acid formation because of being constant during incubation with the ionophore (Fig. 2F).

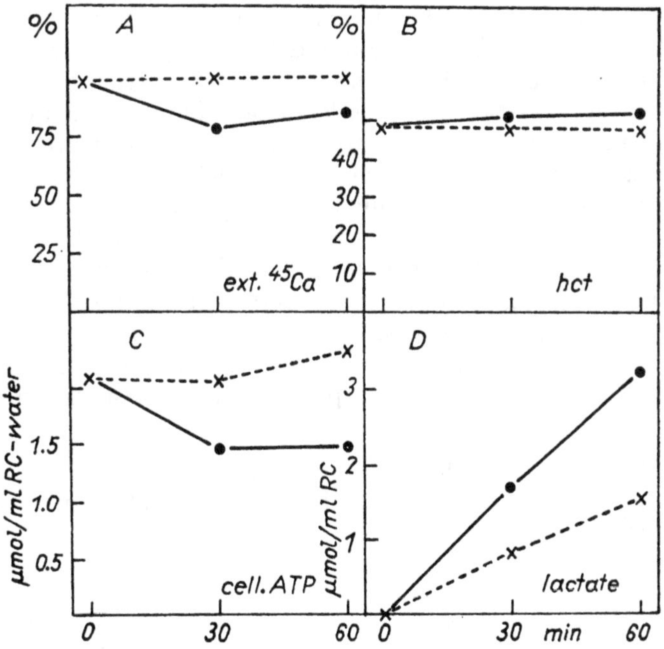

Fig. 4. A 23187 induced alterations in red cells incubated with a K^+ rich medium. Washed red cells were incubated at 37 °C, *p*H 7.4, in saline medium of the following composition: 122 mM KCl, 1 mM Na_2HPO_4, 5.5 mM glucose, 5 mM NaCl, 12 mM $KHCO_3$, 0.16 mM $MgCl_2$, under addition of [^{45}Ca]-$CaCl_2$.

×------× without A 23187; ●——● with A 23187 (5 µM).

A) ext. ^{45}Ca = total radioactivity of extracellular Ca^{2+}; B) hct = hematocrit; C) cell. ATP = cellular ATP concentration; D) lactate = virtual lactic acid concentration per ml RC, recalculated from values determined in the incubate

Adenine nucleotides (Fig. 2G — I): Within 30 min of incubation A 23187 induces a considerable ATP depletion down to a new level of less than 1.0 μmol/ml RC water whereas in untreated cells the level rises to about 2.4 μmoles/ml RC water at the end of incubation. The fall of ATP is accompanied by a slight increase of the concentration and a large increase of the AMP concentration. The sum of the ADP three nucleotides is identical in untreated and ionophore treated cells and nearly constant during the time of incubation. The quotient $\frac{[ATP][AMP]}{[ADP]^2}$ amounts to about 1.2 for both untreated and ionophore treated cells and remains also constant during the time of incubation (range 1.1—1.3).

The dephosphorylation of ATP is in accordance with an enhancement of cellular P_i which leads to an increasing efflux followed by rising extracellular P_i (Fig. 2K, L).

Effects of excess cellular Ca^{2+} on ATP consumption and lactic acid formation in the presence ATPase inhibitors

In the presence of ouabain there is the same A 23187 induced ATP depletion, increased lactic acid formation and K^+ efflux as in the absence of this inhibitor (Fig. 5). During inhibition of the Ca^{2+} pumping system by ruthenium red the Ca^{2+} influx leads to a faster ATP depletion and to a steeper decrease than without the inhibitor (Fig. 2G, broken line). The main ATP depletion takes place during the first 15 min of incubation were the lactic acid formation is about as high as without ruthenium red. After 15 min the lactic acid formation decreases to 1.99 ($n = 3$) μmoles/ml RC · h (Fig. 2M, broken line). The results of more detailed investigations on the action of ruthenium red are given in Fig. 6. 60 min after the addition of A 23187 an equilibrium between extracellular and cellular calcium is established. The addition of ruthenium red at this time does not change this equilibrium (Fig. 6A, C, E, broken lines) whereas the addition of ruthenium red together with EGTA removes excess Ca^{2+} and a considerable part of the initial cellular Ca^{2+} (Fig. 6A, C, E, broken plus dotted lines). Despite different total cellular Ca^{2+} the lactic acid production is identical in both cases (Fig. 6F).

However, ATP remains at the low level after addition of ruthenium red (Fig. 6D, broken line) and rises up to the initial level when cellular Ca^{2+} is removed by EGTA (Fig. 6D, broken plus dotted line).

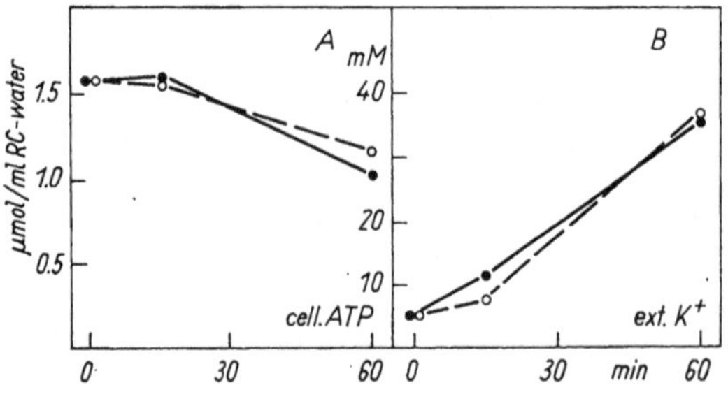

Fig. 5. Effect of ouabain on A 23187 induced ATP depletion and K^+ efflux. Washed red cells were incubated at 37 °C, *p*H 7.4, in saline containing 0.3 mM of ouabain (○– – –○) and without ouabain (●——●) in the presence of A 23187 (5 μM). Cellular ATP concentration (= cell. ATP) and extracellular K^+ concentration (= ext. K^+) are given as mean values of two experiments

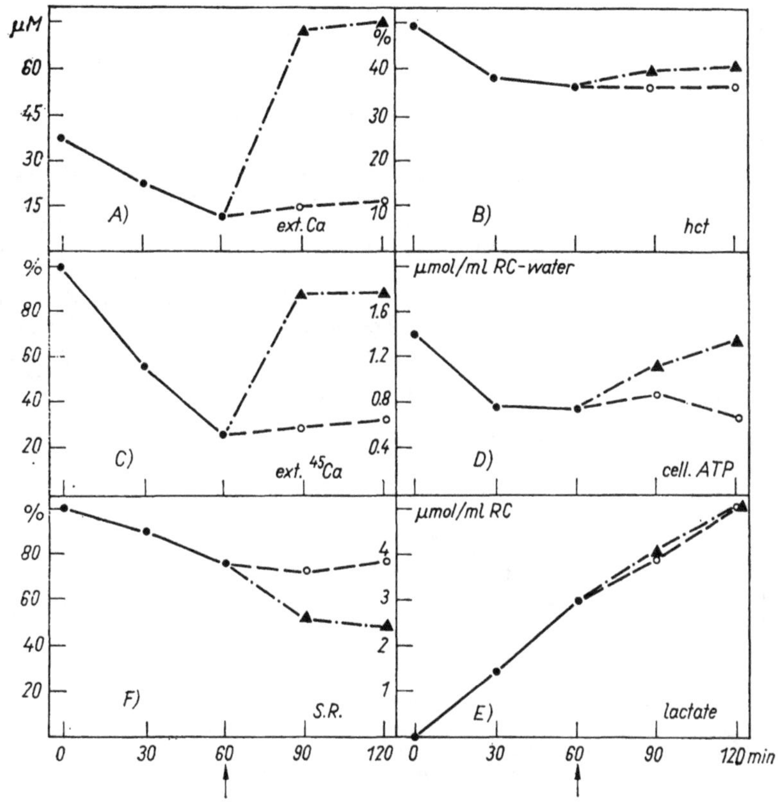

Fig. 6. Action of ruthenium red and ruthenium red plus EGTA on Ca^{2+} loaded red cells. Washed red cells were incubated at 37 °C and *p*H 7.4 in saline in the presence of A 23187 (5 µM) and [^{45}Ca]-$CaCl_2$ (●———●). After 60 min, the incubation mixture was divided into two aliquots for the purpose of different additions (arrow): ○– – –○ with ruthenium red (20 µM); ▲-·-·-·-▲ with ruthenium red (20 µM) plus Mg-EGTA (1 mM). Mean values of two experiments. A) ext. Ca. = extracellular Ca^{2+} concentration; B) hct = hematocrit; C) ext. ^{45}Ca = total radioactivity of extracellular Ca^{2+} in % of the initial value; D) cell. ATP = cellular ATP concentration; E) S. R. = specific radioactivity of extracellular Ca^{2+} in % of the initial value $= \frac{\text{ext. } ^{45}\text{Ca}}{\text{ext. Ca}}$; F) lactate = virtual lactic acid concentration per ml RC, recalculated from values determined in the incubate

Influence of Mg^{2+} on Ca^{2+} loading and accompanying effects

In red cells incubated without Mg^{2+} A 23187 causes the same effects as with 0.16 mM extracellular Mg^{2+}. At increasing Mg^{2+} concentrations above this value the action of the ionophore is reduced. Ca^{2+} loading and accompanying effects are completely abolished at 10 mM extracellular Mg^{2+}. The ATP depletion can be abolished already by 1 mM extracellular Mg^{2+} whereas 10 mM are necessary to abolish the increase of lactic acid formation.

The effects of this Mg^{2+} excess of at least 2 orders of magnitude with respect to the extracellular Ca^{2+} concentration can be explained by Ca^{2+} displacement (Fig. 7): Mg^{2+} removes a part of the accumulated Ca^{2+} from the cells, which leads

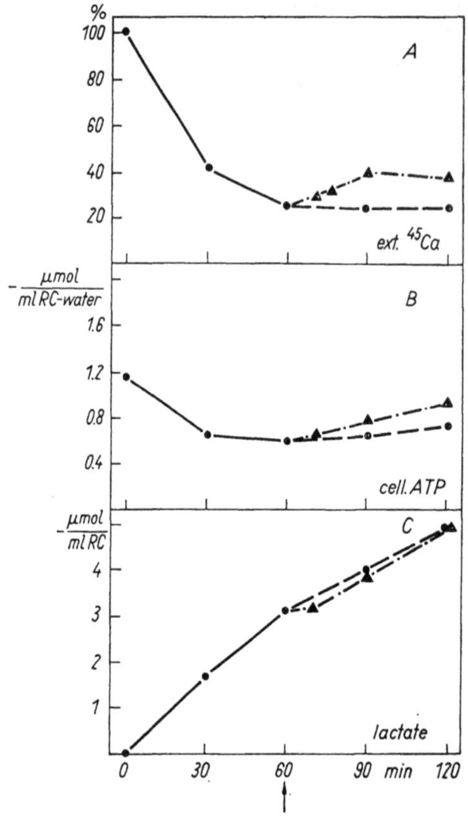

Fig. 7. Action of ruthenium red and ruthenium red plus 10 mM extracellular Mg^{2+} on Ca^{2+} loaded red cells. Washed red cells were incubated at 37 °C and *p*H 7.4 in saline in the presence of A 23187 (5 µM) and [^{45}Ca]-$CaCl_2$ (●———●). After 60 min the incubation mixture was divided into two aliquots with the purpose of further additions (arrow): ○— — - ○ with ruthenium red (20 µM); ▲-·-·-·-▲ with ruthenium red plus 10 mM extracellular Mg^{2+}.
A) ext. ^{45}Ca = total radioactivity of extracellular Ca^{2+} in % of the initial value; B) cell. ATP = cellular ATP concentration; C) lactate = virtual lactic acid concentration per ml RC, recalculated from values determined in the incubate

to a slight increase in ATP concentration and dimishes the lactic acid formation from 3.15 µmoles/ml RC · h to 1.92 µmoles/ml RC · h.

Discussion

Owing to the very low extracellular Ca^{2+} concentration used the A 23187 induced Ca^{2+} loading with 10—25 µM excess Ca^{2+} remains far below the maximum loading (between 5 mM [11] and 45 mM [12]). However, slight Ca^{2+} loading is sufficient to initiate effects on cell volume, ion distribution, glycolysis and adenine nucleotides. Some of these effects are well established in the literature and confirmed in the present paper.

Potassium efflux

From the numerous studies on the possible mechanism of the Ca^{2+} induced K^+ efflux in red cells [2, 7, 11—15, 21, 27—38] controversy remains. With the results of this paper the different views can be restricted:

From Ca^{2+} loading during cold storage it was concluded that the K^+ efflux is a prerequisite for the Ca^{2+} influx [37]. This does not hold for the A 23187 mediated excess Ca^{2+} because here Ca^{2+} loading persists if K^+ efflux is prevented (Fig. 5). The ATP depletion or a competition between (Mg^{2+} + Na^+ + K^+)-dependent

ATPase and (Mg^{2+} + Ca^{2+})-dependent ATPase for ATP as a reason for the K^+ efflux [29—31, 39] seem to be improbable, too:

1. Differences in the time dependence of ATP depletion and K^+ efflux do not correspond with this opinion (Fig. 2E, G.)

2. K^+ efflux is not altered by ouabain (Fig. 4; [11, 15]), ruthenium red (Fig. 2E) or $LaCl_3$ [7].

Therefore, the most probable explanation for the K^+ efflux is as follows: In the presence of A 23187 Ca^{2+} binds to membrane receptors normally not accessible, leading to an increased permeability [1,1 13—15, 21—36]. This is in accordance with the following finding:

The specific radioactivity of the extracellular Ca^{2+} decreases after A 23187 addition by dilution with unlabelled cellular Ca^{2+}. The extent of isotope dilution exceeds that part of total cellular Ca^{2+} which participates in isotope exchange without A 23187 [5].

Adenine nucleotides

The A 23187 induced ATP depletion found is in accordance with the results of most authors [10, 13, 15, 18, 21].

Sarkadi et al. [11] found no ATP depletion after a short time action (2 min) of the ionophore although Ca^{2+} loading took place during this time. This is not contrary to our findings because of the lag phase of ATP depletion (Fig. 2G). During long time action of the ionophore Edmondson and Li [12] found no ATP depletion. This may be due to the high extracellular Mg^{2+} concentration used there. It causes a Mg^{2+} loading of the cells (Fig. 1) which abolishes the Ca^{2+} effect on ATP (see results, influence of Mg^{2+}). The ATP depletion is connected with an increase of ADP and a considerable increase of AMP. The sum of the adenine nucleotides remains constant. The quotient $\frac{[AMP]\,[ATP]}{[ADP]^2}$ (total concentrations) amounts to 1.2 being the same in normal and ionophore treated cells and constant during the incubation. This value indicates thermodynamic equilibrium between the reactants of adenylate kinase (E.C. 2.7.4.3.) under cellular conditions [40—43]. During incubation with ionophore plus ruthenium red the sum of all three nucleotides remain constant, too. However, the ATP decrease and the rise in AMP are much more pronounced (Fig. 2G, I). Contrary to the incubation without ruthenium red the ADP level decreases (Fig. 2H). As a consequence, the quotient $\frac{[AMP]\,[ATP]}{[ADP]^2}$ rises from the initial value of about 1 to about 35 after 30 min and to about 20 after 60 min incubation. The total cellular concentrations of Mg^{2+} and Ca^{2+} as the main cations forming complexes with adenine nucleotides are approximately identical with that of normal cells. It remains to be elucidated whether this is a real or an apparent nonequilibrium of adenylate kinase reactants under those conditions. An apparent nonequilibrium could be present because the predominant part of the remaining ATP (0.2—0.3 mM) will not be available owing to a binding to hemoglobin [44] or complex formation with iron [45].

ATP consuming processes and lactic acid formation

As shown in the results the hexokinase and phosphofructokinase reaction, the (Mg^{2+} + Na^{+} + K^{+})-dependent ATPase and the cell shrinkage are excluded as the reasons for the ATP depletion observed during A 23187 induced Ca^{2+} loading. Our conclusions are supported by findings of an inhibition of the (Mg^{2+} + Na^{+} + + K^{+})-dependent ATPase by Ca^{2+} [28, 35]. Further, only a very small part of ATP will be consumed by the enhanced phosphatidate synthesis [18]. Therefore, the main point of interest should be restricted to the Ca^{2+} pumping system of the membrane which is activated as a consequence of the enhanced Ca^{2+} influx. To calculate the possible ATP consumption by this system the following findings have to be included: the maximum capacity of (Mg^{2+} + Ca^{2+})-dependent ATPase in red cells amounts to 85 μmoles/l cells · min [11] and 150 μmoles/l cells · min [4], respectively. For a rough calculation a value of about 100 μmoles/l cells · min may be used. The most probable stoichiometry of Ca^{2+} moles pumped to ATP moles split seems to be 2:1 [18], being identical with that in the sarcoplasmatic reticulum [30]. Further, in ghosts and membrane fragments, respectively, the enzyme seems to work with the full capacity at small Ca^{2+} loading (less than 10 μM excess Ca^{2+}) [4, 7, 46]. This corresponds to the extent of Ca^{2+} loading in our experiments. A cyclic flux of Ca^{2+} across the membrane during the ionophore induced transport of small amounts of Ca^{2+} is also assumed by Schatzmann and Vincenzi [2], Schatzmann and Rossi [32], Schatzmann [4] and Lee and Shin [47]. The same can be derived from the results shown in Fig. 2C. Taking into account these findings and assumptions one can calculate a maximum ATP consumption of about 3 μmoles/ml RC · h by the Ca^{2+} pumping system. About 2 μmoles/ml RC · h of ATP can be formed by the corresponding enhancement of lactic acid formation (Fig. 2 M). The remaining difference of 1 μmoles/ml RC · h would correspond to the ATP depletion observed within the first hour after A 23187 addition (Fig. 2 G). So far the assumption of Allan and Michell [18] and Ferreira and Lew [48] seems to be supported that the (Mg^{2+} + Ca^{2+})-dependent ATPase theoretically could be responsible for the ATP depletion. However, there are remarkable differences in the time course of ATP depletion and lactic acid formation not being consistent with this assumption (Fig. 2 G, M). Furthermore, after inhibition of the Ca^{2+} pumping system by ruthenium red the ATP depletion is much more pronounced contrary to the expectation (Fig. 2 G). The ATP level decreases within the first 15 min where the lactic acid formation is as high as in the absence of the inhibitor (Fig. 2 M). The inhibiting action of ruthenium red on the Ca^{2+} pumping system under these conditions clearly can be derived from the kinetics of the Ca^{2+} influx (Fig. 2A—C) and corresponds to analogue results with $LaCl_3$ [11, 49]. A further important support are the results of Watson et al. [50] who showed that ruthenium red causes a specific and nearly complete inhibition of the ATP splitting by the (Mg^{2+} + Ca^{2+})-dependent ATPase in human red cell membranes. The concentration of ruthenium red used there was in the same range as in our incubations.

This leads to the conclusion that there is a further Ca^{2+} sensitive ATP consuming process responsible for the ATP depletion observed during Ca^{2+} loading.

The lactic acid formation is increased by excess cellular Ca^{2+} apparently up to a maximal value, since lactic acid formation does not keep pace with the concomitantly increased ATP consumption. Inhibiting the Ca^{2+} pumping system by ruthenium red abolishes the dependence of lactic acid formation on excess Ca^{2+} (Fig. 2 A, M, Fig. 6 F, broken line and Fig. 7 C, broken line). Removal of cellular Ca^{2+} in the presence of ruthenium red does not alter lactic acid formation (Fig. 6 F broken plus dotted line, Fig. 7 C broken plus dotted line). Therefore, the lactic acid formation is controlled either by an active Ca^{2+} pumping system or ruthenium red inhibits the glycolysis as well as the Ca^{2+} pumping system.

After A 23187 high ADP levels with high lactic acid formation are found, after A 23187 plus ruthenium red, however, low ADP levels with low lactate formation (Fig. 2 H, M). This can be taken as a hint that lactic acid formation is governed mainly by ADP. Further data on the relationship between Ca mediated membrane processes and glycolysis will be presented in the accompanying paper [51].

Acknowledgements

We thank Drs. K. WINNEFELD and H. M. DAWTSCHINSKI for carying out the atomic absorption measurements. The skillful technical assistance of Mrs. C. RITTER and Miss. K. BEHRENS is gratefully acknowledged. A 23187 was a gift from Lilli Laboratories, U.S.A.

References

[1] HARRISON, D. G., and C. LONG: J. Physiol., Lond. **199**, 367 (1968)
[2] SCHATZMANN, H. J., and F. F. VINCENZI: J. Physiol., Lond. **201**, 369 (169)
[3] PALEK, J., W. A. CURBY, and F. J. LIONETTI: Am. J. Physiol. **220**, 19 (1971)
[4] SCHATZMANN, H. J.: J. Physiol., Lond. **235**, 551 (1973)
[5] CAMERON, B. F., and P. E. SMARIGA: Biochem. J. **156**, 577 (1976)
[6] REED, P. W., and H. A. LARDY: J. biol. Chem. **247**, 6970 (1972)
[7] REED, P. W., and H. A. LARDY in: Role of Membranes in Metabolic Regulation M. A. MEHTMAN and R. W. HANSON (Eds.). Academic Press, New York 1972
[8] REED, P. W.: Fedn Proc. Fedn Am. Socs exp. Biol. **32**, 635 (1973)
[9] GARDOS, G., B. SARKADI, and I. SZASZ: V^{th} International Congress of Biophysics, Copenhagen 1975. Abstracts p. 100
[10] KIRKPATRICK, F. H., D. G. HILLMANN, and P. L. LA CELLE: Experientia **31**, 653 (1975)
[11] SARKADI, B., I. SZASZ, and G. GARDOS: J. Membrane Biol. **26**, 357 (1976)
[12] EDMONDSON, J. W., and T. K. LI: Biochim. biophys. Acta **443**, 106 (1976)
[13] PLISHKER, G. A., and H. J. GITELMAN: Fedn Proc. Fedn Am. Socs exp. Biol. **35**, 781 (1976)
[14] GOLDINGER, J. M., and J. F. HOFMAN: Fedn Proc. Fedn Am. Socs exp. Biol. **35**, 834 (1976)
[15] REED, P. W.: J. biol. Chem. **251**, 3489 (1976)
[16] WHITE, J.: Am. J. Path. **77**, 507 (1974)
[17] ALLAN, D., and R. H. MICHELL: Nature, Lond. **258**, 348 (1975)
[18] ALLAN, D., and R. H. MICHELL: Biochem. Soc. Transact. **3**, 751 (1975)
[19] ALLAN, D., R. WATTS, and R. M. MICHELL: Biochem. J. **156**, 225 (1976)
[20] ALLAN, M. M., BILLAH, J. B. FINEAU, and R. H. Michell: Nature, Lond. **261**, 58 (1976)
[21] SZASZ, I., and G. GARDOS: FEBS Let. **44**, 213 (1974)
[22] BERGMEYER, H. U. (Eds.): Methoden der enzymatischen Analyse. 2. Ed. Akademie-Verlag, Berlin 1970
[23] ROSE, E. B., and J. LIEBOWITZ: Anal. Biochem. **35**, 177 (1970)
[24] TILL, U., E. BLUME, J. GÜNTHER, J. G. REICH, D. ZAHN, R. KLINGER, K. JAROSZEWICZ, and H. FRUNDER: Eur. J. Biochem. **6**, 373 (1968)
[25] DÜNN, H. J.: Biochim. biophys. Acta **352**, 97 (1974)
[26] ELGSAETER, A., D. M. SHOTTON, and D. BRANTON: Biochim. biophys. Acta **426**, 101 (1976)

[27] Dunham, E. T., and I. M. Glynn: J. Physiol., Lond. **156**, 274 (1961)
[28] Skou, J. C.: Physiol. Rev. **45**, 616 (1965)
[29] Hofman, J. F.: Am. J. Med. **41**, 66 (1966)
[30] Lew, V. L.: Biochim. biophys. Acta **233**, 837 (1971)
[31] Panero, P. J., and R. Whittam: J. Physiol., Lond. **214**, 481 (1971)
[32] Schatzmann, H. J., and G. L. Rossi: Biochim. biophys. Acta **241**, 379 (1971)
[33] Lew, V. L.: Biochim. biophys. Acta **249**, 236 (1971)
[34] Blum, R. M., and J. F. Hofman: J. Membrane Biol. **6**, 315 (1971)
[35] Blum, R. M., and J. F. Hofman: Biochem. biophys. Res. Commun. **46**, 1146 (1972)
[36] Glynn, I. M., and A. E. Warner: Brit. J. Pharmac. **44**, 271 (1972)
[37] Mouat, B., and C. Long: Biochem. J. **142**, 629 (1974)
[38] Gardos, G., I. Szasz, and B. Sarkadi: FEBS Proceedings **35**, 167 (1967)
[39] Gardos, G.: Acta physiol. hung. **15**, 121 (1959)
[40] Rose, J. A.: Proc. natn. Acad. Sci. U.S.A. **61**, 1079 (1968)
[41] Jacobasch G.: Promotin B-Schrift des Wissenschaftlichen Rates der Humboldt-Universität Berlin, 1969
[42] Horn, A., R. Husung, M. Schröder, and H. Böring: Acta biol. med. germ. **27**, 821 (1971)
[43] Bublitz, R.: Promotion A-Schrift der Naturwissenschaftlichen Fakultät der Friedrich-Schiller-Universität Jena, 1973
[44] Gerber, G., H. Berger, G. R. Jänig, and S. M. Rapoport: Eur. J. Biochem. **38**, 563 (1973)
[45] Bartlett, G. R.: Biochem. biophys. Res. Commun. **70**, 1063 (1976)
[46] Wolf, H. H.: Biochim. biophys. Acta **266**, 361 (1972)
[47] Lee, K. S. and B. C. Shin: J. gen. Physiol. **54**, 713 (1969)
[48] Ferreira, H. G., and V. L. Lew: Nature, Lond. **259**, 47 (1976)
[49] Quist, E. E., and B. D. Roufogalis: FEBS Lett. **50**, 135 (1975)
[50] Watson, E. L., F. F. Vincenzi, and P. W. Davis: Biochim. biophys. Acta **249**, 606 (1971)
[51] Brox, D., B. Petermann, and F. Frunder: Acta biol. med. germ. **36**, 611 (1977)

Bezugsmöglichkeiten

Bestellungen sind zu richten:

- **in der DDR** an eine Buchhandlung oder an den Akademie-Verlag, DDR-108 Berlin, Leipziger Straße 3–4
- **im sozialistischen Ausland** an eine Buchhandlung für fremdsprachige Literatur oder an den zuständigen Postzeitungsvertrieb
- **in der BRD und Westberlin** an eine Buchhandlung oder an die Auslieferungsstelle KUNST UND WISSEN, Erich Bieber, 7 Stuttgart 1, Wilhelmstraße 4–6
- **in Österreich** an den Globus-Buchvertrieb, 1201 Wien, Höchstädtplatz 3
- **im übrigen Ausland** an den Internationalen Buch- und Zeitschriftenhandel; den Buchexport, Volkseigener Außenhandelsbetrieb der Deutschen Demokratischen Republik, DDR-701, Leipzig, Postfach 160, oder an den Akademie-Verlag, DDR-108, Berlin, Leipziger Straße 3–4.

www.ingramcontent.com/pod-product-compliance
Lightning Source LLC
LaVergne TN
LVHW081328011026
849161LV00013BB/2872